PURVES, MICHAEL
THE PHYSIOLOGY O...

HCL   QP1.P57.(28)

LIVERPOOL UNIVERSITY LIBRARY
WITHDRAWN
FROM
STOCK

Monographs of the Physiological Society No. 28

The physiology of the
cerebral circulation

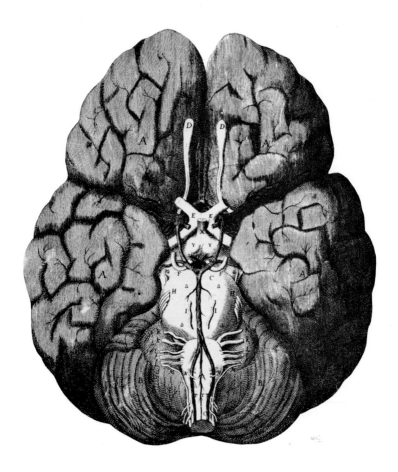

The base of the human brain taken out of the skull, with the roots of the vessels cut off.

From *The remaining medical works of that famous and renowned physician Dr Thomas Willis*. Englished in 1681 by Samuel Pordage, the figure drawn by Dr Christopher Wren and engraved by David Loggan.

Frontispiece

THE PHYSIOLOGY OF THE CEREBRAL CIRCULATION

M. J. PURVES
Department of Physiology
University of Bristol

CAMBRIDGE
At the University Press 1972

Published by the Syndics of the Cambridge University Press
Bentley House, 200 Euston Road, London NW1 2DB
American Branch: 32 East 57th Street, New York, N.Y.10022

© Cambridge University Press 1972

Library of Congress Catalogue Card Number: 70-169577

ISBN: 0 521 08300 1

Printed in Great Britain
at the University Printing House, Cambridge
(Brooke Crutchley, University Printer)

CONTENTS

Preface		*page* vii
1	Some aspects of the anatomy of cerebral blood vessels	1
	Collateral circulations	2
	The fine structure of cerebral vessels	12
	Structures imposed between cerebral vessels and neurones	21
2	Capillary density and oxygen transport in the brain	25
	Capillary density of the brain	26
	Tissue oxygen in the brain	33
3	The innervation of cerebral blood vessels	44
4	Haemodynamic considerations	69
	Factors which regulate cerebral blood flow	72
	The concept of autoregulation	80
	The homeostatic function of c.s.f.	89
	Whole blood viscosity and cerebral blood flow	96
5	The cerebral circulation: experimental approaches	101
	The skull window technique	107
	Measurement of blood flow in vessels	110
	Artificial perfusion of the brain	114
	The use of inert diffusible indicators	123
6	Cerebral blood flow and arterial pressure	156
	Alterations in the relation between pressure and flow	163
	The neural contribution	166

Contents

7	Regulation of cerebral vessels by carbon dioxide	*page* 173
	Factors affecting the vascular response	181
	Carbon dioxide and vascular smooth muscle	187
8	Cerebral blood vessels and pH	200
	Cerebral vascular smooth muscle and pH	201
	The regulation of pH in brain e.c.f. and of c.s.f.	213
9	Regulation of cerebral vessels by oxygen	232
	Oxygen, carbon dioxide and blood pressure	236
	The homeostatic role of the vascular response	243
10	The neural control of cerebral blood vessels	253
	Vasometer nerves and changes in blood gas tensions and pressure	266
11	Cerebral blood flow and metabolism	282
	Measurements of cerebral metabolic rate *in vivo*	288
12	Some aspects of the pharmacology of cerebral vascular smooth muscle	334
References		352
Index		415

PREFACE

The literature concerning the cerebral circulation, its control, its relation to the metabolism of the brain and its disorders is very large and reflects an interest which goes back to Greek and Arabian medicine and which has greatly intensified particularly over the past forty years. Unhappily, the literature also contains much that is repetitious and reveals a very wide range of experimental competence. As in many biological fields, there have been a small number of imaginative and decisive studies; at the same time very many studies merely serve to perpetuate myths while in other areas, it is clear that the proper questions have scarcely been formulated. A further difficulty for the student of the cerebral circulation is that papers on the subject are scattered through a large number of journals, a fact which reflects the background and interests of the authors rather than the breadth of the subject.

The first purpose of this monograph has therefore been to bring together from this literature a coherent account of how our present ideas have developed. In order to keep the text within manageable size, this has meant considerable selection both of topics and references. In this, I have been guided by what seem to me to be the important lines of investigation and by the clarity with which views have been expressed whether or not these have subsequently been shown to be erroneous. Secondly, the evidence from studies which form the basis of our present understanding has been reviewed critically particularly where the methods used are of doubtful validity and, where possible, an attempt has been made to pose present controversies in terms which would admit of direct experimental testing.

A recent and rapid development has been the application of methods for measuring cerebral blood flow in patients with a wide variety of cerebral vascular or metabolic disorders. The results of such studies have been largely omitted except where they shed light on the processes of normal control. This omission is intentional partly in order to limit the size and scope of the present book

Preface

and partly because much of this material has been fully considered in recent reviews and published symposia. To these the present book could be considered complementary.

I am grateful to many colleagues who have read and criticized chapters, to numerous authors who have contributed unpublished material or who have given permission for material to be reproduced, to Miss Lilian Patterson for many of the line drawings, to Mrs Myra Cook who typed the manuscript and to Alistair Purves for translating the Russian texts.

M. J. PURVES

Bristol 1971

1 SOME ASPECTS OF THE ANATOMY OF CEREBRAL BLOOD VESSELS

One of the happier outcomes of the Restoration of the Monarchy in 1660 was the election of Thomas Willis to the Sedleian Professorship of Natural Philosophy at Oxford. Although by this act he exchanged some of the distractions of private practice for those of university life, he was able in the next three to four years to carry out a systematic study of the cerebral vessels in man. He was fortunate in his collaborators; Richard Lower and Thomas Millington as fellow dissectors; Christopher Wren who suggested a new way of preserving post-mortem material and who illustrated many of the dissections and, David Loggan, the engraver. The result, Willis' masterpiece, *Cerebri Anatome*, at one blow disposed of the extravagances of Galen and Vesalius, demonstrated that the rational systems much beloved by the Cartesian philosophers on the continent were no adequate substitute for careful and accurate observation and laid the foundation for all subsequent work on the brain and its blood supply. The tercentenary of the publication of *Cerebri Anatome* has been suitably and magnificently commemorated (Feindel, 1965).

In the intervening years, a large literature on the cerebral vascular bed has accumulated and it is doubtful whether a review of this field which takes into account the embryology, histology and ultrastructure of cerebral blood vessels is now possible. A number of outstanding books appeared in the early nineteenth century dealing particularly with cerebral veins in man (Bock, 1823; Breschet, 1832; Cruveilhier, 1834) and, more recently two restricted but valuable reviews have been published (Kaplan & Ford, 1966; Van den Bergh & Vander Eecken, 1968). These deal solely with the anatomy of cerebral blood vessels in man. The comparative aspects of the cerebral circulation have been considered from the embryological point of view by Padget (1957) and accounts of the cranial circulation in individual species have been given by Ellenberger & Baum (1943), Batson (1944) and

1. Anatomy of cerebral blood vessels

Sagawa & Guyton (1961) for the dog; by Davis & Storey (1943) and Geiger & Magnes (1947) for the cat; by Baldwin (1964) for the ox and sheep and by Hegedis & Shackelford (1965) for the cerebral venous system in various laboratory and domestic animals.

For the physiologist who is a student of the cerebral circulation, there is much in these comprehensive reviews which is of only marginal interest. For him, the anatomical problems are rather more severely practical. In which species, for example, will the arrangement of cerebral blood vessels and the anastomoses between them most satisfactorily allow the vascular isolation of the brain? In what way, if at all, do the intracranial and, in particular, the intracerebral vessels differ from comparable vessels in other vascular beds? Are such differences likely to give rise to different vascular responses to carbon dioxide or pharmacological agents? What are the structures which lie in between the cerebral capillaries and neurones? Are they likely to offer resistance to the diffusion of gases and substances of differing molecular weight and is it possible that they are involved in the intrinsic control of capillary blood flow?

It is a consideration of the evidence which might answer these questions rather than a recapitulation of anatomical fact which forms the basis for the present chapter.

Collateral circulations

The embryological origin of the brain as a tube-like structure explains in part why the vascular supply is so complex. To the leptomeningeal system of arteries, there must be added the arteries of ventricular origin – that is those which originate from subependymal arteries (Van den Bergh, 1961) and the remains of the branchial arch afferent supply. In the adult, there are numerous anastomoses between these systems within and without the skull. Further, there are marked differences between species.

For all these reasons, it has been found particularly difficult to isolate the cerebral afferent supply and the student of the cerebral vascular system may well be forgiven for asking himself from time to time why he did not pick on a structure with a hilum (like the lung) or a pedicle (like the kidney). Certainly, as will be shown in later chapters, it is the failure to obtain a properly isolated

Collateral circulations

cerebral vascular bed which has for so long delayed a proper understanding of the physiology and pharmacology of cerebral blood vessels.

Arterial anastomoses

Extracranial. These arterial anastomoses are of two kinds; those between carotid and vertebral systems and those between internal and external carotid arteries. In the ox and the sheep, the former are of particular importance in maintaining an adequate cerebral circulation (Baldwin, 1964) while in the cat, the vertebro-occipital anastomoses are so great that the respiratory fluctuations of blood gas tensions are transmitted without obvious attenuation while the perfusion pressure is not greatly reduced when the common carotid artery is occluded (Biscoe & Purves, unpublished observations). The anastomoses between internal and external carotid arteries occur principally at the points where the cranial nerves leave the skull, that is around the eye, nose and ear. The evidence, based mainly upon angiographic studies shows that anastomoses occur between the ophthalmic branch of the internal carotid artery and the internal and external maxillary branches of the external carotid artery supplying the face and nasal cavity (Bregeat, David, Fischgold & Talairach, 1952; Decker, 1955; Grino & Billet, 1949; Krayenbühl, 1958; Lazorthes, 1961), between the ethmoidal branches of the ophthalmic artery and the internal maxillary artery (Moniz, 1940) and between the lacrimal and either the superficial temporal or frontal ramus of the middle meningeal artery. In the ear, anastomoses have been described between the tympanic branch of the internal carotid and labyrinthine branch of the basilar arteries (Charachon & Latarjet, 1952) while the persistence of the stapedial artery may also form a collateral channel. Normally present in the embryo, the stapedial artery is formed from both the hyoid artery derived from the second aortic arch as a branch of the internal carotid artery: and the ventral pharyngeal artery which can be regarded as a primitive external carotid artery formed from the third aortic arch. In the embryo, therefore, blood from both internal and external carotid arteries flows in the stapedial artery but, later, it becomes separated from the internal carotid artery while its branches connect between the lacrimal and middle meningeal arteries.

1. Anatomy of cerebral blood vessels

Anastomoses between intra- and extracranial circulations. These occur in man relatively rarely and when present they form in relation to three vessels: the trigeminal artery between internal carotid and basilar arteries (Altmann, 1947; Brea, 1956; Frugoni, 1952; Hinck & Gordy, 1964); the auditory artery (Gros, Minvielle & Vlahovitch, 1956; Maspes, Fasano & Broggi, 1955) and the hypoglossal artery (Murtach, Stauffer & Harley, 1955; Oertel, 1922). Occasionally, large connecting vessels which connect the internal carotid artery to the proximal portion of the basilar artery occur in the region of the vestibulo-cochlear nerve.

In other species, however, arterial anastomoses between intra- and extracranial circulations are the rule and many of these have been considered by Batson (1944) and by Sagawa & Guyton (1961) and are illustrated in Fig. 5.7. Geiger & Magnes (1947) found a similar array of arterial anastomoses in the cat and considered that it was impossible to isolate the afferent blood supply to the brain in this species without gross mutilation and risk of damage to small arteries. Similar conclusions apply to the dog. Instead, most workers have found it easier to try and isolate the cerebral venous drainage.

Intracranial arterial anastomoses. These occur (a) through connections of the large vessels of the brain, (b) through perforating branches to the basal nuclear masses and (c) through the surface vessels on the supra-segmental cell areas. At the base of the brain, in primates and in the rabbit, the arrangement of vessels permits collateral flow of blood between the internal carotid arteries and the vertebro-basilar system through the circle (more strictly, the nonagon) of Willis, Fig. 1.1. Numerous workers have commented on the variability of the size of the vessels which form these anastomoses. For example, Kaplan & Ford (1966) examined fifty adult human brains at autopsy and found that in 44 per cent, the proximal part of the posterior cerebral artery was so fine that it would have been unlikely to have been an effective collateral vessel. However, they rarely found that the anterior and posterior cerebral arteries were small on the same side and they concluded that there would be adequate collateral circulation if one internal carotid artery was occluded provided that both proximal anterior cerebral arteries were large or if the proximal posterior and mesencephalic arteries on the same side were large. The question of how effective these collateral vessels are when one internal carotid artery is occluded has not been extensively studied. Thus Hill & Moore (1894) showed that if the common carotid artery

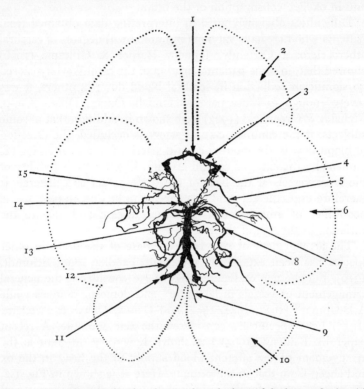

Fig. 1.1. An angiogram showing the vessels at the base of the human brain with the circle of Willis as traditionally presented. 1, Anterior communicating artery; 2, frontal lobe; 3, anterior cerebral artery; 4, internal carotid artery; 5, anterior choroidal artery; 6, temporal lobe; 7, choroidal-diencephalic artery; 8, superior cerebellar artery; 9, anterior spinal artery; 10, cerebellum; 11, vertebral artery; 12, inferior cerebellar artery; 13, basilar artery; 14, mesencephalic artery; 15, proximal stem of posterior communicating artery. (Redrawn from Kaplan & Ford (1966), *The brain vascular system*. Elsevier: Amsterdam, Fig. 118.)

on one side was occluded in normal individuals (themselves), it gave rise to signs of 'irritability' of the pyramidal tract on the same side. The signs could not be ascribed to stimulation of the carotid sinus. On the other hand, Shenkin, Cabieses, Van den

1. *Anatomy of cerebral blood vessels*

Noordt, Sayers & Copperman (1951) measured cerebral blood flow in four patients before and after unilateral internal carotid artery ligation and found that in three, there was no change in blood flow; in the fourth, there was a marked fall in blood flow and in oxygen consumption of the brain.

Difficulties obviously arise in interpreting data obtained from patients who may have varying (and unknown) degrees of cerebral atherosclerosis. Certainly, Jennett, Harper & Gillespie (1966) showed that, in some patients, ligation of the carotid artery caused no significant reduction in cerebral blood flow; in others, it was severe enough to cause paresis. Tindall, Odom, Dillon, Cupp, Mahaley & Greenfield (1963) have shown in addition that in some subjects, if the common carotid artery is occluded, the direction of blood flow in the internal carotid arteries reverses and blood is 'stolen' by the external carotid arteries. The functional ability of the large vessels at the base of the brain to act as collaterals is therefore unpredictable and this appears to be due, in part, to the possibility of atherosclerosis or other disease and also to the variation in the size of the vessels.

The arrangement of vessels at the base of the brain in other species has been extensively reviewed (Tandler, 1899; Schmidt, 1910). Ellenberger & Baum (1943) have considered the general arrangement of vessels in laboratory and domestic animals while Ask-Upmark (1935, 1944, 1953) and Daniel, Dawes & Pritchard (1953) have specifically considered the rete mirabile. A recent paper by Baldwin (1964) has shown how very different is the arrangement of the afferent blood supply to the head in the ox and sheep from that in the primate. Here as is shown in Fig. 1.2, the carotid rete is formed principally by branches of the internal maxillary artery; the internal carotid artery is only rarely patent in the adult and only that portion of it exists as an efferent vessel from the rete carrying blood to the circle of Willis. Further, the basilar artery is more properly thought of as an efferent vessel from the circle rather than, as in the primate, contributing to it. The figure also shows the extensive anastomoses between vertebral and spinal arteries and between the vertebral and carotid arteries. Similar arrangements of vessels have been shown in the cat (Boissezon, 1941; Legait, 1945; Davis & Storey, 1943).

Further studies by Baldwin & Bell (1963a) have clarified some of the properties of the rete mirabile in the calf and sheep. They

Collateral circulations

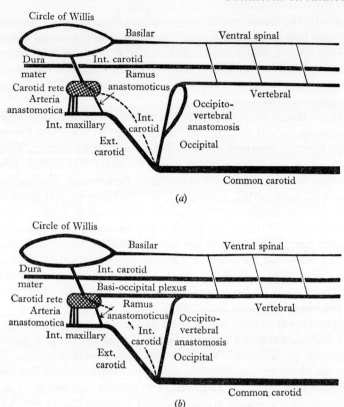

Fig. 1.2. Diagram illustrating the arrangement of the cephalic arteries in (a) the sheep, and (b) the calf. In both species the main blood supply to the circle of Willis is from the external carotid arteries via the carotid rete. The internal carotid artery is vestigial and non-functional except for a short section which carries blood from the extradural rete to the circle of Willis. The vertebral arteries do not contribute to the circle of Willis and the basilar artery carries blood caudally to the ventral spinal artery. In both species the occipital and vertebral arteries show a direct anastomosis but, in addition, the vertebrals in the calf connect with the carotid rete through the basi-occipital plexus. (From Baldwin & Bell (1963b), *Electroenceph. clin. Neurophysiol.* **15**, 465–73, Fig. 1.)

have shown that carotid blood flow is homolateral in distribution and that the rete does not permit admixture of blood from one side to the other. On the other hand it forms a potential anastomosis between left and right sides if one carotid artery is occluded. Although superficially similar, the arrangement of afferent blood supply to the head of the calf and sheep includes one important difference

1. *Anatomy of cerebral blood vessels*

– namely, the existence of a basi-occipital plexus of vessels between the vertebral artery and a rete. Thus when both external carotid arteries in the sheep are occluded, there is virtually complete ischaemia and loss of EEG after 8 s. In the calf, this manoeuvre has no effect because cephalic flow is sustained by flow through the basi-occipital plexus.

These results emphasize that the arrangement of vessels which constitutes the afferent supply to the head has an important bearing upon the way in which cerebral ischaemia can be caused. There have been many experiments in which the effect of tying individual arteries has been observed upon brain function in the dog and rabbit (Cooper, 1836), rabbit, cat, dog and monkey (Hill, 1896), dog (Evans & Saaman, 1936) and sheep (Linzell & Waites, 1957). Specifically, the effect of ligating the carotid and/or vertebral arteries upon EEG patterns has been studied by Meyer & Gastaut (1961), in the cat by Sugar & Gerard (1938), and in the sheep and calf (Baldwin & Bell, 1963*b*). Other less complete studies have been made by Negovski (1945) and ten Cate & Horsten (1952) in the dog; by Leão & Morison (1945) and Van Harreveld & Stamm (1952) in the rabbit; Van Harreveld (1947) in the cat; and Meyer, Feng & Denny-Brown (1954) in the cat and rabbit. Not only do these studies demonstrate in functional terms the arteries which are necessary for adequate cerebral blood flow and oxygenation but they also demonstrate the degree to which collateral blood vessels are effective in maintaining an adequate blood flow. Some further studies have shown that in general, young animals are more resistant to the effects of ischaemia than are adults (Kabat, 1940; Libet, Fazekas & Himwich, 1941).

Anastomoses within the substance of the brain. The leptomeningeal arteries arise from the main trunks at the base of the brain and wrap upwards around the cerebrum as the three cerebral arteries. Caudally, the cerebellar arteries arise from the vertebral and basilar arteries and invest the cerebellum in a similar way. The development of the telencephalon obscures the annular wrapping of the neural tube at the level of the cerebrum but the leptomeningeal arteries can be classified in three groups (Foix & Hillemand, 1925; Lazorthes, 1961). The *paramedian* arteries penetrate into the cerebral parenchyma after a short course (e.g. the branches of the anterior cerebral artery destined for the

Collateral circulations

infundibulum); the *short circumferential* arteries run somewhat further before ending as penetrating arteries (e.g. the rami striati) and the *long circumferential* arteries are the distal arteries which reach the surface of the hemispheres, the cerebellum and the dorsal surface of the brain stem (Van den Bergh & Vander Eecken, 1968).

These arteries may be considered as peripheral arteries whose branches penetrate the substance of the brain towards the centre and supply the cortex (*cortical rami*, Pfeiffer, 1928; Szikla & Zolnai, 1956) and the deeper structures (*rami medullares*) which supply white matter (Rowbotham & Little, 1965; Van den Bergh, 1964); the *rami striati* which supply the grey nuclei and internal capsule (Alexander, 1947; Van den Bergh, 1961); and the *rami perforantes* which supply the greater part of the thalamus and hypothalamus (Foix & Hillemand, 1925; Lazorthes, Gaubert & Poulhes, 1956). The authors quoted above have also described anastomoses between arterioles of these groups at the precapillary level. Van den Bergh and Vander Eecken (1968) have also drawn attention to the existence of arteries of ventricular origin which arise from subependymal arteries which in turn arise from the choroidal arteries and from lateral striate rami. They course away from the ventricles and supply paraventricular grey nuclei and may attain a length of 1.5 cm. These arteries run towards the penetrating branches of the leptomeningeal arteries without apparently any connections with them although Bolonyi (1951) has stated that anastomoses exist between medullary and striate arteries. If these anastomoses can be confirmed, then their presence would ensure that the paraventricular nuclei would escape the cone shaped infarcts which would follow occlusion of the rami striati. On the other hand, the condition known as periventricular leukomalacia which has been described in the newborn (Abramowicz, 1964; Banker & Larroche, 1962) occurs precisely at the junction between centripetal and centrifugal vessels and this might suggest that anastomoses between these systems of vessels was not particularly effective under these conditions.

Venous collateral circulation

Venous collateral vessels are even more abundant. The brain is drained by two series of veins, the deep or great cerebral venous (Galenic) system which functions as a single unit system and

1. Anatomy of cerebral blood vessels

receives tributaries from the entire brain in a neurovascular pattern similar to that provided by the arteries. The second system drains from the suprasegmental cerebral or cerebellar cortices into the overlying venous sinuses. These veins may be considered as meningeal since they lie superficially to the arteries. Extensive collaterals exist between the veins in each system as well as between the veins of the two systems (Kaplan & Ford, 1966).

Three main groups of veins drain the surface of the cerebral cortex, the superior, middle and inferior cerebral veins. They approach the superior sagittal sinus at an acute angle directed rostrally against the flow of blood and the acuteness of this angle is most obvious caudally. There are numerous anastomoses between these groups of veins: some are both constant and prominent. The first of these is the superior anastomotic vein (Vein of Trollard) which connects the superior and medial groups of veins. In the medial group of veins, the deep middle cerebral vein which drains the cortex of the insula drains into the inferior cerebral veins and also connects with the superficial medial cerebral vein. The latter, in turn, drains medially into the cavernous sinus. Yet another vein extends inferiorly from the superficial medial cerebral vein and connects with the transverse sinus. The veins which drain the medial aspect of the cerebral hemispheres connect with the great cerebral vein while the veins on the ventral surface of the brain connect with those on the lateral surface and with the deep venous system.

The pattern of blood flow in the dural sinuses into which the superficial and deep venous systems empty is equally varied. Batson (1944) describes three patterns. (1) The superior longitudinal sinus continues as either the left or right transverse sinus – usually the right. The straight sinus to the opposite side. (2) The superior sinus divides and is equally distributed between right and left transverse sinuses, and (3) there is a true confluence of all three sinuses. The importance of these varying patterns of blood flow in the sinuses lies in the fact that sampling of blood in the internal jugular veins could lead to erroneous conclusions as to the area of brain which was being drained. This question is dealt with in detail in Chapter 11.

From corrosion studies of the venous system, it is clear that venous blood can leave the skull by a number of routes other

Collateral circulations

than the internal jugular veins. There are, for example, numerous venous plexuses at the base of the skull which have been called the basilar or cavernous sinuses and which communicate with similar plexuses within the bones of the skull, the pterygoid plexuses and the veins of the orbit. A further widespread system of venous connections exists between the superficial veins of the cortex, the dural sinuses and the veins within the bones of the calvarium on the one hand and, on the other between these diploic veins and the extracranial veins. This means that first, the veins which truly drain the bones of the skull contaminate dural sinus blood with blood from other than cerebral sources. Secondly, there is the possibility that cerebral venous blood may drain via these emissary veins into the extracranial veins of the scalp. Ordinarily, the flow of blood in these veins is likely to be small: certainly, Batson (1944) was able to show that only in a small proportion was the flow great enough to show the passage of contrast material. Occasionally, however, and especially in the mastoid emissary veins, the flow of blood may be considerable.

For the experimentalist who attempts to isolate the cerebral venous drainage with the intention of making quantitative measurements of blood flow, the most serious difficulty arises in all species from the anastomoses outside the skull between the jugular and vertebral venous systems. These anastomoses have been particularly well described by Geiger & Magnes (1947) in the cat. In this and other species which have been described (Breschet, 1832; Hill, 1896; Batson, 1944), at least three pairs of vertebral veins extend longitudinally for the whole length of the spinal column from the plexus of veins which originate from the dural sinuses in the posterior fossa. These series of vertebral veins anastomose with each other at each vertebral segment and also with systemic veins in the thoracic and abdominal cavities. Geiger & Magnes (1947) also report that, in the cat, there is an additional large connection between the vertebral and posterior facial veins so that even if the external jugular veins are tied off, blood from extracranial sources will enter the vertebral veins by this route.

The size of the vertebral veins varies considerably but from corrosion studies (Batson, 1944), it appears that the total cross-section of these veins is not less than that of the internal jugular veins. Precisely what proportion of intracerebral blood drains into the vertebral veins under normal conditions is not certain but it

1. *Anatomy of cerebral blood vessels*

is known that if both internal jugular veins are tied, the vertebral plexus can drain the whole cranial cavity (Hill, 1896; Evans, 1942).

It therefore seems to be essential that the vertebral veins be ligated if quantitative measurement of cerebral blood flow is attempted and this argument becomes more powerful the greater the resistance imposed on venous flow by cannulation. Geiger & Magnes (1947) have described a method for occluding the vertebral plexus in the cat and in their hands this was successful since they were able to obtain values for the rate of cerebral blood flow which were similar to that obtained by other methods. However, the importance of occluding the vertebral veins may vary between species since Reivich (1964) was able to obtain satisfactory measurements of cerebral blood flow in the rhesus monkey together with normal responses to carbon dioxide merely by measuring blood flow from cannulated jugular veins.

But the difficulties do not end with the vertebral veins. A comprehensive account of the different types of cerebral venous drainage has been given for the dog, monkey, sheep, pig, cat, rabbit, horse and ox by Hegedis & Shackelford (1965) and these workers also list the number and extent of venous communications in the dog. This indicates that the measurement of blood flow in the dog by isolating the cerebral venous drainage is as formidable a procedure as isolating the afferent blood supply to the brain (Sagawa & Guyton, 1961): and it may well explain why workers who have used this method, as will be discussed in later chapters, have obtained unusually low values for cerebral blood flow.

If all the factors considered in this section are taken into account, it is clear that a satisfactory isolation of the afferent or efferent blood supply of the brain can only be carried out in primates. While Geiger & Magnus (1947) have shown that it is feasible to isolate the venous drainage of the brain in the cat, the technique is a formidable one and it may be extremely difficult to establish conclusively that cerebral venous blood is not being diverted through anastomotic channels. These arguments apply even more strongly in the dog.

The fine structure of cerebral vessels

The comprehensive study of cerebral vessels using conventional methods of staining and light microscopy has not shown any

The fine structure of cerebral vessels

obvious difference between the vessels of the brain and those in other organs.

These studies have been extended using electron microscopy in order to examine the fine structure particularly of intracerebral vessels and in order to determine how far these vessels are innervated. This latter question is considered in detail in Chapter 3: in the present section, the evidence concerning the fine structure is reviewed together with any differences which have been observed between cerebral and other vessels.

Extracerebral vessels

The fine structures of extracerebral vessels have been studied in the cat and monkey (Pease & Molinari, 1960), in the rat (Samarasinghe, 1965; Sato, 1966) and in man (Dahl & Nelson, 1964; Dahl, Flora & Nelson, 1965). The arteries consist of the usual three coaxial coats, the tunica intima, tunica media and tunica adventitia. The *tunica intima* includes the endothelium and internal elastic lamina. The features of the endothelium are similar to those described in arteries of other vascular beds (Buck, 1958; Karrer, 1965; Moore & Ruska, 1959). Mitochondria, endoplasmic reticulum, occasional osmophilic structures resembling lysosomes (Novikoff, 1959) and pinocytic vesicles are seen. The nucleus is enclosed in a double membrane with peripherally arranged chromatin. The endothelial cells often overlap (Fig. 1.3) and annular attachments are shown by increased osmophilia of portions of cell membranes of abutting cells (Bennett, Luft & Hampton, 1959).

The *internal elastic lamina* consists of two layers which can be distinguished by staining with phosphotungstic acid. The layer which is adjacent to the endothelium consists of a finely flocculent material and is only slightly affected by staining: that layer adjacent to smooth muscle appears dense and finely granular with staining and is thought to represent elastic tissue which is deposited in poorly organized matrix or ground substance (Pease & Molinari, 1960; Dahl, Flora & Nelson, 1965).

The internal elastic lamina is fenestrated, the interruptions being filled with undifferentiated matrix. Parker (1958) observed that in coronary arteries, the endothelium dipped into these fenestrations: this has not been confirmed in pial vessels. The regular occurrence of these fenestrations has suggested that the

1. Anatomy of cerebral blood vessels

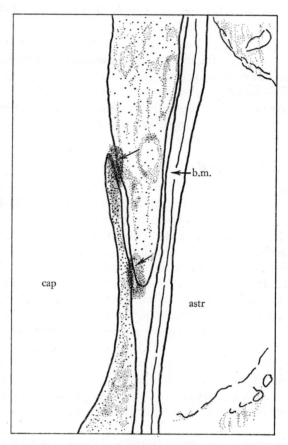

Fig. 1.3. The lateral junction between two adjacent epithelial cells at high magnification. The lumen of the vessel (cap) is to the left; astrocytic cytoplasm (astr) is to the right. The endothelium rests on a basement membrane (b.m.). The endothelial margins overlap slightly, the overlapping here amounting to 330 nm. Structures resembling adhesion plates are present at the beginning and end of the overlap as indicated by the arrows. × 60000. (From Maynard et al. (1957), *Am. J. Anat.* **100**, 409–34, Fig. 3.)

internal elastic membrane is not in fact a laminated but is a net-like structure which would allow greater resilience during contraction and relaxation.

The tunica media of cerebral vessels consists of smooth muscle cells which are defined by a cytoplasmic membrane and invested with a basement membrane of variable thickness: there is no

The fine structure of cerebral vessels

evidence of a syncytial arrangement (Moore & Ruska, 1959; Caesar, Edwards & Ruska, 1957). Occasionally, collagen is observed between smooth muscle cells. The variations and internal arrangement of smooth muscle cell nuclei have been described by Dahl et al. (1965). The size and arrangement of myofibrils in smooth muscle cells as well as the numerous vesicles arranged at the surface of the cell which are a feature of smooth muscle (Caesar et al. 1957) have been observed in cerebral vascular smooth muscle (Pease & Molinari, 1960).

The *tunica adventitia* has been less completely studied because of the poor penetration of osmium tetroxide. By light microscopy, the adventitia appears as collagen interspersed with fibrous tissue cells. With electron microscopy, the outer border of the artery is seen to be sharply defined and to consist of elongated spindle shaped cells that are similar to fibroblasts (Duff, McMillan & Ritchie, 1957). Embedded in the collagen matrix deep to this layer of fibroblasts are the myelinated and unmyelinated nerves which, except at their termination, are surrounded by Schwann cell cytoplasm. The appearance of these nerves is no different from that described in adventitia of other arteries (Parker, 1958; Caesar et al. 1957; Hess, 1955). Dahl et al. (1965) have also reported the presence of cells within the adventitial layer which are neither fibroblasts nor Schwann cells and which have irregular nuclei with dense cytoplasm in which prominent granular endoplasmic reticulum and numerous vesicles are seen. These cells are similar to the interstitial cells of Cajal which have been seen in relation to the autonomic innervation of the intestinal wall (Ruska & Ruska, 1961; Thaemert, 1963; Taxi, 1958).

The relationship between cerebral vessels and leptomeninges

A number of studies have shown the close relation between pial vessels – arterioles and vesicles – and the pia-arachnoid membranes (Dahl et al. 1965; Samarasinghe, 1965) and the fine structure of the meninges in the rat has been described in detail (Pease & Shultz, 1958). Similarly there is now ample evidence that, as branches of the pial vessels penetrate the neuropil, they are invested with a leptomeningeal sheath, Fig. 1.4. This sheath forms one boundary of a perivascular space, the meningeal sheath on the neuropil forming the other. The space extends for a variable

1. Anatomy of cerebral blood vessels

distance and is terminated at approximately the arteriolar level by the fusion of the leptomeningeal membranes (Maynard et al. 1957). The perivascular space was thought by earlier workers to extend certainly to capillary level but a review of the evidence by Woollam & Millen (1954) has made a good case for these spaces

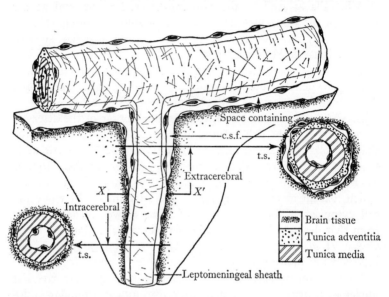

Fig. 1.4. A diagram representing the changes in the relation between a penetrating vessel, the leptomeninx and the neuropil. The perivascular space is in communication with the space containing cerebrospinal fluid and terminates at the level marked XX'. The vessel above this level is defined as extracerebral and below this level as intracerebral. Transverse sections of the extracerebral and intracerebral portions of the vessel are shown. (From Samarasinghe (1965), *J. Anat.* **99**, 815–28, Fig. 8.)

being either artifactual or else confused with astrocyte extensions. The perivascular space, then, appears to be limited to the comparatively large vessels, i.e. $> 100 \mu$m in diameter.

Intracerebral arterioles

Samarasinghe (1965) from whom Fig. 1.4 is taken, has classified the penetrating vessel in extra- and intracerebral portions, the limit being the fusion of the meningeal membranes. This classification is somewhat arbitrary and was developed by the author as a matter of convenience to describe the portions of artery

The fine structure of cerebral vessels

being examined for the presence of nerves, see Chapter 3. A more sensible classification is that proposed by Rhodin (1967) for arterioles in general and which can be applied to intracerebral vessels.

Arterioles have a diameter ranging from 50 to 100 μm and have two or more layers of smooth muscle and a well developed elastic interna, Fig. 1.5. The *terminal arteriole* has a diameter of less

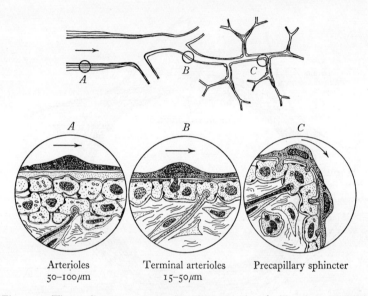

Fig. 1.5. These diagrams summarize the analysis of dilated mammalian arterioles and precapillary sphincters by electron microscopy with particular reference to the location, frequency and general organization of the myo-endothelial junctions. For a description of the arterioles (*A*), terminal arterioles (*B*) and precapillary sphincters (*C*), see text. (From Rhodin (1967), *J. Ultrastruct. Res.* **18**, 181–223, Fig. 37.)

than 50 μm, a single muscle layer, frequent membranous contacts between endothelium and muscle cells and a scanty or absent elastic interna. Rhodin (1967) for the first time demonstrated the presence of precapillary sphincters with electron microscopy (in the thigh muscle) although it had previously been described by Tannenberg (1926) Sandison (1931), Chambers & Zweifach (1944) and others. The *precapillary sphincters* are to be found in the wall of the terminal arteriole when it reaches an inner diameter of about 30 μm and at the point where branches with an average diameter of 10–15 μm are given off at right angles. The branches

1. Anatomy of cerebral blood vessels

are constricted to a diameter of about 7 μm for a distance of 50–100 μm and at the branching point, the smooth muscle cells of the terminal arteriole are arranged in a circular fashion.

It should be noted that precapillary sphincters have not yet been demonstrated in intracerebral arteries. It is of some importance that their presence should be sought in future studies since, in

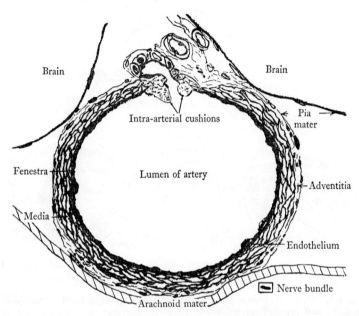

Fig. 1.6. A reconstruction of a transverse section of the basilar artery of the rat which shows the relation of the vessel to the leptomeningeal membranes, the intra-arterial cushions and the disposition of nerve bundles in the adventitia. (From Samarasinghe (1965), *J. Anat.* **99**, 815–28, Fig. 2.)

other vascular beds, they appear to play a crucial role in the regulation of capillary blood flow and hence of vascular resistance.

An analogous structure has been described in pial arteries by Hassler (1962 a, b) and by Samarasinghe (1965), Fig. 1.6. These are found in larger arterioles, i.e. > 100 μm diameter and occur in the wall of the parent artery as a side branch is given off. According to the light microscope studies of Hassler, these 'cushions' consist of aggregations of longitudinally oriented smooth muscle fibres and elastic tissue and are richly supplied by nerves. Their function is obscure but it should be borne in mind that the

The fine structure of cerebral vessels

precise point at which the major resistance changes in the cerebral vascular bed occurs is still in doubt. As will be discussed in later chapters, some evidence, e.g. Mchedlishvili, Ormotsadze, Nikolaishvili & Baramidze (1967) points to the larger arterioles rather than to the precapillary sphincters and terminal arterioles as in other vascular beds. If this is confirmed, then, clearly the internal cushions described by Hassler require more careful consideration.

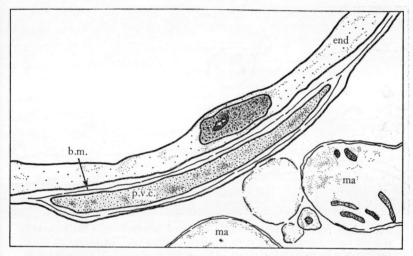

Fig. 1.7. A portion of a transversely sectioned capillary at high magnification. The vessel lumen is to the upper left and astrocytic cytoplasm is to the lower right. The endothelium (end) shares a basement membrane with a second layer of cytoplasm of perivascular cells (p.v.c.). The latter is sheathed by basement membrane. The lamina densa of the basement membrane (b.m.) is shown as are two astrocytic mitochondria (ma). × 36,000. (From Maynard et al. (1957), Am. J. Anat. **100**, 409–34, Fig. 6.)

Intracerebral capillaries

The principal features of intracerebral capillaries are illustrated in Figs. 1.3. and 1.7. The endothelial layer is continuous, there being no evidence of fenestration seen in capillaries elsewhere (Pease, 1955). The endothelial cells bulge into the lumen of the capillary while the endothelium rests upon a basement membrane which it shares with the astrocyte sheath. The basement membrane is 3–5 μm thick and consists of a central dense lamina sandwiched between cement layers which form the means of attachment of cells to the basement membrane. Donahue & Pappas (1961) have

1. Anatomy of cerebral blood vessels

shown that the basement membrane increases in thickness and the endothelial cells become attenuated with age in the rat. Fig. 1.7 illustrates a perivascular cell described by Farquhar & Hartmann (1956, 1957) and by Maynard, Schultz & Pease (1957) which is completely surrounded by basement membrane and which is orientated longitudinally. This last observation together with the nature of its cytoplasm serves to differentiate it from a typical

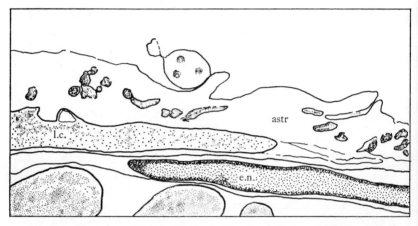

Fig. 1.8. A portion of longitudinally sectioned venule in rat's brain. A well defined astrocytic perivascular sheath (astr) is present. The elongated dense nucleus of a leptomeningeal cell (l.c.) and some of its flattened cytoplasm is to be seen forming a sheath over the venule. An endothelial nucleus (e.n.) extends to the lower right. × 5100. (From Maynard et al. (1957), Am. J. Anat. 100, 409–34, Fig. 11.)

smooth muscle cell. Nevertheless the fact that it is enveloped by a basement membrane suggests that it may be a poorly developed or primitive form of smooth muscle cell.

Intracerebral venules

Maynard et al. (1957) have shown that intracerebral venules may become quite large, i.e. up to 30 μm in diameter before they acquire a muscular coat. In general, the venules are indistinguishable from the smaller capillaries having an endothelial layer resting upon a basement membrane. Perivascular cells similar to those seen in the walls of arterioles have been described and a leptomeningeal sheath extending into the neuropile has also been reported, Fig. 1.8.

The capillary: brain tissue interface

Structures imposed between cerebral vessels and neurones
Perivascular spaces

As mentioned in the previous section, penetrating intracerebral arterioles carry with them layers of arachnoid and pial meningeal membranes and which persist for a variable distance into brain substance. The perivascular space is presumably filled with cerebrospinal fluid (c.s.f.) and this means that the total surface area of blood vessels which are exposed to c.s.f. is greatly increased. It is unlikely that exchange of substances other than gases, oxygen and carbon dioxide takes place through the arteriolar walls but, as will be discussed in Chapter 10, the intimate relation of arterioles (and venules) with c.s.f. could (*a*) enable perivascular c.s.f. pH to be affected by intravascular P_{CO_2} and (*b*) allow cerebrovascular smooth muscle to be affected by changes in c.s.f. pH. In general, the smaller the vessel whose smooth muscle contractility is affected, the greater will be the changes in vascular resistance offered to blood flow: hence the strategic importance of the arrangement whereby c.s.f. bathes the small vessels of the brain.

As a result of numerous studies, it now seems certain that these perivascular spaces are strictly limited to arterioles of $> 50\ \mu$m in diameter. The spaces seen by earlier workers around capillaries, neurones and neuroglial nuclei were almost certainly artifactual. Woollam & Millen (1954) have reviewed the literature exhaustively and Cammermeyer (1960) has given clear technical instructions as to how these artifacts may be avoided during fixation of post-mortem material. It would seem then, that extensions of the subarachnoid space play no part in the exchange processes between cerebral capillaries and neurons and cannot be invoked to account for the resistance to diffusion of certain substances, the so-called blood–brain barrier.

Glial cells

Astrocytes. Particular difficulty has been encountered in identifying astrocytes because of the 'watery' nature of their cytoplasm and the fact that only traces of precipitated protein remain after osmic acid fixation. Features of the cytoplasmic and nuclear arrangement of astrocytes have been well described and illustrated by Schultz, Maynard & Pease (1957), by Maynard *et al.* (1957) and by Cammermeyer (1960). An important feature of the astro-

1. Anatomy of cerebral blood vessels

cytes are their perivascular end-feet which extend to invest the smaller intracerebral vessels, at first interspersed with leptomeningeal cells and at the capillary level, over approximately 84 per cent of the surface area (Maynard et al. 1957). The watery appearance of their cytoplasm is of course itself artifactual as a result of fixation for electron microscopy and other substances, e.g. lipids and carbohydrates are also likely to be affected in preparation of the material. For this reason, it is difficult to assign any function to these cells from their histological appearance beyond the obvious fact that their cytoplasm constitutes a 'space' or compartment which is interposed between capillaries, extracellular fluid and neurones and whose contents are presumably in some form of equilibrium with those of the other three compartments.

Oligodendrocytes. According to Del Rio Hortega (1922, 1928), oligodendrocytes are to be found in three distinctive positions – next to neurons, between myelin sheaths and alongside blood vessels. Their precise function is unknown but their position suggests that they be involved in the metabolism of neurones and myelin sheaths.

An exhaustive study of the ultrastructure of oligodendrocytes by Cammermeyer (1960) showed that their primary relation was with small intracerebral vessels, principally arterioles and capillaries. In gray matter, they appear as clusters of considerable size and as rows of unusual length in white matter. The clusters are particularly numerous in the spinal cord and the perivascular arrangement is best seen in longitudinal sections.

Histiocytes and microglial cells. Microglial cells are scattered widely through the substance of the central nervous system and it is difficult to isolate one type of tissue with which they are in the closest relation. This wide distribution of histiocytes and microglia has led a number of workers to conclude that (a) they are responsible for the elimination of necrotizing tissue and the products of cellular metabolism (Del Rio Hortega, 1930, 1932; Penfield, 1928; Scholz, 1933) and (b) they are involved in the transport of substances for neuronal metabolism (Neissing, 1952) or insoluble material (Lumsden, 1955; Pomerat, 1960). When cells die, the microglia, histiocytes and pericytes or adventitial cells (Schlote, 1959) are mobilized in order to eliminate debris. Their mobilization is a function of blood flow: if flow is arrested, the microglia and histiocytes do not survive, autolysis cannot take place, and tissue is transformed to a coagulative necrosis (Spielmeyer, 1922; Adams, 1958; Cammermeyer, 1953, 1960).

The relation between neurones and blood vessels

The precise way in which blood vessels are related to neurones in the central nervous system is still controversial. Cammermeyer (1960) has pointed out that observations made on single sections can be erroneous and that relationships between tissues can only be defined by partial or complete reconstruction of an area by examining a series of companion sections.

According to one view, the neurone is situated within a capillary meshwork at some distance from a capillary (Ramon y Cajal, 1913; Woollam & Millen, 1954, 1955; Tschirigi, 1958). The distance has been given as between 5 and 15 μm (Friede, 1953). Schiebel & Schiebel (1955) emphasize that there is no spatial orientation of neuronal cell bodies in respect to blood vessels while Dempsey & Luse (1958) hold that separation of the neurone from blood vessels by neuroglia is typical of adult nervous tissue.

A second group of workers have reported that neurones bear a much more intimate relation to blood vessels (Adamkiewicz, 1900; Bielschowsky, 1928; Yamada & Maie, 1954). So close indeed that blood vessels perforate neuronal perikarya. Some studies with the electron microscope have shown that occasionally neurones are in direct contact with blood vessels without imposition of glial tissue (Maynard et al. 1957; Luse & Harris, 1960).

This occasional juxtaposition of neurone and blood vessel is extended to be the rule, according to a third view (Cammermeyer, 1960); that is, with the exception of certain areas, e.g. dentate gyri, and cochlear nuclei, the cytoplasmic membrane of neurones is in contact with blood vessels over wide areas and at these junction points, there are no PAS-stained granules which are usually described as being distributed over the surface of neurones and in between synaptic endings (Palay, 1956). Similar observations have been made in the newborn mouse (Dempsey & Luse, 1958).

These discrepancies in interpretation can no doubt be partly explained by differences in preparation of tissue and in staining methods and partly by the inevitable selection of areas studied. Thus it does not seem unreasonable that all of the views quoted

1. *Anatomy of cerebral blood vessels*

above could be correct, that is that in some areas or even over the length of a single vessel, neuronal cytoplasm is in direct contact with the capillary basement membrane. At other points, glial cells are interposed between the two. Obviously, if one or other of these viewpoints is chosen exclusively, the anatomical basis for the blood–brain barrier could vary considerably. If the juxtaposition of capillary and neurone is treated as the norm, it could explain the exclusive pigmentation of neurones after administration of tellurium (Pentschew, 1958) the incrustation of acutely damaged neurones by blood constituents after damage to the vascular wall following ischaemia (Bakay, Hueter, Ballantine & Sosa, 1956).

Cammermeyer (1960) has drawn attention to the potential importance of oligodendrocytes not only as part of the glial tissue which could form part of a blood–brain barrier and as concerned in the metabolism of neurones but as regulators of capillary blood flow. Their strategic position beside blood vessels and particularly at their junctions suggest that some sort of intrinsic control of vascular lumen is possible. This idea receives some support from two characteristics of oligodendrocytes in tissue culture: their spontaneous rhythmic pulsatile activity (Canti, Bland & Russell, 1937; Lumsden & Pomerat, 1951; Pomerat, 1952) and their responsiveness to serotonin (Benitez, Murray & Woolley, 1955; Lumsden, 1959). The overall function of neuroglia still remains unclear both with respect to support for neurones and transport between brain capillaries and neurones. Certain characteristics of glial cells have now been firmly established and these are considered in the authoritative review by Kuffler & Nicholls (1966). The merit of this review and of the papers quoted in this section is that they tend to dispose of the idea that glial cells are inert and merely spacers. On the contrary, there is evidence that they are concerned with transport of ions and in the context of this book, may well be concerned in the distribution of H^+ and HCO_3^- and in the local regulation of capillary flow. They must also be considered as forming part of the observed barrier between blood and brain but the inhomogeneity of the glial–vascular and neuronal–vascular relationships suggests that this barrier is a complex one.

2 CAPILLARY DENSITY AND OXYGEN TRANSPORT IN THE BRAIN

Krogh's cylinder

One of the most fundamental requirements of the blood supply to the brain or other tissues is that it should meet the oxygen demand of the tissue in terms of quantity (ml 100 g^{-1} min^{-1}) and partial pressure (mmHg) .The cylinder model envisaged by Krogh (1919) assumed that oxygen flows from the capillary to cell or mitochondrion by a process of diffusion and that according to the Fick law of diffusion, the flow of particles (molecules of oxygen) is proportional to their concentration gradient. Krogh's original model was that of a cylinder of tissue of radius R with a central capillary of radius r and Erlang's equation which took account of the rate of metabolism p, the product of the diffusion coefficient for oxygen and solubility of oxygen d was:

$$T_0 - T_R = p/d\,[1 - 15R^2 \log(R/r) - (R^2 - r^2)/4], \tag{1}$$

where T_0 and T_R are the oxygen tensions at the centre of the capillary and R cm away respectively. From this equation, it is possible to predict that if T_0-T_R diminishes as in hypoxia and if p remains constant, then either R must diminish or r must increase. R can diminish if the intercapillary distance shortens which can be accomplished by the opening up of capillaries: r may increase as a result of capillary dilatation. Similarly, from equation (1), it can be predicted that if capillary P_{O_2} (T_0) remains constant and p, the metabolic rate of the cells increases, in order to maintain an adequate level of P_{O_2} at the periphery of the cylinder, R will have to diminish or r increase.

In terms of the vascular system, Krogh's model can be validated if it can be shown that in response to arterial hypoxia or an increase in metabolic rate, the surface area available for gas exchange increases either by an increase in the cross-sectional area of the capillary bed or in the length of capillaries or both. Further, a distinction has to be made with respect to the length of time

2. Capillary density and oxygen transport

for which the stimulus is applied. Thus, as will be shown there is clearer evidence that the response to transient changes in P_{a, O_2} or in metabolic rate involves an increase in r rather than a reduction in R. But for long term changes, e.g. in response to hypoxia at altitude, there is more impressive evidence for changes in intercapillary distance. This evidence will be discussed in the context of the brain.

Capillary density of the brain

Krogh (1919) concluded that 'in most glands and in the central nervous system, the capillary network is even richer than in striated muscles'. However, Cobb & Talbott (1927) showed from a number of sources that resting skeletal muscle is twice as rich in capillaries as grey matter and heart muscle ten times as rich while Sjostrand (1935) found that resting skeletal muscle is twice as rich, liver 2.8 times heart 3.5 times and kidney 4 times. Lorente de No (1927) on the other hand found that the capillary network in the cerebral cortex was richer than in any other tissue in the body. It is possible that some of these discrepancies can be accounted for by differences in technique. Craigie (1920) has catalogued these and emphasized the errors which can be caused by shrinkage and other artifacts.

It has been known for a long time that there are considerable differences in capillary density within the brain. Guyot (1829) and Magendie (1836) commented on the rich vascularity of grey matter and the virtual absence of vessels in white matter. Sterzi (1904) found that the vascularity of grey matter increases in spinal cords of vertebrates from the lowest classes to mammals both absolutely and relatively to white matter and Hoche (1899) confirmed that in the dog, capillaries are two or three times as dense in grey matter as in white.

In the first edition of his book, Kolliker (1856) observed that the diameter of capillaries was nearly inversely proportional to their numbers and added, in a later edition (1896), that the capillaries in white matter are larger than those in grey matter. The first systematic and quantitative study of cerebral capillaries was carried out by Craigie and in a remarkable series of papers (1920, 1921, 1924, 1930), he established not only the relative density in different parts of the brain but also the relation

Capillary density of the brain

between the capillary density and age. He also covered a number of comparative aspects of the subject.

Similar studies were carried out by Cobb & Talbott (1927) and by Dunning & Wolff (1936). A summary of their findings with respect to the brain of the cat is shown in Fig. 2.1 and this

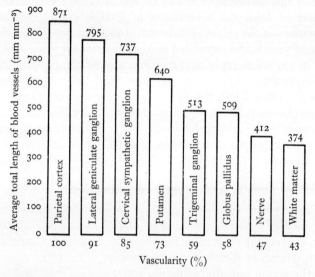

Fig. 2.1. Diagrammatic representation of the relative vascularity of the various parts of the nervous system of the cat. The average total length of blood vessels expressed as mm mm^{-3} is indicated by the columns and the appropriate vascularity is expressed per cent at the foot of the columns. Ten to fifteen fields in seven cats. (From Wolff (1938), *Res. Pub. Ass. Res. nerv. ment. Dis.* **18**, 29–68, Fig. 22.)

emphasizes how greatly capillary density varies between grey and white matter. Similar results have been obtained in man (Lierse, 1961, 1963; Diemer, 1964, 1965a, b). The observation that the capillary density of the superior cervical ganglion is more similar to that of the parietal cortex than to the trigeminal ganglion raises the question as to which component of neural tissue is related to the density of capillaries. The obvious difference between the superior cervical and trigeminal ganglia is the presence of synapses in the former and virtual absence of them in the latter. Further evidence that the presence or absence of synapses may be related to the degree of vascularity is to be found in the observation by

2. Capillary density and oxygen transport

Lorente de No (1927) that the outermost layer of the cerebral cortex is richer in blood vessels than a deeper layer containing many more cell bodies, and he pointed out that the first layer, as demonstrated by Ramon y Cajal (1909), consists of a plexus composed of dendrites of the few cell bodies contained therein and of the terminal arborizations of axons originating in cell bodies lying in the deeper layer of the cortex. A further correlation can be derived from the observation of Craigie (1921) that of the laminations of the cortex (Brodmann, 1909), lamina IV was richest in blood vessels and contained more numerous synaptic structures than in any other layer of the striate area of the occipital cortex (Poljak, 1927).

On the other hand, the data of Dunning & Wolff (1936) show no correlation between capillary density and either the number or mass of cell bodies in laminae I–VI in parietal cortex, cervical sympathetic ganglia or trigeminal ganglia.

Capillary density and oxygen consumption

The reason why capillary density is more closely related to the frequency of synapses rather than of cell bodies is a matter for speculation at the moment. There have been only a few satisfactory measurements of the oxygen consumption of neurones associated with the action potential (Ritchie, 1967) and no comparable data for the oxygen consumption associated with transmission at synapses. If it is assumed that there is a close relation between capillary density and the metabolic rate of the tissues they perfuse, of which oxygen consumption is used as an index, then the correlations outlined above would suggest that the oxygen consumption at synapses was higher than in cell bodies. As will be shown later, it is technically very difficult to measure the oxygen consumption of small areas of tissue *in vivo*. There is evidence from studies *in vitro* that the oxygen consumption of different parts of the brain varies considerably but there is also evidence that such values substantially underestimate values obtained *in vivo*, Table 2.1. The values shown in this table confirm that between grey and white matter, oxygen consumption varies as does capillary density but data for more local correlations do not exist. Lierse (1963) has summarized the position thus: 'Up to now, capillary density remains the only anatomical indicator

Capillary density and oxygen consumption

Table 2.1. *Respiratory rate of the human brain* in vivo *and of preparations from it* in vitro. (Data from Kety & Schmidt (1948a), McIlwain (1953), Korey & Orchen (1959) and sources there quoted)

Method of preparation	Part of brain	Respiration (μmol O_2 g^{-1} h^{-1})
In vivo		
Arterio-venous difference and blood flow	Whole	90
Estimate*	Grey matter	120
	White matter	60
In vitro		
Slice: glucose–phosphate saline, no applied pulses	Grey matter	60
	White matter	30
Slice: same saline, with pulses	Grey matter	120
	White matter	50

* Estimate, regarding the brain as of equal volumes of grey and white matter, the grey of twice the respiratory rate of the white.

of the oxygen consumption of limited regions of the brain.' There have been no recent developments which have altered this position.

Capillary density and age

Rather more complete data on the changes of regional metabolism are available in young animals and during maturity (see Chapter 11) and it is possible to make a more definite correlation between capillary density during this period. In Fig. 2.2, the changes in capillary density and mean intercapillary distance in the frontal cortex are shown from birth until the fourth year and in adult life in man (Diemer, 1968). Allowing for differences in maturity at birth, this relation is very similar to that found by Craigie (1925) in the albino rat and by Diemer (1965b) in the guinea pig. In man, the capillary density at birth is about one third of that in the adult and is even lower in the premature infant. Capillary density increases at approximately the same rate as in prenatal life for the first three months but from the fourth to sixth months, the rate of increase approximately doubles. Thereafter, the rate of increase again falls so that by the end of the first year, capillary density is approximately twice that at birth and by the end of the fourth year it is almost that of the adult. This period of

2. Capillary density and oxygen transport

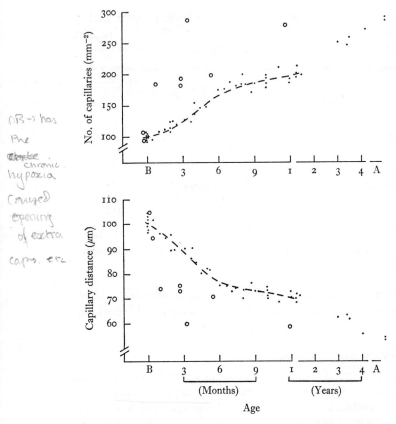

Fig. 2.2. The relation between above, the number of capillaries per square millimetre and below, the mean intercapillary distance (μm) in the frontal cortex and the age of children from birth (B) until the fourth year. Corresponding values for adults (A) are included for comparison. Dots represent values from children dying from acute illnesses: open circles from children dying after prolonged hypoxia. (Redrawn from Diemer (1968), in *Oxygen transport in blood and tissues*. Thieme: Stuttgart, pp. 118–23, Fig. 4.)

intense increase in capillary density occurs between the tenth and twenty-fifth day after birth in the rat and between the 35th and 50th day of gestation in the guinea pig. In the rat, this period coincides with the peak increase in oxygen consumption (Tyler & van Harreveld, 1942), see Fig. 11.2, and corresponds to the 'critical phase' (Flexner, 1952) of cerebral development during which several processes, e.g. the differentiation of nerve cells

Capillary density and age

formation of the neuropile, enzymatic differentiation of the brain and development of cortical function are taking place. These circumstances would in general support the thesis that capillary density and rate of oxygen consumption are related.

Capillary density and hypoxia

The second prediction of the Krogh equation is that if the oxygen pressure in capillary blood falls, there should be either dilatation of capillaries or a reduction in the intercapillary distance and this might be accomplished by an increase in capillary density. There is evidence that both these factors are involved in the vascular response to hypoxia in the brain. The acute vasodilatation is considered in Chapter 9: in the present chapter, evidence which indicates that capillary density increases with hypoxia is discussed.

Morphological studies have confirmed the anticipated increase in capillary density in the retina (Opitz, 1951), in cardiac muscle (Toth, 1965) and in skeletal muscle (Cassin, Gilbert & Johnson, 1966). An alternative approach has been to obtain an index of increased vascularity. Thus an increase of 15–20 per cent in pulmonary diffusing capacity in normal human subjects at altitude has been observed by West (1962) but this could to a large extent be accounted for by the increase in pulmonary blood flow and haematocrit. Tenney & Ou (1970) measured the carbon monoxide uptake from subcutaneous gas pockets in rats and found that after three months at 5600 m, this uptake almost doubled. When corrections had been made for the secondary polycythaemia and certain other factors, the authors deduced that the capillary density had increased by 50 per cent above control at sea level. The effect of this increase in capillary number was to ensure that venous P_{O_2} was only 10 mmHg below sea level values and still well above zero.

In the brain, Opitz & Palme (1944), Merker & Opitz (1949), Opitz (1951) and Diemer (1968) have shown morphologically that an increase in capillary density occurs in rats and rabbits exposed to hypoxia for long periods. The results of Diemer (1968) are representative and are shown in Table 2.2. Further results given by Diemer (1968) are also of interest since they afford comparison between the brain capillary density in children dying after long periods of hypoxia due to the presence of cardiac

2. Capillary density and oxygen transport

Table 2.2. *Capillary density in the parietal cortex of rats under chronic hypoxia* (from Diemer, 1968).* Capillarization and oxygen supply of brain. In *Oxygen transport in blood and tissues* (ed. Lübbers D. W., Luft, U.C., Thews, G. & Witzleb, E. Thieme: Stuttgart)

Control animals (760 torr)		Animals under lack of oxygen	
No. of capillaries (mm^{-2})	Mean capillary distance (μm)	No. of capillaries (mm^{-2})	Mean capillary distance (μm)
644	39.4	1020	31.3
627	39.9	1012	31.4
630	39.8	1020	31.3
626	40.0	968	32.1
630	39.8	—	—

* Not allowing for shrinkage of the histological tissue.

defects and those children who died without significant hypoxia during acute illnesses, Fig. 2.2. This confirms that under hypoxic conditions, the intercapillary distance is less than control certainly during the first year of life.

There is thus some evidence that the vessels of the brain participate in the general increase in vascularity in response to chronic exposure to hypoxia. As in other vascular beds, the mechanism of this response remains unknown. It is possible that the increase in vascularity is caused by a substance akin to erythropoetin which appears to be involved in the production of secondary polycythaemia under similar conditions. The increase in vascularity could be caused by the increase in tissue flow or it could be due to some direct action of prolonged alterations in blood and tissue gas tensions. Whatever mechanism is involved, it is to be presumed that the increased vascularity of the brain operates to maintain a venous P_{O_2} at a higher level than could be achieved by an increase in blood flow and haematocrit alone. It must therefore be accounted an important factor in enabling animals to live at high altitude without compromise to brain function.

Tissue oxygen in the brain

It is generally accepted that oxygen is supplied to tissue from capillaries by a process of diffusion alone. This is an economic form of transport which depends upon the kinetic energy of the gas molecules themselves and although the laws of diffusion were stated rather more than a hundred years ago by Fick (1855) and subsequently applied, it is only recently that the factors which affect diffusion of oxygen in the brain have been considered either theoretically or experimentally.

Theoretical considerations

The tissue oxygen tension in brain depends upon the oxygen tension of capillary blood, tissue oxygen consumption, the diffusion coefficient of oxygen in tissue and the capillary density. The first measurements of cerebral arterial and venous oxygen tensions were made by Opitz & Schneider (1950) and of oxygen consumption of the brain in man by Kety & Schmidt (1948a) and by Schneider (1953). The diffusion coefficient of oxygen in brain was calculated from experimental determinations by Thews (1960a) and found to be 1.6×10^{-5} cm^2 s^{-1} and the oxygen solubility coefficient to be 2.25×10^{-2} ml ml^{-1} atm. In addition to these constants, the estimation of tissue oxygen supply requires a knowledge of the area to be supplied around a single capillary. The simplest model which has been used since Krogh (1919) is the cylinder surrounding the central capillary (see also Hill, 1928; Thews, 1953; Roughton, 1952; Kety, 1957a) and this cylinder is supposed to follow the windings of the capillary. The arrangement of capillaries in the brain, Plate 1, suggests that this model only occasionally describes the possible capillary–tissue relationships and other possible arrangements are considered later. From measurements of the intercapillary distance, values for radius R of Krogh's cylinder in cerebral grey matter of various mammals were estimated by Hortsmann (1960) and these are given in Table 2.3. For the theoretical estimate of tissue oxygen tension in man, a value of 30 μm for R has been used.

A further factor which has to be considered is the level of tissue oxygen uptake. A figure of 3.3 ml O$_2$ 100 g^{-1} min^{-1} has been given for the human brain (Kety & Schmidt, 1948a) but as will be discussed in more detail in Chapter 11, this figure

2. Capillary density and oxygen transport

Table 2.3. *Radius R of Krogh's cylinder in cerebral grey matter of warm blooded animals* (from Horstmann, 1960. In *Structure and function of the cerebral cortex*, ed. Tower, D. B. & Schadé, J. P. Amsterdam: Elsevier; and Lierse, 1961. *Z. Zellforsch.* **54**, 199–206)

	R (μm)
Mouse	22
Rat	25
Guinea-pig	26
Man (gyrus precentralis)	29
Man (visual cortex)	32
Cat	27–31
Horse	36

represents mean cerebral oxygen uptake and there is some evidence which suggests that cortical oxygen uptake is not less than 5 ml 100 g^{-1} min^{-1} and may be as high as 7.0 ml 100 g^{-1} min^{-1} (Gleichman, Ingvar, Lassen, Lübbers, Siesjö & Thews, 1962). Finally it is necessary to define boundary conditions in the capillary. The diffusion rate for oxygen is calculated from the value of oxygen tension at the vessel wall and this can be computed for any point in the capillary from the arterial end (90–100 mmHg) to the venous end (30–37 mmHg) by means of the oxygen dissociation curve for haemoglobin. But these average values may be modified at the capillary wall for two reasons. First, the reaction time of deoxygenation of haemoglobin may form a significant proportion of the time occupied by the process of diffusion (Gibson, 1959). Secondly, it is possible that a significant gradient for oxygen may exist within the capillary. Thus Thews (1960b) has calculated that if axial streaming of red cells occurs in capillaries or other small vessels, P_{O_2} at the capillary wall could be 5 mmHg lower than at the centre of the stream.

Using these data and the Krogh cylinder as a model, modified by Opitz (1948) and illustrated in Fig. 2.3, Thews (1960b) has calculated the oxygen gradients in brain tissue under normal and hypoxic conditions. The gradients which would be expected at the venous end of the capillary are shown in Fig. 2.4. Under normal conditions, P_{O_2} at the periphery of the cylinder is 17 mmHg and this contrasts with the earlier calculations of Opitz & Schneider

Tissue oxygen in the brain

(1950) which led them to believe that 'there was an oxygen surplus everywhere and at any time'.

Noell & Schneider (1944) had demonstrated that certain reactions to reduced oxygen supply were associated with repro-

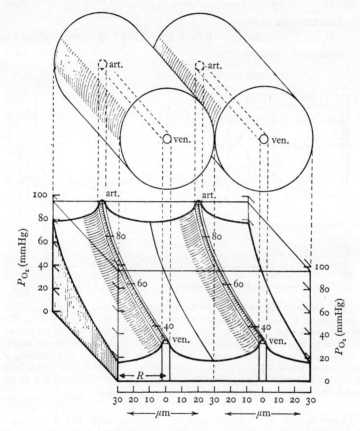

Fig. 2.3. Model of two adjoining cylinders in the grey matter of the human brain (above) with the corresponding relief of oxygen tension values (below). For reasons of clarity, the overlapping of the cylinder which is necessary for complete supply have been omitted. (From Thews (1963), in *Selective vulnerability of the brain in hypoxaemia*. Blackwell: Oxford, pp. 27–35, Fig. 3.)

ducible values of oxygen tension in venous blood and this has since been shown to be a valuable correlation since hypoxaemia will first show its effects at the venous end of the capillary and then at the periphery of the cylinder. Three characteristic threshold-

2. Capillary density and oxygen transport

values of venous oxygen tension have been defined (control venous P_{O_2} = 34–38 mmHg).

(a) P_{v,O_2} = 25–28 mmHg: 'reaction threshold' at which brain capillaries start to dilate in response to hypoxaemia.

(b) P_{v,O_2} = 17–19 mmHg: 'critical threshold' at which humans lose consciousness.

(c) P_{v,O_2} = 12 mmHg: 'lethal threshold' at which there is immediate danger to life.

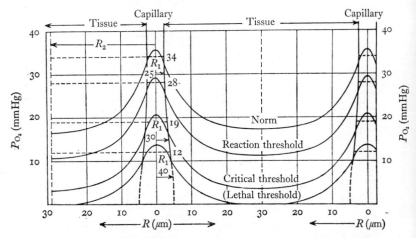

Fig. 2.4. Values of oxygen tension at the venous ends of two adjoining cylinders in grey matter of human brain at the thresholds of venous oxygen tension. The fact was allowed for that the capillaries become larger with decreasing oxygen tension in venous blood. The oxygen tension in the region of the lowest supply is approximately 17 mmHg: under conditions of hypoxia it is 11 mmHg at the 'reaction threshold', 4 mmHg at the 'critical threshold' and 0 mmHg at the 'lethal threshold'. (From Thews (1963), in *Selective vulnerability of the brain in hypoxaemia*. Blackwell: Oxford, pp. 27–35, Fig. 4.)

These levels of venous oxygen tension and the associated calculated levels of P_{O_2} at the periphery of the cylinder are of further interest because at the 'reaction threshold' the tissue P_{O_2} of 17 mmHg is that at which the saturation of cytochrome oxidase just becomes incomplete and the ensuing dilatation of capillaries ensures an improvement of oxygen supply. At P_{v,O_2} of 17–19 mmHg, the lowest level of P_{O_2} in tissue is 4 mmHg and this corresponds to a reduction of cytochrome oxidase to 83 per cent. At the lethal threshold, the lowest level of P_{O_2} in tissue is 0 mmHg.

These calculations of Thews may well have to be modified in

Tissue oxygen in the brain

view of a number of theoretical and experimentally determined factors which have since been raised. In the first place, the simple cylinder model proposed by Opitz & Schneider (1950) and used by Thews assumes parallel flow in the same direction in capillaries and a diffusion coefficient, solubility coefficient and tissue oxygen uptake which are homogeneously distributed along the tissue. Although the model has the merit of simplicity, it cannot account for the presence of diffusion shunts which have been described by Lübbers (1968a). An alternative model has been proposed by Diemer (1965a, b) in which flow was still parallel but in which capillaries were perfused in opposite directions. Since this model allows for some adjacent capillaries being perfused in the same direction, a true counter-current system is not envisaged. Diemer's model implies a cone rather than a cylinder of supply and although a supply along longer capillary distances is possible, certain parts of the tissue would be supplied to excess. In contrast to the cylinder model, the minimal P_{O_2} would be at a point in the tissue at mid- rather than at end-capillary.

However, neither of these models take into account the clearly established fact – namely that capillaries are asymmetrically distributed. An attempt has been made to construct a model on such a basis by Grunewald (1968) but which assumes equal distribution of oxygen consumption, solubility and diffusion coefficients along the tissue. Although this model has many of the same restrictions as those of Thews and Diemer, it allows computation of the likely range of oxygen consumption of the cortex, 6 to 8 ml 100 g^{-1} min^{-1} unless, with hypoxia, there is a reduction in oxygen consumption. It is to be hoped that this kind of approach will lead in due course to the unravelling of the system of capillary arrangement and blood flow which ensures adequate oxygen supply with the maximum economy.

Experimental considerations

It has been possible to test some of Thews' calculations of mean cortical P_{O_2} by making surface measurements with modified Clarke electrodes (Gleichmann, Ingvar, Lübbers, Siesjö & Thews, 1962). The closeness of the correlation between calculated and experimentally determined values is remarkable. It should be borne in mind however that the values for surface P_{O_2} are always likely to underestimate the true P_{O_2} since it is technically very

2. Capillary density and oxygen transport

difficult to measure P_{O_2} with relatively large cathodes (100 μm) incorporated into flat discs without interfering with the local microcirculation. Certainly, a wide variety of measured values for cortical P_{O_2} have been obtained (Davies & Brink, 1942; Davies & Bronk, 1957; Ingvar, Lübbers & Siesjö, 1960) and these almost certainly can be explained by differences in electrode size and techniques of applying them to the surface of the brain.

But the measurement of mean tissue P_{O_2} in the cortex is only one step away from measuring mean tissue P_{O_2} of the brain in jugular venous blood. Such a method gives little idea of the oxygen tension gradients involved and of the way in which these are affected by manoeuvres which affect blood flow or tissue oxygen uptake. Ideally, the electrodes should measure P_{O_2} of an extremely small volume of tissue and at the same time should neither consume much oxygen nor damage tissue or its blood supply. The greater number of electrodes which could be placed locally, the more completely would the pattern of P_{O_2} distribution in tissue be known.

Early attempts to measure P_{O_2} on the cortical surface using small noble metal electrodes have been made by Cooper, Crow, Walter & Winter (1960), Cooper, Moskalenko & Walter (1964), Meyer & Denny-Brown (1955), Meyer & Gotoh (1961), Meyer, Gotoh & Tazaki (1961) and Sonnenschein, Stein & Perot (1953). The usefulness of these studies was limited by the fact that the electrodes were not calibrated and the changes in P_{O_2} were therefore qualitative. In view of this, it is surprising that the convention of describing the current output of these electrodes as 'available oxygen' (aP_{O_2}) was adopted since such measurements would give only the most indirect index of the state of oxygen supply to the tissue in question.

A very considerable advance towards the ideal was the introduction by Silver of the oxygen microelectrode – at first, an Araldite-covered, gold plated stainless steel electrode and later, the glass covered platinum electrode with tips of between 0.5 and 5 μm (Cater & Silver, 1961; Cross & Silver, 1962). These electrodes have been shown to be stable over many hours, they can be easily calibrated and so give quantitative measurements of tissue P_{O_2}, their oxygen consumption is small and there is evidence that they can be thrust into brain or other tissue without significant disturbance of the microcirculation. This type of electrode has been

Tissue oxygen in the brain

used in two different types of experiments. In the first, the oxygen tension at one point in tissue has been measured under control and again under a variety of experimental conditions. This approach has been extended in a second type of experiment in which the distribution of oxygen tension in tissue has been determined either by moving the single electrode or by the use of multiple electrodes. An example of the changes in tissue P_{O_2} which can be sensed along a single track in brain is shown in Fig. 2.5. It will be seen that as the electrode tip passed over two capillaries, the P_{O_2} was high, while that at the furthest point from the capillaries

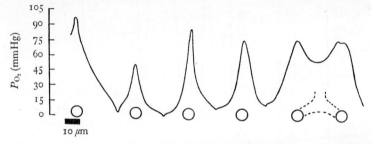

Fig. 2.5. Record of oxygen tension obtained by moving an oxygen electrode across the surface of the cerebral cortex of a rat. Circles indicate the positions of capillary blood vessels in the pia-arachnoid. On the right of the record, the electrode passed over the junction of two vessels. Note that the lowest P_{O_2} is not always midway between vessels. (From Silver (1965), *Med. Electron. Biol. Engng.* **3**, 377–87, Fig. 4.)

was near to zero. The pattern is similar to that calculated by Thews (1963). It will be noted however that, under control conditions, oxygen tension fell to lower levels than predicted by Thews and this is confirmed in Fig. 2.6 from Lübbers which in addition shows that the gradients for oxygen tension are large in brain as is shown by the large changes in P_{O_2} which occur when the microelectrode is moved by small amounts. The irregularity of Lübbers' trace can of course be accounted for by the arbitrary nature of the electrode track with respect to the brain capillaries: but the fact that relatively low values for tissue P_{O_2} were obtained suggests that the values for cortical oxygen consumption is in fact higher than that used by Thews, i.e. 5 ml 100 g^{-1} min^{-1} or the intercapillary distance is greater or both.

The fact that rather low values for P_{O_2} have been found in brain tissue also raises the question as to the value for critical

2. Capillary density and oxygen transport

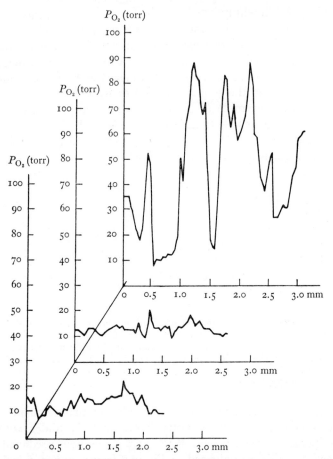

Fig. 2.6. Oxygen pressure fields in the guinea-pig brain. With a membrane covered needle electrode (diameter of tip about 2–5 μm), the P_{O_2} of brain tissue has been measured by pushing the electrode perpendicularly into brain. Every 40 μm, a reading was made. Abscissa, depth of needle electrode (mm): ordinate, local P_{O_2} (torr). (From Lübbers (1968a), in *Oxygen transport in blood and tissues*. Thieme: Stuttgart, pp. 124–39, Fig. 4.)

oxygen tension for brain mitochondria respiring at their maximum rate. In brain, there is evidence that there is a higher concentration of respiratory enzymes than is necessary for maximum respiration (Wodick, Schwickardi & Lübbers, 1966). The distribution of respiratory enzymes in brain has been determined by a number of workers (Tolani & Talwar, 1963; Hess & Pope, 1960; Elliot &

Table 2.4. *Content of respiratory enzymes of the brain* (from Wodick, Schwickardi & Lübbers, 1966. *Pflug. Archiv. ges. Physiol.* **291**, R 25)

		Guinea-pig nmol g^{-1} (fresh brain)		Man nmol g^{-1} (fresh brain)
Total brain	cyt. *a*	12.6	$F^* = a/c = 1.34$	—
	cyt. *b*	17.5		—
	cyt. *c*	14.7		—
Brain cortex	cyt. *a*	14.4	$F = 1.29$	22.2
	cyt. *b*	21.6		—
	cyt. *c*	19.5		28.8
White matter	cyt. *a*	11.3	$F = 1.25$	17.5
	cyt. *b*	15.6		—
	cyt. *c*	14.2		21.7
Thalamus	cyt. *a*	9.75	$F = 1.26$	—
	cyt. *b*	12.6		—
	cyt. *c*	12.3		—

* F = ratio of cytochrome *a* to cytochrome *c* in the guinea-pig.

Heller, 1957). Representative data are given in Table 2.4 and this shows that there is a rough correlation between the concentration of enzymes and the known level of blood flow and capillary density and presumed levels of oxygen uptake. Further confirmation of this correlation has been given by the observation that as with capillary density, the highest levels of enzymes are to be found in layers II, III & IV of the human cerebral cortex (Hess & Pope, 1960; Elliot & Heller, 1957).

From these concentrations of enzymes it has been possible to compute the effective mitochondrial P_{O_2} under normal conditions and Lübbers (1968*a*) has given values of 1–2 mmHg. This level has been found experimentally in brain tissue and in this respect therefore, Thews' calculations could be modified.

The use of oxygen microelectrodes has also confirmed that P_{O_2} varies considerably in different parts of the brain. Thus Cross & Silver (1962) found that the mean tissue P_{O_2} for all parts of the brain studied was 12.5 mmHg and ranged between 3.8 in the hypothalamus to 25.4 in the cortex and 45.3 in tissue immediately adjoining the IIIrd ventricle. There was close resting level of

2. Capillary density and oxygen transport

P_{O_2} and both the height to which and the rate at which it rose when the animal was given a high level of oxygen to inhale. In dead tissue, the P_{O_2} was uniformly low. Tissue P_{O_2} was raised by the inhalation of carbon dioxide and by the administration of adrenaline though in the latter case, interpretation is difficult because of the accompanying rise in arterial pressure, Fig. 10.5. Hypotensive agents such as amyl nitrite or haemorrhage caused a fall in tissue P_{O_2} as did stimulation of the cervical sympathetic nerves or hypothalamus.

This study raises a number of points of interest. In the first place, it emphasizes that even when the electrode is calibrated isolated measurements of tissue oxygen tension are of limited physiological value unless (a) the relation of the electrode tip to a capillary is known or (b) the change in P_{O_2} at that point is measured in response to some particular manoeuvre. As mentioned previously, some workers have used such isolated measurements as an index of oxygen supply to the tissue in question. Presumably, at any point the oxygen tension will be determined by the rate of blood flow and oxygen consumption but it is by no means certain which of these two factors is more important. Thus the data of Cross & Silver (1962) suggest a close relation between the level of P_{O_2} in tissue, the level of capillary density and the level of blood flow obtained by autoradiographic techniques (Sokoloff, 1961). It would be expected that, if other factors remain constant the higher the level of oxygen consumption of tissue, the lower would be the local P_{O_2}. In fact, the opposite is the case for the highest P_{O_2} is to be found in the cortex, the lowest in white matter of the thalamus and in dead tissue where oxygen consumption is presumably zero, P_{O_2} is also zero. The closest relation therefore appears to be between tissue P_{O_2} and blood flow and evidence which supports this is the observation by Cross & Silver (1962) that the inhalation of carbon dioxide causes a rise in P_{O_2}. As far as is known, this does not affect brain oxygen consumption but it does increase blood flow and by causing vasodilatation, increases the surface area available for the exchange of gases. When arterial pressure is raised or lowered, it is probable that tissue P_{O_2} follows changes in the distribution of blood flow as additional capillaries open and close. A similar explanation would account for the fall in P_{O_2} which occurs when the cervical sympathetic nerves are stimulated.

Tissue oxygen in the brain

This last observation is of particular interest since it provides some evidence that the sympathetic nerves can affect cerebral blood flow. It is also possible that, as will be discussed in more detail in Chapter 11, the effect of an increase in sympathetic activity by causing a redistribution of blood flow may also affect the rate of tissue respiration.

All these points indicate how important cerebral blood flow is in determining the supply of oxygen to brain tissue and also, possibly, in determining the rate of tissue respiration. For this reason, it is important to consider the factors which affect the volume of blood flow and the mechanisms whereby the supply of oxygen and other nutrients is maintained under adverse conditions, e.g. hypotension and hypoxia. These factors are considered in detail in the chapters which follow.

3 THE INNERVATION OF CEREBRAL BLOOD VESSELS

Although nerves were seen to accompany cerebral vessels in the seventeenth century, their presence was only accepted, and that grudgingly, at the beginning of the present century. This was due in part to the lack of good staining techniques and in part to the difficulty of reconciling independent vasomotor activity with the so-called Monro–Kellie doctrine (for further discussion, see Chapter 4). Subsequently, the cerebral vessels have been shown to be richly supplied with nerves using light microscope, electron microscope and histochemical techniques and, as will be shown in the present chapter, these nerves terminate on vascular smooth muscle in the same way as in other vascular beds. There is now compelling evidence that many of these nerve fibres are adrenergic and derive from the superior cervical ganglion. They most probably have a vasoconstrictor action. Other fibres have non-adrenergic type terminals upon smooth muscle. There is some evidence that they are cholinergic and these may well correspond to the vasodilator fibres which are carried in the VIIth cranial nerve. The least convincing evidence is for the existence of sensory fibres from cerebral blood vessels.

The basis for controversy has now shifted and the question now posed is: what physiological function, if any, do these nerves have? This question is discussed in Chapter 10. To demonstrate that nerves accompany blood vessels is one thing; to show that they are vasomotor and exert their effect by releasing transmitter substances is another. The morphological evidence is considered in the present chapter and the pharmacological aspects in Chapter 12.

Innervation of cerebral arteries
Early studies to determine the course and origin of vasomotor nerves to cerebral vessels

Willis (1664) described nerve fibres on the anterior and posterior cerebral arteries and showed that they were branches of the 'intercostal' nerve – a name which both he and Reid (1616) had given to the prevertebral plexuses. Although Willis erred in thinking that the intercostal nerve originated in the brain, his description was otherwise accurate and his findings though the subject of controversy in the nineteenth century have now been abundantly confirmed. Further studies by Benedikt (1874) and Aronson (1890) confirmed that these fibres accompanying blood vessels of the brain originated in sympathetic plexuses around the major vessels at the base of the brain and that the vessels of the choroid plexus were similarly innervated while Köllicker (1893) showed that these fibres could be seen on pial vessels down to a diameter of 90 μm or less.

In a paper published jointly with Bayliss & Hill (1895), Gulland reported that he had been unable to find any nerves supplying pial vessels and this afforded morphological confirmation for the contention of Bayliss and Hill that cerebral vessels were without vasomotor reflex control. In 1898, however, Gulland published a further paper in which he reported that as a result of using a new staining method, he had seen numerous nerves on cerebral vessels and their distribution conformed to that described by previous workers. Further studies by Bochenek (1899) and by Hüber (1899) confirmed the sympathetic origin of many of these nerves, although Hüber also observed myelinated fibres which reached the pial vessels from bundles which joined the middle cerebral artery at its junction with the posterior communicating artery and secondly, from nerves which accompanied the vertebral and basilar arteries. The unmyelinated fibres were supplied from plexuses on the vertebral and internal carotid arteries. Both types of fibres could be followed along pial vessels and their branches and just before their termination, the unmyelinated fibres could be seen as delicate fibrils lying parallel to the long axis of the muscle fibres. The myelinated fibres branched repeatedly, the final ramifications being unmyelinated and varicose and were therefore indistinguishable from the terminations of the unmye-

3. The innervation of cerebral blood vessels

linated fibres. The myelinated fibres however ended clearly in the adventitia and the course of the fine fibrils was parallel to the long axis of the artery rather than to the muscle fibres.

Hüber was unable to find nerve fibres which followed the intracerebral arteries which penetrated brain tissue. The first reports of the finding of such nerves were by Robertson (1899) and by Rohnstein (1900). Their illustrations are extremely difficult to interpret and in retrospect could equally well be of artifacts, a possibility which the authors admit. Hunter (1900) stained a plexus of nerves on vessels in the cerebral cortex but he was not certain whether or not they were myelinated. He could not trace these fibres into the white matter of the cerebrum but he stated that 'in the cerebellum, midbrain, pons and medulla and spinal cord, similar examples of this plexus are found'.

Stohr (1922 a, b) using Schultze's modification of Cajal's reduced silver chloride method made an extensive study of the innervation of vascular smooth muscle in many organs of the body and these included what was up till that time the clearest demonstration of the innervation of pial vessels. He was not able to confirm that nerves followed the intracerebral arteries and in a later publication (Stohr, 1928) he summarizes his own work and the position at that time:

Es wäre denkbar, dasz von der Pia her feine Nerven gleichzeitig mit den Gefäszen ein Stück weit in die Gehirnsubstanz hineinzögen, eine Angabe, die sich schon bei Köllicker (1896) vorfindet. Ich habe mich hiervon nur in ganz vereinzelten Fällen überzeugen können und halte die Gefäsze der Substanz des Zentral nervensystems für nervenlos, ein allerdings negatives Ergebnis, zu dem auch Berger (1924) und audere Autoren gelangt sind.

Hassin (1929) also used Schultze's staining technique and largely confirmed Stohr's findings both with respect to pial artery innervation and the lack of nerves to intracerebral vessels. He also commented that the arrangement of perivascular neuroglial pedicles made it unlikely that such nerves existed. In the same year, the results of two series of experiments were published in which there was clear evidence for the existence of nerves supplying the intracerebral arteries. Clarke (1929) used the pyridin–silver method of Ranson and impregnated the brainstem and portion of spinal cord of cats and dogs. Serial sections were prepared and separate examination was made of pia mater and medullary vessels. Clarke found the same general arrangement of myelinated and unmyelinated fibres on the pial vessels as Hüber (1899) had done but he was also able to show that both types of nerves branched to give fibres which accompanied the intracerebral arteries for the greater part of their course. The unmyelinated fibres were more commonly seen and they terminated with varicose endings near the media. The myelinated fibres usually lost their myelin before they terminated and various types of endings were observed, either as a thin fibril or as a plexus of fibres with fusiform swellings or, in one case, as an end bulb. These types of endings had been seen by Stohr (1922a) in the pia mater. Virtually identical findings were made by Kurusu & Hamada (1929) using the same technique in the cortex of dogs and monkeys.

At about this time, Chorobski & Penfield (1932) and Cobb & Finesinger (1932) were investigating vasodilator pathways to cerebral vessels carried by various cranial nerves, in particular, the VIIth facial nerve. In view of these additional pathways and the discrepant results obtained by Stohr (1922a) and Clarke (1929), Penfield (1932) reinvestigated the problem of innervation of the intracerebral arteries. He records that he studied some of Stohr's sections and the alternative possibility occurred to him that as the nerves branched to join the intracerebral arteries, their staining character altered. He therefore first confirmed Stohr's findings using Schultze's technique and then evolved a method based on the Gros–Bielshowsky silver technique. With this method, he was able to show in the cortex of dog, man, monkey and cat, that there were indeed nerves accompanying the intracerebral arteries and as Clarke (1929) had found using a different staining technique, these nerves were continuous with those observed on the pial vessels. The arteries most commonly studied were between 10 and 60 μm in diameter and the nerve fibres either formed a net or else ran as long threads along the arterial walls. In the smaller arteries, the fibres were found between adventitia and media. As Hüber had found, these nerves were both myelinated and unmyelinated and varied between 0.5 and 2.5 μm in diameter. Penfield also confirmed by degeneration that only a proportion of these fibres originated in the cervical sympathetic nerves.

A further and equally convincing study was undertaken by Humphreys (1939) on pial and intracerebral arterial material removed at autopsy in man. As Penfield had done, Humphreys teased the vessels away from the cortex and dissected off the

3. *The innervation of cerebral blood vessels*

perivascular sheath and initially used the Gros–Bielschowsky staining technique. More consistent results were obtained by impregnation with strong protein silver activated by copper. Humphreys also showed that when the arteries were fixed with formaldehyde–acetic acid, there was little change in the arterial diameter. He was therefore able to make a roughly quantitative estimate of the number of fibres which were related to arteries of diameter 10–250 μm. He was also able to confirm previous findings that the nerves accompanying intracerebral arteries were to be found in two principal layers, in the adventitia and on the media, and he gives one example of the way in which a fibre branches in a horse-tail fashion as it disappears into the media.

In Plate 4 an example, from Penfield (1932), of such a nerve on an intracerebral artery, 10 μm in diameter is given and the summary of Humphreys' findings with respect to the frequency of fibres found with pial and intracerebral vessels is shown in Fig. 3.1 (*b*) together with the cortical areas studied, Fig. 3.1 (*a*). This shows that as the diameter of the intracerebral artery diminishes, so the frequency falls. Neither Humphreys nor Penfield were able to find nerves supplying cerebral capillaries or veins.

A survey of the literature has not revealed any further studies on the innervation of pial vessels using the silver staining methods cited above and light microscopy. As in other fields, the earlier controversies and difficulties were largely overcome by the development of better techniques, specifically in the preparation, fixing and staining of the vessels. The later studies, in particular those of Clarke, Penfield and Humphreys show clearly the presence of nerves accompanying the intracerebral nerves and that these nerves originate from and share most of the characteristics found in the myelinated and unmyelinated fibres of the pia mater and its vessels. These studies however by themselves give little information about the types of ending which may be found – in particular the characteristic types of neuromuscular junctions and specific types of terminal which may be associated with afferent nerves. The question as to whether some or all of the myelinated fibres are afferent has not been satisfactorily solved: some evidence which suggests that the pial arteries do have sensory endings close by is considered in a later section.

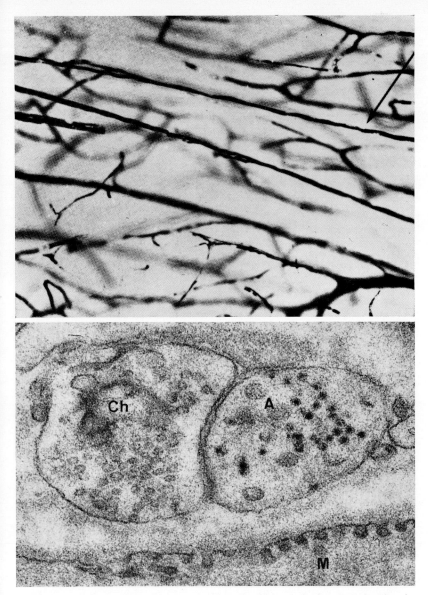

PLATE 1. Photomicrograph of a section of rabbit cerebral cortex showing capillaries filled with Indian ink. Two capillaries (in focus) lie almost parallel: in the background (blurred) is the more typical arrangement of vessels. The arrow indicates the electrode track. Section 25 μm thick. ×174. (From Silver (1965), *Med. Electronic. Biol. Engng* **3**, 377–87, Fig. 3.)

PLATE 2. Electron micrograph of the anterior cerebral artery of the cat, $KMnO_4$-fixation. A nerve terminal, provisionally characterized as cholinergic (Ch) with empty synaptic vesicles and an adrenergic (A) nerve terminal containing dense-cored synaptic vesicles both approach a smooth muscle cell (M) of the arterial media at a distance of 110 nm which is the distance characterizing the neuro-muscular contacts in arterial walls in other peripheral vascular beds in the body. ×72000. (From Nielsen, Owman & Sporrong (1971), *Brain Res.* **27**, 25–32, Fig. 3.)

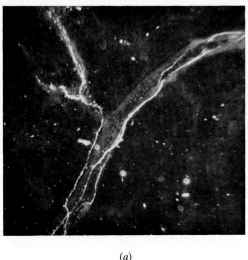

(a)

PLATE 3. (a) Whole-mount of pial membrane from a cat, taken from the cerebral convexities. The adrenergic nerves surrounding or accompanying the pial arteries can be followed as far as arteries having a diameter down to approximately 15 μm, which is the diameter of the smallest vessel branchings in the illustration being accompanied by adrenergic nerve terminals. × 170.

(b) Fluorescence photomicrograph of cerebral cortex from a cat. The pial membrane with some pial vessels are seen at top. One of the pial vessels is supplied with green-fluorescent adrenergic nerves. This pial vessel issues an arterial branch which radiates into the brain parenchyma. This penetrating brain vessel is seen to be surrounded by a small number of varicosed nerve terminals forming a wide-meshed plexus in the adventitia of the vessel. × 212.

(c) Whole-mount of pial membrane from a cat taken at the cerebral convexities. Adrenergic nerve fibres are seen to follow a number of small arteries in the pial membrane. Autofluorescent granular material is seen in the pial membrane. × 133. (Unpublished records: K. C. Nielsen and Ch. Owman.)

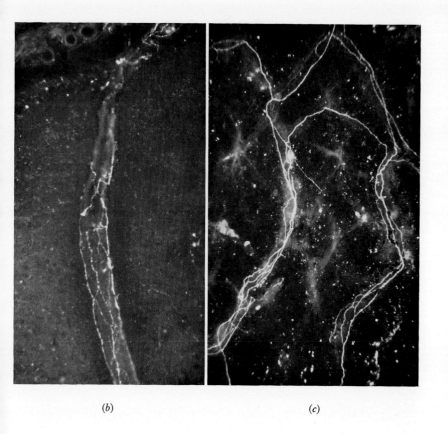

(b)　　　　　　　　　(c)

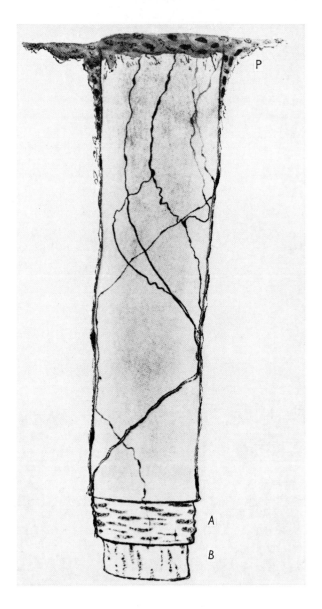

PLATE 4. Drawing of an artery 10 by 70 μm entering the basal ganglion of a monkey. The nerve fibres are drawn as they appeared on all sides of the vessel and therefore appear more numerous than in individual photomicrographs. The nerves are situated directly on the outer surface of the muscular layer and the two layers of muscle are shown (*A* and *B*). Cells are seen coming down from the pia, P, about the vessel but the perivascular sheath is not indicated. (From Penfield (1932). *Archs Neurol. Psychiat., Chicago* **27**, 30–44, Fig. 5. Copyright 1932, American Medical Association.)

Innervation of cerebral arteries

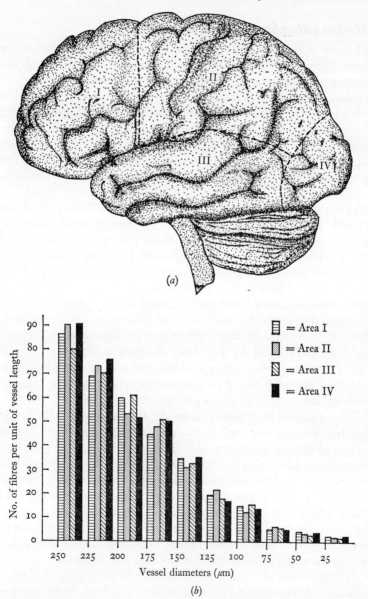

Fig. 3.1. (a) Areas of cortex from which vessels were removed for fibre counts. (b) Graph illustrating mean fibre counts from various portions of the brain. (From Humphreys (1939), *Archs Neurol. Psychiat.*, *Chicago* **41**, 1217–21, Figs. 9, 10. Copyright 1939, American Medical Association.)

3. The innervation of cerebral blood vessels

Studies using electron microscopy

Electron microscopy has confirmed the results of studies using light microscopy and has extended these by demonstrating the pattern of innervation of vascular smooth muscle. From further studies, it has been possible to compare this type of innervation with that seen in other vascular beds (Devine & Simpson, 1967; for pancreatic vessels, Lever, Spriggs & Graham, 1968; for the aorta, Pease & Paule, 1960 and for the pulmonary vessels, Verity, Bevan & Oström, 1966), and to determine how and how fast degeneration takes place following section of the sympathetic trunks.

Nerves which accompany cerebral vessels. Pease & Molinari (1960) described nerves which accompanied cerebral vessels in the cat and monkey but because these nerves were confined to the adventitia, they were regarded as being of little, if any, functional significance. A more extensive study was carried out by Dahl & Nelson (1964) on human post-mortem material and numerous myelinated and unmyelinated fibres were observed on the superior cerebellar and posterior inferior cerebellar arteries. Essentially similar findings were made by Samarasinghe (1965), by Sato (1966) and by Iwayama, Furness & Burnstock (1970) and Iwayama (1970); the typical arrangement of these nerves is identical to that reported on comparable vessels in other vascular beds (Gasser, 1958; Elfvin, 1958; Richardson, 1960).

These studies have been principally directed towards the larger cerebral arteries but both Samarasinghe (1965) and Sato (1966) observed that the frequency of both myelinated and unmyelinated fibres diminished with the calibre of the vessel. Nerve fibres could be distinguished on pial vessels having a diameter of 15–20 μm but neither worker was able to show that the smaller vessels, i.e. those which were without smooth muscle cells, were accompanied by nerve fibres. All the studies have confirmed that as the fibres terminate on or near smooth muscle cells, the axons are denuded of Schwann cell and basement membrane investments.

The question as to whether the intracerebral arteries are innervated has been studied by Samarasinghe and Sato. Samarasinghe proposed that the course of the intracerebral arteries should be divided into two portions – extra- or intracerebral, depending on whether there was a perivascular space and adventitial covering

Fig. 1.4. However, in neither of these portions was he able to demonstrate the presence of nerve fibres accompanying the vessels (twelve arteries in five rats). Sato, however, confirmed the occasional presence of fibres accompanying intracerebral arterioles. There thus appears to be a discrepancy between the results obtained by light microscopy and, as will be discussed in the next section, by fluorescent histochemistry and those obtained by electron microscopy. It is most probable that this discrepancy merely reflects the relatively few studies carried out with electron microscopy to determine whether intracerebral arteries are accompanied by nerve fibres. It is also probable that in the two studies so far carried out, the method used of studying the vessels in transverse section may have made the task of identifying fibres more difficult, for Kajikawa (1969) has demonstrated how easy it is to miss the fluorescence in such nerves because of their spiral relation to the arteries. In consequence, it is easy to gain the impression that the vessels are either not innervated or that they are only poorly supplied by nerves.

Other workers, also using the fluorescent technique for identifying nerve fibres on these vessels, have emphasized that a more satisfactory approach is to examine whole mounts of small vessels and if necessary to reconstruct from serial sections (Nielsen & Owman, 1967); they also draw attention to the fact that the frequency of innervation of cerebral vessels and their branches may vary considerably. For all these reasons, therefore, the question as to whether intracerebral arteries are innervated requires further attention particularly in studies with electron microscopy.

The neuro-muscular junctions. The use of electron microscopy has greatly clarified the position with respect to the neuromuscular junctions on cerebral vessels. Some preliminary observations by Dahl & Nelson (1964) have shown that the axon terminals are to be found at variable distances from the smooth muscle cells – the range given in this and in a later paper (Nelson & Rennels, 1970) being 78 to 300 nm. A similar range of values has been found by Nielson, Owman & Spurrong (1971) and by Iwayama, Furness & Burnstock (1970) and Iwayama (1970). All these workers have confirmed that as the axons terminate, the Schwann cell plasmodium is partially or completely absent. They also describe the presence of a moderately electron-dense material in

3. The innervation of cerebral blood vessels

the neuromuscular cleft without obvious specialization of the axonal membrane facing the cleft.

The absence of a Schwann cell sheath gives rise to a varicose appearance of the terminal itself and all the authors quoted above have been able to distinguish two types of axonal terminal on the basis of the type of vesicles contained. The commonest size of these vesicles is 50 nm but a range of 30 to 100 nm have been described. In any given terminal, the vesicles are predominantly granular or non-granular showing as electron dense or translucent cores, Plate 2. In view of the accumulating evidence from similar studies in other vascular beds that the granules represent the storage site for noradrenaline (Parker, 1958; Lever, Graham, Irvine & Chick, 1965; Grillo, 1966; Devine & Simpson, 1968), axons containing the granular type of vesicle have been designated as adrenergic.

Rather more direct evidence that these axons are adrenergic is provided from a study by Iwayama (1970) who was able to show that degeneration of only the axons with granular vesicles occurred after removal of the superior cervical ganglion; the axons with non-granular vesicles, myelinated fibres and a further group of fibres without synaptic vesicles but which terminated as close as 1 μm to smooth muscle cells were unaffected.

The position with respect to the axons containing predominantly non-granular vesicles is more controversial. Whittaker (1965) has observed such vesicles in synaptosomes while Verity, Bevan & Ostrom (1966) have observed them in nerve terminals on the pulmonary artery and in both cases, there has been some collateral evidence to suggest that the nerves were cholinergic. However, before such nerves can be accepted as being cholinergic, it is essential that (1) there should be additional evidence, e.g. the demonstration of choline acetylase or cholinesterase which would suggest that acetyl choline was being synthesized and hydrolysed and that (2) the possibility of artifacts be borne in mind. Thus it has been shown that granules can be depleted after treatment with osmic acid (Richardson, 1960; Hökfelt, 1968; Roth & Richardson, 1969) and, further, if the sections have been inadequately fixed or if the photographs are slightly out of focus, the vesicles will appear as non-granular. In this connection, the method of permanganate fixation used by Iwayama, Furness & Burnstock (1970) appears to have been the most satisfactory since these workers

Innervation of cerebral arteries

had no difficulty in distinguishing between the two populations of vesicles having electron-dense and lucent cores.

The finding that some fibres accompanying small pial vessels contained cholinesterase has been reported by Lavrentieva, Mchedlishvili & Plechkova (1968) using a modification of Koelle's technique. By itself this finding cannot be taken as proof of the existence of cholinergic fibres though it is suggestive. If further studies show that they are cholinergic, then it would be important to see whether they corresponded to the dilator group of fibres carried by the VIIth cranial nerve and this could be accomplished by carrying out appropriate degeneration studies. Further, if these fibres are shown to be cholinergic, their presence would explain the observation by Mchedlishvili & Baramidze (1967) that the ability of pial vessels to dilate in response to a fall in arterial pressure is almost completely abolished following the administration of atropine.

Studies using electron microscopy, then, have shown that the fine structure and organization of nerves and their terminals on cerebral vessels conform in almost all particulars with that described in other vascular beds. The evidence is strongest that the axons having predominantly small granular vesicles are adrenergic and that these are sympathetic post-ganglionic fibres whose degeneration is readily and quickly demonstrated after removal of the superior cervical ganglia. The evidence which suggests that the axons having non-granular vesicles are cholinergic is, at the moment, circumstantial: but they may well correspond to the vasodilator fibres carried by the VIIth cranial nerve. These studies have yielded very little evidence concerning the presence of receptors on or near smooth muscle whose afferent fibres might be represented by the myelinated nerves. It is possible that some kind of sensory function may be found for the terminals described by Iwayama, Furness & Burnstock (1970) which have no synaptic vesicles and which are unaffected by sympathectomy. A further problem which requires elucidation is the function of the arrangement of the vasomotor nerves into 'peri-adventitial' and adventitial bundles as described by Iwayama *et al.* (1970).

Histochemical studies

Additional evidence for the existence of adrenergic fibres which supply pial arteries and their branches has been provided by

3. The innervation of cerebral blood vessels

Nielsen & Owman (1967), Falk, Mchedlishvili & Owman (1965), Falck, Nielsen & Owman (1968), Ohgushi (1968) and Kajikawa (1969) together with Iwayama (1970) and Iwayama *et al.* (1970) who used fluorescent techniques in parallel with their electron microscope studies. Using the formaldehyde fluorescent technique developed by Falck, Hillarp, Thieme & Torp (1962), these workers have demonstrated the presence of adrenergic fibres on intracranial vessels in the rabbit, cat, dog and man. The most dense distribution was seen to be on the main arteries forming and leaving the circle of Willis, being most marked on the intracranial portion of the internal carotid arteries and least marked on the middle cerebral and basilar arteries.

The pial arteries receive a moderate number of adrenergic nerves and these can be followed down to vessels having a diameter of 15–20 μm, Plate 3. An adrenergic nerve supply to intracerebral arteries has been demonstrated by Ohgushi (1968) and Kajikawa (1969) and has been confirmed by Nielsen & Owman (personal communication). Using this method, it has also been possible to demonstrate that following denervation or pretreatment with reserpine, the fluorescence in fibres on pial and intracerebral vessels disappears (Falck, Nielsen & Owman, 1968; Kajikawa, 1969) and that the pattern of this disappearance indicates that the sympathetic supply to cerebral blood vessels is mainly homolateral although the supply to the midline vessels – the anterior cerebral, the posterior cerebellar and basilar arteries – is bilateral. Secondly, there appears to be an interconnecting plexus of adrenergic nerves on the pia mater between formations on adjacent arteries. Thirdly, some fibres from the pial vascular plexus appear to enter the cortex and mingle with adrenergic fibres which run in the outer layer of the neocortex. Falck *et al.* (1968) raise the interesting and novel possibility that true synapses may occur between adrenergic and non-adrenergic cell bodies. After treatment with reserpine, all the fluorescence seen in both cortical and vascular nerves was abolished. After bilateral cervical sympathectomy, fluorescence in the vascular nerves was abolished while that in the adrenergic cortical nerves was unaffected.

Vasodilator pathways

The function of the depressor nerve in the neck was discovered by Cyon & Ludwig (1866). They thought that the afferent nerves were derived from receptors in the heart but they showed that stimulation of the central end of the nerve caused a fall in blood pressure and they concluded that the nerve was inhibitory to the vasoconstrictor centre. Bayliss (1908) confirmed these results and elaborated the concept of reciprocal innervation to vasomotor nerves. Afferent stimulation of this nerve brings about vasodilatation of the mucous membrane of the nose (Martin & Mendelhall, 1915; Fofanow & Tschalussow, 1913) and this effect is unimpaired if the sympathetic efferent fibres are cut, indicating that the vasodilator reflex is direct and not merely inhibitory to sympathetic vasoconstrictor fibres. Forbes & Wolff (1928) also demonstrated that afferent stimulation of the vagodepressor trunk in the neck causes pial artery dilatation. Ranson & Billingsley (1916) had provided evidence that the afferent circuit entered the medulla oblongata and part of the reflex arc could be reached by stimulation of the area postrema.

By analogy with the fact that vasodilator fibres to blood vessels of the tongue and salivary glands were carried with the cranial nerves, Cobb & Finesinger (1932) and Chorobski & Penfield (1932) undertook a study of the effects of stimulating cranial nerves upon the pial circulation.

The results are summarized in Table 3.1, and the points at which faradic stimulation was carried out are shown in Fig. 3.2. When the results of experiments carried out by Forbes & Wolff (1928) were included, the following points emerged. Stimulation of the cervical sympathetic nerve in the neck gave rise to ipsilateral pial artery constriction; central stimulation of the vago-depressor trunk caused bilateral pial artery vasodilatation; stimulation of the peripheral cut end of the VII cranial nerve intracranially caused ipsilateral vasodilatation; stimulation of the III, V, VIII, IX, XI and XII were without effect upon the calibre of pial arteries; cranial nerves IV and VI for technical reasons were difficult to expose and therefore were not satisfactorily studied; preliminary evidence was obtained that the vasodilator pathway passed through the geniculate ganglion of the VIIth nerve.

3. The innervation of cerebral blood vessels

Table 3.1. *Effect of stimulating various nerves upon pial artery diameter* (from Cobb & Finesinger, 1932. *Arch. Neurol. Psychiat.*, Chicago **28**, 1243–56)

Locus of stimulation	Number of animals	Number of trials	Effect		% change in diameter of pial artery		
					Low	High	Average
Sympathetic in neck:							
Ipsilateral	58	172	Constriction	132	1.3	20	8.03
			No change	40	—	—	—
Contralateral	10	33	No change	33	—	—	—
Vagus:							
Ipsilateral	48	137	Dilated	106	3.0	50	14.5
			No change	31	—	—	—
Contralateral	9	29	Dilated	21	4.0	36	14.0
Stimulate at 2 to 1	4	22	Dilated	10	3.0	22	15.5
			No change	3	—	—	—
Peripheral to cut at 1	5	20	No change	20	—	—	—
With VII nerve cut	5	16	No change	16	—	—	—
IIIrd nerve:	3	10	No change	9	—	—	—
			Slight constriction	1			
Vth nerve:	13	38	No change	38			
VIIth nerve:							
Total	25	147	Dilated	122	4.0	60	16.0
			No change	25			
Point 3 with VIIth nerve intact	21	124	Dilated	104	4.0	60	17.0
			No change	20			
Point 3 with cut at 4	5	10	No change	10			
Point 4 with cut at 3	6	18	Dilated	16	4.0	20	12.0
			No change	2			
Point 7 with cut at 6	4	14	No change	14			
VIIIth nerve:	8	25	No change	25			
IXth nerve:	6	14	No change	14			
XIth nerve:	10	42	No change	42			
XIIth nerve:	5	17	No change	17			

Vasodilator pathways

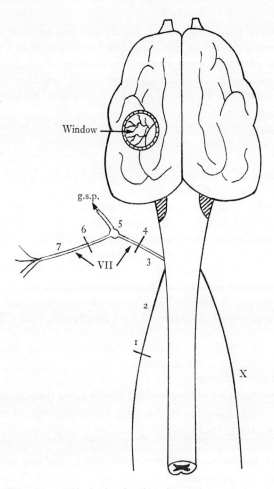

Fig. 3.2. Diagram of cat's brain showing the position of the skull window used to view the pial arteries. X, the vagus nerve, was stimulated and again at position 2 after section at 1. VII, the facial nerve. This nerve was stimulated at positions 3 and 7 and again after section at points 4 and 6 respectively. 5, the geniculate ganglion: g.s.p., the greater superficial petrosal nerve. The effects of stimulating the vagus and facial nerves are summarized in Table 3.1. (From Cobb and Finesinger, (1932). *Archs Neurol. Psychiat.*, Chicago **28**, 1243–56, Fig. 4. Copyright 1932, American Medical Association).

3. The innervation of cerebral blood vessels

Origin of the nerve plexus on pial and intracerebral arteries

The origin of the nerves which supply the pial and intracerebral arteries was reviewed by McNaughton (1938) and it is a measure of the lack of any progress in this field that the diagram which he used to summarize the known connections together with the question marks he employed can be reproduced today without modification (Fig. 3.3).

The source of sympathetic vasoconstrictor fibres is the plexus of nerves on vertebral and carotid arteries (Hüber, 1899; Clarke, 1929).

According to Williams (1936) who studied the innervation in four human brains, there is little or no connection between these plexuses, the posterior cerebral artery and its branches being innervated from the internal carotid plexus. However this may be an arrangement peculiar to man since McNaughton (1938) reported that in the Macacque monkey, there was clear evidence of innervation of posterior cerebral vessels from the basilar artery nerve plexus and only a few fibres were derived from the internal carotid plexus. A similar continuity of vertebral and carotid plexuses was reported in the cat by Christensen, Polley & Lewis (1952).

The vertebral plexus

Langley (1900) traced the vertebral nerves from their origin in the inferior cervical ganglion to the third cervical nerve. In an earlier paper, Langley (1894) had observed that the vertebral nerves ran past the third cervical nerve, but he had not followed them. He concluded that the vertebral nerves were grey rami communicating vasomotor and pilomotor nerves to the third, fourth and lower cervical nerves. These observations were confirmed by Chistensen, Polley & Lewis (1952) and they extended them in two further ways. First, they distinguished between the vertebral nerves and the plexus of nerves around the vertebral artery which was most apparent in the rostral part and which was continuous with the intracranial vertebral and basilar plexus. They found no evidence of continuity between vertebral nerves and plexus. Secondly, they showed that the fibres of the vertebral nerves were sympathetic

The vertebral plexus

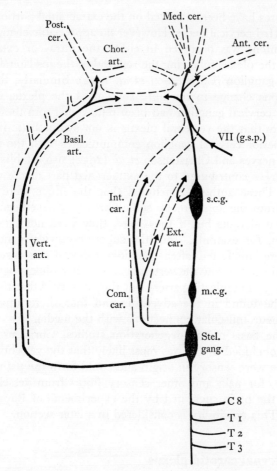

Fig. 3.3. A diagram summarizing the known neural pathways to cerebral blood vessels. m.c.g. and s.c.g., middle and superior cervical ganglia; VII (g.s.p.), greater superficial petrosal nerve carrying vasodilator fibres from the intracranial portion of the facial nerve. (Redrawn from McNaughton (1938), *Res. Publs Ass. Res. nerv. ment. Dis.* **18**, 178–200, Fig. 88.)

in origin by degeneration studies after the inferior cervical ganglion had been removed.

The plexus of nerves around the vertebral artery has been described in the cat (Hüber, 1899; Christensen *et al.* 1952) and in man (Dowgjallo, 1932; Lazorthes & Reis, 1942). The fibres in the intracranial portion are principally unmyelinated although

3. The innervation of cerebral blood vessels

some fibres have been observed on the extracranial portion rostral to the IIIrd cervical nerve. However no appreciable change in the fibre content was observed in either the intra- or extracranial parts of the plexus following the removal of the ipsilateral inferior cervical ganglion (Christensen *et al.* 1952). Similarly, there was no obvious change in the fibre content of the plexus after the superior cervical ganglion had been removed. From these results it appears that the vertebral plexus is not sympathetic in origin. The superior cervical ganglion communicates with the first two cervical nerves and Christensen *et al.* (1952) also established that these nerves contributed to the extracranial part of the vertebral plexus. These authors concluded that the fibres were derived mainly from the dorsal roots: some fibres may have been derived from ventral roots but if they were, they were not sympathetic in origin, for example, from the superior cervical ganglion for they were unaffected after ganglionectomy. Christensen *et al.* (1952) did not follow the vertebral and basilar plexus beyond the posterior communicating artery but they observed that the plexus was to be found in the adventitia and that there appeared to be few neuromuscular connections with the media.

On the basis of the degeneration studies, Christensen *et al.* (1952) concluded that it was most likely that the vertebral plexus of nerves were sensory in origin and could form the pathway, for example, for pain and other sensory fibres from vessels at the base of the brain suggested by the experiments of Ray & Wolff (1940). This possibility is considered in a later section.

The internal carotid plexus

The principal sympathetic supply to blood vessels of the brain is derived from the internal carotid plexus and nerve which is the largest branch of the superior cervical ganglion. Ranson & Billingsley (1918) showed that the majority of fibres in the internal carotid nerve were unmyelinated but a number were myelinated and of relatively small size, $3-4 \mu$m. There has been some doubt as to the significance of these fibres. Langley (1896) showed that if the branches of the superior cervical ganglion were cut, the myelinated fibres connected to the ganglion were normal; those separated from the ganglion degenerate. Ranson & Billingsley (1918) also showed that if the sympathetic trunk caudal to the ganglion was

The internal carotid plexus

cut, the myelinated fibre content of the internal carotid nerve was unchanged from control. These studies would indicate that the myelinated fibres originated from cells in the superior cervical ganglion and acquired sheaths (Langley, 1896; Michailow, 1911). On the other hand there is evidence from a number of experiments that some of these myelinated fibres in the superior cervical ganglion are preganglionic (Kuntz, Hoffman & Napolitano, 1957) since they observed degeneration of myelinated fibres in the internal carotid nerve after dividing the sympathetic trunk caudal to the superior cervical ganglion.

A number of workers have described numerous additional ganglion cells or aggregations of cells in the plexuses of both external and internal carotid arteries (Chorobski & Penfield, 1932; Hoffman, 1958; Lazorthes, 1949) in man and the cat. It is thus possible that a number of post-ganglionic fibres originate from these ganglia rather than from the superior cervical ganglion and in turn this may explain why Chorobski & Penfield (1932) were unable to detect much evidence of degeneration in pial vessel nerves after extirpation of the superior cervical and stellate ganglia. In addition, there appears to be a limited number of preganglionic fibres from rami communicating with C1 and 2 which do not pass through the superior cervical ganglion and which contribute to the external and internal carotid plexus.

An alternative approach to the study of the origin of intracranial vascular nerves has followed the introduction of the fluorescent technique of identifying adrenergic nerves (Falck *et al.* 1962). There have been a number of recent studies which have all shown that the nerves which supply the vessels at the base of the brain (Falck, Neilson & Owman, 1968; Ohgushi, 1968; Kajikawa, 1969) are derived from the superior cervical ganglion and not from the stellate ganglion. This immediately suggests that the pial artery innervation is not significantly supplied from the vertebral nerve or plexus or at least that part derived from the stellate ganglion. However Kajikawa (1969) found that although removal of the stellate ganglia did not affect the innervation of all basal arteries, including the vertebral and basilar arteries, section of the trunk caudal to the superior cervical ganglion, in a proportion of experiments affected the innervation of the vertebral and basilar arteries. He therefore proposed that the rostral part of the basilar artery was supplied by post-ganglionic fibres in

3. *The innervation of cerebral blood vessels*

the internal carotid nerve while the caudal part of the basilar and upper vertebral arteries are supplied by caudal branches from the superior cervical ganglia.

The results of experiments cited so far are reasonably consistent with two important exceptions. First, there is still doubt as to the origin of the vertebral and basilar plexus. Christensen *et al.* (1952) claim that it is largely afferent while Kajikawa (1969) provides evidence that it is largely sympathetic in origin and Ohgushi (1968) observed degeneration of this plexus following removal of the stellate ganglia. No suitable reconciliation of these findings seems possible at the moment. Secondly, we have the remarkable discrepancy between the findings of Chorobski & Penfield (1932) who found no evidence of degeneration of pial nerves following removal of the superior cervical ganglion and the three groups of workers who found that fluorescence from adrenergic fibres in intracranial vascular nerves was abolished after ganglionectomy.

The most obvious question which arises is: how reliable are the silver impregnation staining methods used by Stohr (1922a) and by Chorobski & Penfield (1932) in measuring degeneration of nerves with sufficient precision that at least semi-quantitative estimates can be given. This topic has been reviewed in some detail by Richardson (1960).

Afferent nerves from blood vessels of the pia

Hüber (1899) and Stohr (1922a, b) demonstrated the existence of myelinated and unmyelinated fibres supplying the blood vessels of the pia mater and although there is some evidence that postganglionic sympathetic fibres may be myelinated, as given in the previous section, an alternative possibility is that these myelinated fibres are sensory in function. This possibility is strengthened by the demonstration by Stohr (1922a, b), Penfield (1932) and Humphreys (1939) of endings on or near pial blood vessels which appear similar to Meissner and other terminals in other vascular beds.

Further evidence that the pia and its blood vessels may have a sensory innervation was given by experiments in cats, immobilized with 'bulbocapnine' and given local anaesthetic, in which the exposed pia was stimulated with a unipolar electrode (Levine &

Afferent nerves from blood vessels of the pia

Wolff, 1932). They used a change in skin resistance brought about by reflex changes in blood flow as an index of effective stimulation in the same way that the effectiveness of somatic painful stimuli or psychological stress cause reflex vascular changes. With random testing of pial vessels but excluding the pia over the motor cortex, these workers obtained skin responses in some 39 per cent of occasions when vessels were stimulated and 18 per cent of occasions when the intervening pia was stimulated.

In these experiments, the possibility of spreading cortical excitation cannot be ruled out although the weighting of responses in favour of vessel stimulation makes this unlikely. If these do represent true sensory responses, the modality of sensation remains obscure.

In a further study, additional evidence was given for pial vessels being the site of sensory receptors which if stimulated either mechanically by handling or crushing or faradically could give rise to painful sensations referred to specific areas of the head (Ray & Wolff, 1940). The study was undertaken in thirty human patients given only local anaesthetic. The observations can be summarized as follows. The dural arteries and principal venous sinuses, the extracranial arteries and pial arteries all gave rise to painful responses when stimulated. The pia mater was relatively insensitive except near the arteries at the base of the brain as was the parenchyma of cerebrum and cerebellum. In general, the pial and dural arteries were sensitive only in their proximal portions.

This further study again does not precisely define the nature of the sensory modality involved since the method of stimulation was fairly crude and it is possible that a number of different types of receptor, e.g. temperature, stretch, may have been supramaximally stimulated to give rise to a sensation of pain. The elaborate hypothesis proposed by Ray & Wolff (1940) on the basis of these experiments, that headache is produced by traction or dilatation of the sensitive arteries, veins or dura is plausible and is buttressed by clinical studies in which drugs, e.g. histamine or carbon dioxide which are known to cause vasodilatation, also are associated with headache. Nonetheless, these results, suggestive as they are, should not obscure the fact that very little is known about the types of sensory innervation of intracranial blood vessels and structures, and although pain was induced in the experiments

3. The innervation of cerebral blood vessels

cited above, the physiological function of the sensory structures remains obscure. The possibility that the sensory receptors initiate reflex activity which affects cerebral blood vessels themselves should not be overlooked.

Afferent pathways

There is only scanty evidence as to the pathways taken by the myelinated fibres described by Hüber (1899) and which are presumed to be sensory in function. Following the studies on referred pain from manipulation or stimulation of pial and dural blood vessels, Schumacher, Ray & Wolff (1940) examined the afferent pathways involved. Their results may be summarized as follows. Structures which are pain-sensitive and which lie above the tentorium cerebelli are supplied by fibres from the fifth cranial nerve. Sub-tentorial structures receive few if any fibres from the fifth nerve. Lushka (1850) described the 'nervus spinosus' which accompanied the middle meningeal artery and its branches and was derived from the third division of the trigeminal nerve while Arnold (1851) described the 'nervus meningens medius' as a branch principally of the second division of this nerve. McNaughton (1938) has also described fine unnamed branches of the artery which accompany the nerve. Thus it is probable that all the principal branches of the fifth nerve contribute to the afferent innervation of the supra-tentorial structures.

Dowgjallo (1932) has offered some evidence that the nerve plexuses around the large arteries around the base of the brain have connections with various cranial nerves, the IIIrd, Vth, VIth, VIIth, VIIIth, IXth, Xth, XIth and XIIth: but the evidence does not indicate whether the fibres involved were sensory. Other observations have been made by Princeteau (quoted by Hovelacque, 1927) who described connections between Vth cranial nerve and internal carotid nerve in the carotid canal; Rosenstein (1935) who mentions a branch from the ophthalmic branch of the trigeminal nerve to the periarterial plexus of the carotid; and Rauber (1872) who described branches of the IXth cranial nerve to the internal carotid nerve by way of the tympanic plexus and carotico-tympanic nerves.

The question of an afferent pathway through branches of the vagus through the superior cervical ganglion and internal carotid

Afferent pathways

nerve was raised by Kuntz (1934). He showed that after entire denervation of the cervical sympathetic trunk and including the ganglion, section of the internal and external carotid nerves showed intact fibres which he concluded were of vagal origin. Hinsey (1935) has made the comment that these fibres could equally well be vagal preganglionic fibres and that their sensory nature was by no means certain. Ranson & Billingsley (1918) found no evidence for such fibres after similar degeneration experiments.

The evidence therefore for the existence of afferent nerves from intracranial structures, notably from pial and dural blood vessels, is suggestive but largely circumstantial and virtually nothing is known either about their function, beyond the fact that various types of stimulation can cause pain, or the afferent pathways which are involved.

Innervation of dural vessels and sinuses

Dural vessels

The principal nerves which accompany the middle meningeal artery and its branches described by Lushka (1850) and Arnold (1851) have already been mentioned. Many of the features of these nerves in the Macacque and in man have been reviewed by Hovelaque (1927) and by McNaughton (1938). The nerves, according to McNaughton (1938) are principally derived from the ophthalmic division of the trigeminal nerve rather than the second or third divisions and arise from the ganglion itself. The nerves have traditionally been described as vascular or 'nervi proprii' which travel along the vessels but take no part in their innervation and pass off at different levels to end in the dura. Unmyelinated fibres which branch in the tunica media have been described by Hüber (1899) and by Traum (1925) and these appear to be vasomotor in function. Stohr (1928) was not able to demonstrate any true sensory endings in man nor did McNaughton (1938) demonstrate them in the Macacque using the Gros–Bielchowsky staining method. Dowgjallo (1932) however claimed to have described three different kinds of ending using intravital methylene blue methods:

(a) 'Büschelschen' of fine varicose fibres in the adventitia of the larger vessels, usually connected with myelinated fibres.

3. *The innervation of cerebral blood vessels*

(b) 'Knopfschen – similar to those described by Stohr in the pia.

(c) 'Endigungen mit Plättchen' – a network of thin and thick fibres with small swellings where the fibres cross one another were found in the adventitia of the dog but these are not illustrated.

Dural sinuses

Very little is known about the innervation of the dural sinus. McNaughton (1938) and Ray & Wolff (1940) report the existence of nerves which are to be found on the sagittal sinus derived from the ophthalmic division of the trigeminal nerve. At its anterior end, the fibres are derived from the anterior ethmoidal nerves while the posterior third of the sinus has a plexus of nerves derived from fibres of the falx, the torcular region and occipital dura being in turn derived from the two tentorial nerves. The tentorial nerves also run to the transverse, straight and superior petrosal sinuses. McNaughton (1938) describes twigs from the IXth and Xth cranial nerves which run to the sigmoid sinuses and branches of the first trigeminal nerve have been seen to run to the cavernous sinuses (Hovelaque, 1927).

Although the staining methods were apparently adequate to define the course of these nerves, no study of neuromuscular junctions has apparently been carried out and, in consequence, the function of these nerves remains obscure. It will be noted also that all these references are to work carried out thirty years or more ago. The literature does not reveal any more recent references in this field.

Innervation of cerebral and other vascular beds compared

The accumulated evidence from light microscope, electron microscope and fluorescent histochemical studies indicates that the cerebral blood vessels have a system of vasomotor innervation which is fully comparable with that in other vascular beds. The evidence is most complete for the adrenergic group of fibres whose axons show the same type of fluorescence as in other parts of the body, the fluorescence disappearing after pretreatment with reserpine or after cervical sympathectomy: and whose neuromuscular terminals correspond in all particulars with those described on other vessels. The distribution of adrenergic fibres has not been systematically studied but the evidence so far available suggests that there is considerable difference in frequency of fibres to be seen on the principal cerebral arteries and that the frequency of fibres diminishes with the calibre of arterioles. No fibres have been seen supplying capillaries or venules.

One obvious difference between cerebral and other arterioles has emerged – namely, the apparent absence of cerebral arterioles and precapillary sphincters which can be grouped as 'resistance' vessels. There does not therefore appear to be the intense innervation of such vessels as has been described in other vascular beds (Folkow, 1955). The morphological description of vasomotor innervation therefore yields no direct information about the situation of resistance vessels in the cerebral circulation.

The existence of a vasodilator pathway seems to be clearly established and vasodilator fibres may well be represented by the cholinergic axons and neuromuscular terminals with non-granular vesicles described in preceding sections of this chapter. The physiological function and pharmacological properties of this pathway have been much less completely studied in the cerebral than for example in the skeletal vascular system (Folkow & Uvnäs, 1948, 1949): the evidence suggests that these fibres are genuinely vasodilator – that is, they act independently and not by inhibiting sympathetic vasoconstrictor activity.

The evidence presented in this chapter contains some internal inconsistencies and emphasizes certain lines of investigation which it would be profitable to pursue. Notable among these is the question of the sensory innervation of cerebral blood vessels. As has been pointed out, the conclusions arrived at concerning the physiological role of such sensory fibres and their possible role in the pathogenesis of headache rest on singularly indirect evidence: it would be important to establish the existence of sensory terminals rather more clearly than was possible in early studies with light microscopy and it would also be important to establish the presence of afferent neural discharge in what might be presumed to be mixed nerve trunks inside the skull and its elimination following application of local anaesthetic. It would then be necessary to establish which sensory modality was involved, whether for example as has been suggested, the receptors respond to stretch or distortion of the vascular wall. When these points have been

3. The innervation of cerebral blood vessels

established, it would then be possible to study the reflex consequences of stimulation of these receptors.

It is also clear that discrepancies exist between Penfield's observation that section of the cervical sympathetic nerves leads to no obvious degeneration in post-ganglionic sympathetic fibres and the observation by Falck and his colleagues (1968) that sympathetic section leads to virtually complete abolition of fluorescence in adrenergic vasomotor fibres on the same side. Similarly, there is still some uncertainty about the frequency of fibres supplying intracerebral arteries. Such fibres were seen with light microscopy and in a limited number of studies using histochemical techniques but they have not been consistently observed with electron microscopy. Both these discrepancies may be explained in terms of technique but they remain incompletely resolved and deserve further study.

4 HAEMODYNAMIC CONSIDERATIONS

The Monro–Kellie doctrine

The apparently unyielding nature of the skull led Alexander Monro the younger (1783) to suppose that expansion of cranial contents was severely limited and that active changes in the calibre of cerebral vessels was unlikely.

> For, as the substance of the brain, like that of the other solids of our body, is nearly incompressible, the quantity of the blood within the head must be the same, or very nearly the same, at all times, whether in health or disease, in life or after death, those cases only excepted, in which water or other matter is effused or secreted from the blood vessels: for in these a quantity of blood, equal in bulk to the effused matter, will be pressed out of the cranium.

This observation was one of several which Monro made from post-mortem examination and it will be apparent that Monro in company with other anatomists of his time recognized neither the origin nor significance of cerebrospinal fluid: hence the somewhat ambiguous reference to 'water or other matter (is) effused or secreted from the blood vessels.' Monro's general concept was confirmed by Kellie partly from the post-mortem examination of men who had died from exposure (1824a) and partly from some experiments on dogs (1824b) and this concept, which subsequently became known as the Munro–Kellie doctrine, exerted a powerful influence over the interpretation of experimental results in the nineteenth century and for some years of the present century.

Some workers, failing perhaps to realize the assumptions which were implicit in Munro's statement, applied it literally. Of these, the most articulate was Hill (1896) who concluded that 'the whole circulatory system of the brain will have assimilated itself into a system of rigid tubes' and 'In every experimental condition, the cerebral circulation passively follows the change in the general arterial and venous pressures.' These conclusions were particularly unfortunate for they were made in the face of contrary evidence from his own experiments (Bayliss & Hill, 1895), they directed attention away from evidence which might have indicated independent vasomotor activity in cerebral vessels (Donders, 1850,

4. *Haemodynamic considerations*

1859; Roy & Sherrington, 1890). These results came perilously close to being dismissed as artifacts.

Other more perceptive workers pointed out some of the flaws in Monro's arguments. Burrows (1846), for example, while recognizing that the contents of the skull must be nearly constant realized the importance of cerebrospinal fluid (c.s.f.) and its relation to the veins in the cranio-spinal hydraulic system and emphasized that although the brain was incompressible, this was not as important as the fact that changes in brain volume could occur at the expense of other fluids. This view of Burrows is close to that which is accepted today. However, further consideration of the matter has suggested that the bony and membranous coverings of the skull and spinal cord are not as rigid as previously thought. Although the dura mater is closely applied to the inner surface of the skull, in the spinal region this is not so for the epidural space enclosing an abundant venous plexus and fatty areolar tissue is interposed between dura and the bony vertebral canal and segmental ligaments. The plasticity or compliance of this covering has not been formally measured but is unlikely to be negligible. Magendie (1836) and many subsequent workers have demonstrated, for example, that the occipito-atlantoid ligament possesses sufficient elasticity to transmit pulsations of the c.s.f. in the same way as does the membranous covering of the anterior fontanelle in young children. More recently pressure–volume measurements have been made by Langfitt, Weinstein & Kassell (1965) and a representative example illustrated in Fig. 4.1 indicates that the volume of an intracranial balloon can be increased or saline added to the subarachnoid space before pressure starts to rise. One further possibility requires consideration – namely the escape of c.s.f. around the cranial and segmental nerves. The presence of such perineural channels has been demonstrated by injection (Key & Retzius, 1876; Weed, 1914) but their physiological significance is uncertain.

In addition to these anatomical considerations, it is important to recall the unusual features of the venous drainage of the brain and other intracranial structures and their relationship to pressure changes in c.s.f. The veins are of two sorts, those which traverse the subarachnoid space and the cranial sinuses. The latter being enclosed in thick dural membranes can exert little effect upon c.s.f. pressure nor are they likely to be much affected by this

The Monro–Kellie doctrine

pressure. The pial or subarachnoid veins, on the other hand have no valves, they are thin walled and lack the supporting action of muscles and they empty into the sinuses against the direction of flow. These features not only help to ensure that flow of blood is smooth and that the brain is protected against the effects of sudden changes in pressure occasioned by, e.g.

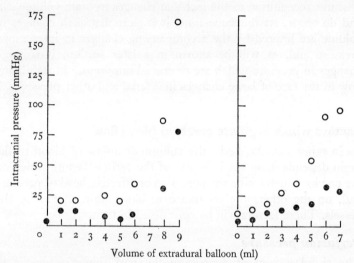

Fig. 4.1. Intracranial volume–pressure curves in two rhesus monkeys. Pressure was measured in the subarachnoid space and volume refers to that injected into an extradural balloon. Closed circles represent the final pressure following each increment of water added to the balloon and open circles, at each point, the peak pressure produced by rapid injection of 0.5 ml of saline into the cerebral subarachnoid space. Interval between injections about thirty minutes; each saline injection made about 10 min before the balloon injection; balloon injection time 50 s; volume of deflated balloon about 1.5 ml in each experiment. (From Langfitt, Weinstein & Kassel (1965), *Neurology* 15, 622–41, Fig. 1A, B.)

changes in posture, but they also ensure that changes in c.s.f. pressure are readily and completely transmitted to the veins. In a hydraulic sense therefore the c.s.f. and systemic venous system can be considered as one, and there is now abundant experimental evidence that a linear relationship between jugular venous and c.s.f. pressure exists over a wide range of pressure above and below atmospheric (Forbes & Nason, 1935; Williams & Lennox, 1939; Courtice, 1940; Ferris, 1941; Britton, Corey & Stewart, 1946; Britton, Pertzoff, French & Kline, 1947; Rushmer, Beckman & Lee, 1947; Kety, Shenkin & Schmidt, 1948*b*).

4. Haemodynamic considerations

Because of all these factors, it is clear that substantial changes in volume of the vascular compartment of the brain can take place. Quite how great this change can be has not been determined but it is sufficient to render the Monro–Kellie doctrine if not invalid, at least misleading. For this doctrine by its emphasis on the fact that the skull is rigid and the brain incompressible obscures recognition of the fact that changes in brain volume can and do occur. At the same time, it is clear that such changes in volume are limited by the accompanying changes in intracranial pressure and, as will be shown in a later section, it is these changes in pressure which are of crucial importance in stabilizing flow in the face of large changes in arterial and other pressures.

Factors which regulate cerebral blood flow

As in other vascular beds, the volume of inflow of blood to the brain depends upon the interplay of the various factors which on the one hand determine the perfusion or effective head of pressure and, on the other, the resistance to blood flow offered by the vessels. These factors will be considered in turn.

Perfusion pressure

The simplest way of expressing the perfusion pressure is to consider it as the net head of pressure forcing blood through the cerebral vessels and it is therefore the difference between the mean arterial and venous pressures. Because the latter is very small by comparison being some 20–40 mmH$_2$O positive with respect to the right auricle, it is often disregarded and perfusion pressure is assumed to be equivalent to mean arterial pressure. This form of treatment ignores the fact that both pressure and flow are phasic: this matter is dealt with in a later section in the present chapter.

Perfusion pressure may be altered either by changes in arterial or venous pressure. The effect of alterations of arterial pressure upon cerebral blood flow is considered separately in Chapter 6; but the relation between perfusion pressure and flow, illustrated in Fig. 4.3 in diagrammatic form and in Fig. 6.4 from experimental data, is such that flow is independent of perfusion pressure over the physiological range and for a variable range below it. If perfusion pressure is reduced below a critical level which is approximately 40–50 per cent of control, blood flow falls rapidly to very low levels.

Factors which regulate cerebral blood flow

This has been confirmed by Jacobson, Harper & McDowall (1963a) who altered sagittal sinus pressure in the dog by graded compression of the superior vena cava and found that if venous pressure was raised and perfusion pressure reduced to not less than 60 mmHg, cortical blood flow was unaffected: to be precise, the changes in blood flow fell within the errors of their method of measuring flow. This finding is of some historical importance for it disposes of the argument used by both Roy & Sherrington (1890) and Bayliss & Hill (1895) that the changes in volume of the brain (which they used as an index of cerebral blood flow) in asphyxia were caused by changes in venous pressure.

A similar type of study was carried out by Moyer, Miller & Snyder (1954) in man. They raised jugular venous pressure by inflating a cuff around the neck and even after jugular pressure had been raised to 300 mmH$_2$O causing a reduction in calculated perfusion pressure to 60 mmHg, there were no significant changes in cerebral blood flow.

This type of experiment has obvious limitations: in man, for ethical considerations and in both dog and man because the system of venous drainage of the brain which ensures alternative pathways protected from externally applied pressure limits the height to which sagittal or jugular venous pressure can be raised.

An alternative approach has been to study the effects of altering c.s.f. pressure since this will cause proportionate changes in venous and hence in perfusion pressure. The effect of raising c.s.f. pressure upon the calibre of pial vessels is shown in Fig. 4.2: as c.s.f. pressure is raised, the pial vessels dilate without obvious change in femoral artery pressure. Similar studies were carried out by Williams & Lennox (1939) who measured blood flow (using the A–V oxygen difference as an index) in patients with raised intracranial pressure due to cerebral tumours. Blood flow was found to be within normal limits. In comatose patients, blood flow appeared to have increased (Kety, Woodford, Harmel, Freyhan, Appel & Schmidt, 1948). In small and heterogeneous groups of patients or subjects, Courtice (1940) and Ferris (1941) thought that cerebral blood flow fell as intracranial pressure increased. Their methods of measuring blood flow were, however, indirect and the first satisfactory study was by Kety, Shenkin & Schmidt (1948b) in which it was shown that (a) c.s.f. pressure increased linearly with mean arterial pressure and that (b) cerebral

4. Haemodynamic considerations

blood flow was unaffected by increments of c.s.f. pressure up to approximately 450 mmH$_2$O (33 mmHg). Above this level, cerebral blood flow fell progressively.

Several points of interest arise from this study. First, it largely confirms the early observations of Cushing (1901, 1902) that

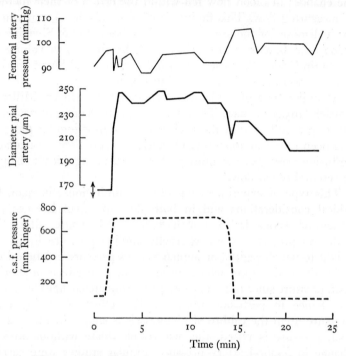

Fig. 4.2. Effect on pial vessels of a sudden moderate increase in intracranial pressure. The arrows indicate the probable error of measuring pial artery diameter. C.s.f. pressure was raised to 750 mm Ringer over 1 min and held at that level for 12 min. Within 45 s, the pial vessels started to dilate and after a further 45 s, had dilated by 47 per cent of control. Flow in pial veins slowed as c.s.f. pressure was first raised and then increased to control levels. (From Wolff & Forbes (1928), *Archs Neurol. Psychiat., Chicago* **20**, 1036–47, Fig. 1 Copyright 1928, American Medical Association.)

a rise in intracranial pressure is associated with a rise in arterial pressure. The mechanism for this is unknown. Cushing proposed that it was due to 'anaemia of the vasomotor centre'. However it is clear that mean arterial pressure rises before there is any significant fall in blood flow so, in order to substantiate Cushing's proposal, it would be necessary to suppose that blood flow fell

Factors which regulate cerebral blood flow

selectively. There is certainly some evidence that ischaemia of brain tissue does occur with raised intracranial pressure since Kety, Shenkin & Schmidt (1948b) showed that, as mean arterial pressure increased, cerebral venous P_{CO_2} and hydrogen ion concentration increased while oxygen saturation fell.

The second point is that, as c.s.f. pressure increases, so the exposed pial capillaries and veins tend to collapse and the resistance to blood flow increases. Further, the rise in c.s.f. pressure is transmitted to the veins so that at a c.s.f. pressure of 450 mmH$_2$O, the perfusion pressure would have been reduced to approximately $85 - 35 = 50$ mmHg had not mean arterial pressure increased to 100 mmHg. In this sense therefore, the rise in arterial pressure can be considered as an important compensating mechanism which maintains perfusion pressure and hence flow at normal levels. Above a certain critical intracranial pressure, this means of compensation no longer holds and it is possible that the failure of this mechanism may also explain the fall in estimated flow with raised intracranial pressure in comatose patients (Williams & Lennox, 1939; Courtice, 1940).

A reduction in intracranial pressure should increase perfusion pressure and hence blood flow providing that arterial pressure is not affected. Forbes & Nason (1935) found that intraperitoneal injections of hypertonic urea or sodium chloride caused an immediate dilatation of pial arteries with increased oxygenation of pial venules and this was followed by a more prolonged phase of venous engorgement, cyanosis and some evidence of more sluggish blood flow. Reduction of intracranial pressure by removal of c.s.f., that is to -300 mm Ringer caused changes in pial vessels similar to the second phase of response to hypertonic solutions. A similar difference in response was shown by Shenkin, Spitz, Grant & Kety (1948) since a reduction of c.s.f. pressure by removal of c.s.f. caused no change in cerebral blood flow while the intravenous injection of hypertonic glucose caused a substantial rise in blood flow and fall in calculated vascular resistance. These workers also noted that with the latter manoeuvre, there was a significant fall in arterial oxygen content which they attributed to haemodilution. The increase in flow and fall in resistance may therefore have been due to a fall in viscosity of blood. In neither group of subjects did a reduction in c.s.f. pressure by at least half its control value materially affect mean

4. Haemodynamic considerations

arterial pressure. The possible ways in which blood flow is held constant with a rise in perfusion pressure are discussed in a later section.

Cerebrovascular resistance

The volume of cerebral blood flow is also determined by the resistance offered by blood vessels and by the viscosity of blood. The factors which affect vascular resistance include intracranial and tissue pressure, organic changes in the vessel wall together with the state of contraction of vascular smooth muscle particularly in the small resistance vessels. Vascular smooth muscle responds to changes in the chemical environment both in the lumen and in the perivascular space, to certain drugs and hormones and to changes in reflex activity. These aspects of the regulation of cerebral vascular smooth muscle are considered in later chapters in this book (7–10 and 12). In the present section, the concept of vascular resistance is discussed together with the effects of changes in tissue pressure and blood viscosity.

The problems associated with the calculation of vascular resistance have been reviewed on a number of occasions (Green, Lewis, Nickerson & Heller, 1944; Burton, 1951, 1952, 1954; McDonald, 1960; Daly & Hebb, 1966). The relationship between pressure and flow in the cerebral vascular bed differs in certain important respects from other parts of the body and this raises particular problems. Traditionally, vascular resistance (R) is calculated from the ratio of perfusion pressure (ΔP) to flow (F) and is thus analogous to Ohm's law. Under certain conditions, the relation can be extended from Poiseuille's equation:

$$R = \Delta P/F = 8\eta l/\pi r^4,$$

where η is blood viscosity (cP), l the length of the tube of radius r. This relation holds however only if, as in Ohm's law, R is independent of either ΔP or F, as is shown in Fig. 4.3, curve A. This assumption is unlikely to be met in the vascular system because of problems posed by passive distension or active constriction of blood vessels and the anomalous viscosity of blood. Moreover, the phasic nature of both pressure and flow make calculations based on 'mean' arterial pressure a crude approximation. Under these circumstances, measurements of vascular impedance are more appropriate. Although attempts have been

Factors which regulate cerebral blood flow

made to measure total vascular impedance or that of local vascular beds (Bergel, McDonald & Taylor, 1958; Bergel, Caro & McDonald, 1960), these have been only partially successful and the formidable experimental and theoretical difficulties have been discussed by Defares, Osborn & Hara (1963), Defares & Van Der Waal (1969) and Noordergraf (1964). Comparable studies have not been attempted in the cerebral vascular bed: one study however has shown that the pressure–flow curves obtained during

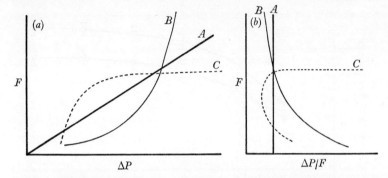

Fig. 4.3. The relation between (a) blood flow (F) and perfusion pressure (ΔP) and (b) the relation between blood flow and the calculated vascular resistance ($\Delta P/F$). In each diagram, curve A represents conditions where Poiseuille's law holds; curve B, a vascular bed which is distensible or where blood exhibits anomalous viscosity and curve C, the cerebral vascular bed showing in particular the range over which flow is independent of perfusion pressure. (Modified from Daly & Hebb (1966), *Pulmonary and bronchial vascular systems*. London: Arnold, Fig. 4A, B.)

pulsatile and non-pulsatile flow to the brain are not the same, Fig. 4.4 (Held, Gottstein & Niedermayer, 1969). These authors have also shown that if at constant perfusion pressure, steady flow is abruptly switched to pulsatile flow, the volume of inflow rises by 25 per cent of control: thereafter, blood flow returns to control levels while pulsatile pressure rises by some 20 per cent with a commensurate rise in vascular resistance. It is possible that this study demonstrates that vascular smooth muscle responds to the rate of change of pressure as well as to a steady mean level.

If as these points suggest, calculation of vascular resistance is the only practicable method of relating flow and pressure, the question arises as to what physiological meaning such calculated values have. Clearly, isolated measurements of resistance are

4. Haemodynamic considerations

potentially misleading since apparently normal values could be obtained from proportional changes in both flow and pressure. A better approach is to determine resistance from experimentally determined pressure–flow curves. If these curves or part of them are linear as is the case in the pulmonary vascular bed of foetuses or newborn lambs (Cassin, Dawes & Ross, 1964) then the resistance can be calculated simply from the slope of the curve and the effect of any given manoeuvre can be estimated by its effect on the slope.

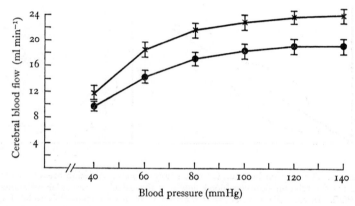

Fig. 4.4. The relation between sagittal sinus blood flow and mean arterial pressure. The upper curve describes the relation with non-pulsatile perfusion of cerebral vessels, the lower curves with pulsatile perfusion. Points on each curve, the mean values for ten experiments ±s.e. of the mean. The curves differ significantly. $P < 0.01$. (From Held, Gottstein & Niedermayer (1969), *Cerebral blood flow*. Berlin: Springer, Fig. 1.)

But in the majority of vascular beds, the pressure–flow relation is not linear and in other situations it is practically difficult, e.g. in man, to obtain formal and extended pressure–flow curves. Such a curve is shown (B) in Fig. 4.3(a) and the same values for flow plotted against $\Delta P/F$ in Fig. 4.3(b). It is clear that $\Delta P/F$ is not constant as it is in curve A, Fig. 4.3(b), being high at low rates of perfusion, and low at high rates. Now the effect of common physiological manoeuvres, e.g. the infusion of adrenaline or sympathectomy will be to shift both position and shape of these curves. It should be clear that to compare the control value for resistance on one curve with the experimental value on another curve is without meaning, and it is certainly unjustifiable to infer

Factors which regulate cerebral blood flow

that changes in vascular resistance are due to 'active' or 'passive' vasomotor changes.

The situation however in the cerebral circulation is in some respects more simple. The typical curve relating pressure and flow is shown (curve C) in Fig. 4.3(a) and shows that above a certain pressure, flow is independent of pressure and that in contra-distinction to most other vascular beds, over the physiological range $\Delta P/F$ varies only with pressure. If therefore any stimulus, e.g. the inhalation of carbon dioxide causing a rise in $P_{a,\,CO_2}$, is found to cause an increase in cerebral blood flow, irrespective of the accompanying changes in pressure, it is justifiable to infer an active vasomotor response. On the other hand, if two groups of subjects, e.g. normotensive and hypertensive, are compared and, as is shown in Table 4.1 they are found to have the same level of blood flow, it is not possible to infer any vasomotor mechanism.

Table 4.1. *Comparison of cerebral blood flow, vascular resistance and oxygen consumption in a group of normal subjects and hypertensive patients* (from Kety & Schmidt, 1948a. *J. Clin. Invest.* **27**, 476–83 and Kety, Shenkin & Schmidt, 1948a. *J. Clin. Invest.* **27**, 511–14)

	No.	B.P. (mmHg)	$P_{a,\,CO_2}$ (mmHg)	pH (units)	c.b.f. (ml 100g^{-1} min^{-1})	c.m.r. O$_2$*	c.v.r.† (units)
Normal subjects	14	86	42	7.40	54	3.3	1.6
Hypertensive patients	13	159	41	7.41	54	3.4	3.0

* Mean cerebral oxygen consumption.
† Calculated cerebral vascular resistance: units, mmHg ml^{-1} 100 g^{-1} min^{-1}.

This raises a difficulty of interpretation of values for resistance at high pressure. In this range, i.e. > 140 mmHg, the relation between pressure and blood flow has been obtained either in patients with some form of hypertension, e.g. essential hypertension (Kety, Shenkin & Schmidt, 1948a) or eclampsia (McCall, 1953) or following the administration of pressor drugs. It is not therefore certain whether the independence of flow and pressure holds in this range of pressure although Yoshida, Meyer, Sakamoto & Handa (1966) found that negligible changes occurred in flow when the thoracic aorta was clamped and the animals were made

4. Haemodynamic considerations

hypertensive. Because of the paucity of evidence on this point, some caution is recommended in interpreting the significance of changes in resistance at high levels of perfusion pressure.

Ideally, vasomotor responses can only satisfactorily be estimated when the vascular bed is perfused at constant pressure or flow. Even here there are difficulties. Green, Lewis & Nickerson (1943) have shown that with the type of pressure–flow curve (B) shown in Fig. 4.3(a), perfusion at constant pressure results in a ratio of values for flow under control and experimental conditions which is not constant at all perfusion pressures. On the other hand, if the vascular bed is perfused at constant flow, the ratio of control to experimental pressures has been found to be constant over a wide range of flow. Burton & Stinson (1960) have suggested that the explanation for this discrepancy lies in the fact that with constant pressure perfusion, increasing vascular tension of vessels should cause 'critical closure' of vessels; with constant inflow volume such closure could not occur since, with increasing vascular tension, the perfusion and transmural pressure also increase.

No completely satisfactory method for perfusing cerebral vessels at constant volume or pressure has yet been devised. Sagawa & Guyton (1961) attempted such perfusion in monkeys but the near linear relation between pressure and flow which they obtained suggested that the reactivity of the blood vessels had been seriously affected, possibly as a result of the extensive surgery required to prepare the perfusion circuit. Therefore at the moment, we have no experimental basis for assuming that comparison of ratios of flow at constant pressure or ratios of pressure at constant flow can yield reliable information about changes in vasomotor activity. The development of a satisfactory method of perfusing the isolated but innervated cerebral vascular bed is therefore urgently needed.

The concept of autoregulation

The observation by Fog (1937, 1939b) that pial vessels responded to abrupt changes in arterial pressure by altering their calibre to oppose the change and the subsequent demonstration by Kety & Schmidt (1948b) that over the physiological range of pressure cerebral blood flow remained virtually constant indicates that there is a remarkable mechanism in the cerebral circulation which

The concept of autoregulation

stabilizes blood flow. Because a neural component has been considered to be slight or negligible, it has for many years been believed that the control of cerebral blood vessels is essentially intrinsic. Cerebral blood vessels are thus said to exhibit autoregulation whether this is defined as the capability of an organ to regulate its blood supply in accordance with its needs or as the intrinsic tendency of an organ to maintain constant blood flow despite changes in arterial perfusion pressure. The response of

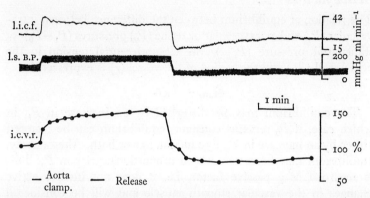

Fig. 4.5. The effect of a sudden rise in pressure in the arteries supplying the head (measured in the left subclavian artery, l.s.) upon blood flow in the left internal carotid artery (l.i.c.f.) and upon the calculated internal carotid vascular resistance (i.c.v.r.). At the marks, the thoracic aorta was clamped and released. The flow and pressure records indicate a larger initial rise in flow than pressure and hence a rise in calculated vascular resistance. Thereafter, blood flow slowly falls although pressure is held constant and vascular resistance rises proportionately. (Redrawn from Yoshida, Meyer, Sakamoto & Handa (1966), *Circulation Res.* **19**, 726–38, Fig. 4.)

flow and of calculated vascular resistance to abrupt changes in pressure is illustrated in an experiment of Yoshida *et al.* (1966) Fig. 4.5. As pressure is raised and held constant, flow at first rises passively but subsequently falls slowly and the calculated vascular resistance rises. It is to be presumed that if, in this experiment, pressure were to be maintained at the higher level for 5–10 minutes, flow would fall to control levels so that the steady-state pressure–flow curve in which flow is independent of pressure is obtained. Insofar as this degree of stability of flow is achieved, the cerebral vessels must be accounted as among the most active in the body.

What mechanisms are involved in this regulation? It must be

4. Haemodynamic considerations

said that in this respect the cerebral vessels possibly because of their inaccessibility have been less precisely studied than those in other vascular beds. The problem has however been extensively studied in peripheral vessels generally (Mellander & Johansson, 1968) and the factors involved have been classified as active, passive and extrinsic. The role of these factors with respect to cerebral vessels will be assessed in the following sections.

Active factors

The relation at equilibrium between the difference between intravascular (P_1) and extravascular or tissue (P_0) pressures ($P_1 - P_0$) = transmural pressure ($P_{t.m.}$), the circumferential tension in the vessel wall (T_c) and the radius of the vessel (r) is expressed as:

$$P_{t.m.} = T_c/r.$$

The equilibrium may be disturbed by an increase in P_1, in which case, if P_0 remains constant, equilibrium can be restored either by an increase in T_c, a reduction in r or both. Alternatively, equilibrium may be restored by a proportionate rise in P_0. This is regarded as a passive factor, i.e. it does not involve active changes in the vascular smooth muscle and will be considered later. The active changes in vascular smooth muscle as defined above may be described as myogenic: and the myogenic hypothesis has been proposed to explain the phenomenon of autoregulation in its restricted sense, i.e. the property of blood vessels to respond to changes in transmural pressure so as to maintain flow constant.

But the equilibrium relation, $P_{t.m.} = T_c/r$ can be affected in other ways, notably by drugs or chemical agents which primarily affect the tension of smooth muscle in the vessel wall, T_c without affecting $P_{t.m.}$. As an example of this class of agents we may consider the local action of carbon dioxide which may act directly or more probably (see Chapter 8) by altering the pH of the extracellular fluid ($pH_{e.c.f.}$). In either case, the effect of raised carbon dioxide or $[H^+]$ is to reduce T_c. If equilibrium is to be maintained we would predict either a reduction in r or a proportionate reduction in $P_{t.m.}$ and this could be achieved either by a reduction in P_1 or an increase in P_O. This readjustment of equilibrium where the disturbing factor is chemical, forms the basis of the 'chemical' hypothesis which seeks to explain autoregulation of blood in its less restricted sense, i.e. the ability of

The concept of autoregulation

an organ to regulate blood flow to meet its metabolic requirements. How far either of these two hypotheses (*a*) fit the known properties of vascular smooth muscle and (*b*) explain the phenomenon of autoregulation will be considered.

Responses of vascular smooth muscle to changes in transmural pressure. In 1902, Bayliss reported that if arterial pressure were reduced in the cat by stimulating the depressor nerve, the volume of the hind limb diminished and when normal pressure was restored, the volume of the limb increased above control; opposite results were obtained when arterial pressure was raised. Bayliss concluded that blood vessels relaxed when pressure was reduced or constricted when pressure was raised and since the limbs were denervated, that this was not a reflex but a local reaction of blood vessels. Further, Bayliss reported that he had observed similar responses in isolated perfused vessels but no evidence was given. These results were criticized by Anrep (1912) on a number of grounds; that the time course of events could be great enough for the changes to have been accounted for by circulating catechol amines or metabolites; that the changes in volume of the limb could be accounted for by other factors, e.g. changes in venous volume, rather than changes in cross-sectional area of muscular vessels and, finally, that he, Anrep, was unable to confirm Bayliss' findings that isolated vessels responded to changes in pressure in the isolated perfused preparation.

Although subsequent workers (Fog, 1937; Nicoll & Webb, 1955) have been able to demonstrate changes *in vivo* in small arteries in response to changes in pressure, the relatively slow response (see, for example, Fig. 6.1) renders an interpretation solely in terms of a local myogenic response susceptible to some of Anrep's criticisms.

Bayliss' hypothesis was revived by Folkow (1949) in order to explain the results of experiments in which blood-borne and neural factors were excluded and the calibre of small vessels was shown to oppose changes in pressure. Folkow (1964) has subsequently reviewed in detail the evidence which supports or contraverts the hypothesis. The two most serious obstacles to the hypothesis are (1) the fact that, in response to distension, the contractile elements would shorten and so abolish the stimulus

4. Haemodynamic considerations

(Gaskell & Burton, 1953; Johansson & Bohr, 1966) and (2) the mechanism implies a positive feedback which should lead to ever-increasing vascular resistance and arterial pressure. Folkow (1964) attempts to resolve these difficulties by considering the effect on blood flow of phasic contractions of smooth muscle rather than the sustained contraction in response to distension assumed under steady state conditions. Such phasic contractions would be analogous to those accompanying the spontaneously generated depolarization seen in intenstinal smooth muscle, the frequency of which is proportional to tension (Bülbring, 1962). Certainly, such action potentials have been observed in vascular smooth muscle (Funaki, 1961) and certainly capillary flow is known to be intermittent (Wearn, Ernstene, Bromer, Barr, German & Zchiesche, 1934; Vogel, 1947; Lee & Dubois, 1955; Wagner & Filley, 1963). But whether the properties of intestinal smooth muscle can be assumed in vascular smooth muscle and whether the pressure–flow relationships assumed by Folkow's model are valid remains to be determined. In this sense therefore there are some grave theoretical objections to the myogenic hypothesis and the decisive experiments which would test it under conditions of no flow have not yet been devised.

Is it possible to make any predictions about the behaviour of blood vessels in terms purely of the myogenic hypothesis and how far are these predictions borne out experimentally in the cerebral vascular bed? The myogenic hypothesis would predict that a fall in transmural pressure would cause dilatation of vessels and that this dilatation should be pre- rather than post-capillary. The transmural pressure could be reduced by lowering arterial or by raising tissue pressure. Certainly as is shown in Fig. 4.6, an abrupt reduction in arterial pressure (B) causes a passive reduction in flow followed by a slow rise, suggesting vasodilatation in response to sustained reduction in transmural pressure. It is also noteworthy that a more modest reduction in pressure (D) was without effect which suggests, in the absence of any change in venous or c.s.f. pressure that the myogenic response only occurs to a large stimulus, that changes in blood flow are an insensitive method of deducing (possibly) small changes in vessel calibre or that the change in (B) was due to some mechanism other than the myogenic. Johnson (1964) has suggested two further ways in which the myogenic response can be distinguished from a response

The concept of autoregulation

due to an altered chemical environment, or due to altered tissue pressure. If venous pressure is raised, the accompanying changes in tissue pressure will cause precapillary vasoconstriction; the accompanying changes in the chemical environment, precapillary

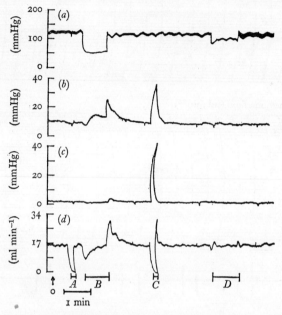

Fig. 4.6. (*a*) Carotid artery pressure, (*b*) spinal fluid pressure, (*c*) venous outflow pressure and (*d*) cerebral venous outflow in a dog whose head was perfused as shown in Fig. 5.6. (*A*) Position of zero flow determined by shunting the blood around the flowmeter. (*B*) The effect of reducing carotid and vertebral artery pressure to 60 mmHg. (*C*) The effect of occluding the cerebral venous outflow downstream from the pressure transducer. (*D*) Effects of reducing carotid pressure from 125 to 110 mmHg and vertebral pressure from 108 to 100 mmHg. (From Rapela & Green (1964), *Circulation Res.* **14**, 1, 205–11, Fig. 2.)

vasodilatation. As is shown in Fig. 4.7, abrupt increases in c.s.f. pressure, which would cause similar increases in venous pressure are without effect upon blood flow and it is just possible that, in this situation, the rise in tissue and venous pressure cancel each other's effects upon precapillary resistance. The example shown in Fig. 4.6 of a sudden rise in venous pressure (*C*) is too short for any interpretations to be made.

Johnson's (1964) second suggestion is that the myogenic response

4. Haemodynamic considerations

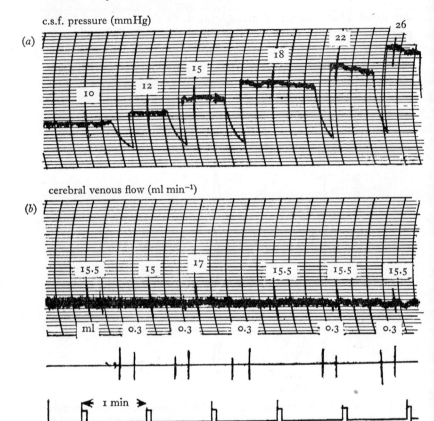

Fig. 4.7. (a) Cerebrospinal fluid pressure (mmHg) and (b) cerebral venous outflow (ml min^{-1}). Between the vertical marks below the flow record, 0.3 ml of saline was injected into the cisterna magna. The figures above the c.s.f. pressure and blood flow records indicate the steady levels achieved after each saline injection. (From Rapela & Green (1964), *Circulation Res.* **14**, 1, 205–11, Fig. 3.)

can be distinguished from that due to changes in tissue pressure by 'paralysing' the smooth muscle. One form of 'paralysis' of smooth muscle is to apply adrenaline (Bülbring & Kuriyama, 1963) which causes hyperpolarization and a reduction in the number of spontaneous spikes (Shanes, 1958). These manoeuvres are more suitable for conditions *in vitro* rather than *in vivo* because of the difficulty of administering these drugs while avoiding systemic changes. This is particularly true of adrenaline and almost certainly explains why it is that the action of adrenaline upon cerebral vessels is not known with any certainty (Sokoloff, 1959). One further possibility is that the known direct effects of acetylcholine upon isolated taenia coli (Burnstock, 1958; Bülbring & Kuriyama, 1963), lowering of the resting potential, increase in frequency and duration of spike could be abolished by atropine. If this is so, then the observation by Mchedlishvili & Nikolaishvili (1970) that atropine abolishes the response of cerebral vessels to a reduction of perfusion pressure could be otherwise interpreted, i.e. in terms of the myogenic rather than a neural hypothesis.

In summary, therefore, it can be said that some of the observed responses of cerebral vessels are consistent with predictions from the myogenic hypothesis. But these responses can be otherwise interpreted and the evidence is at present too slight and circumstantial for this matter to be resolved.

Responses of vascular smooth muscle to changes in the chemical environment. Autoregulation of blood flow has been explained in terms of the 'chemical' hypothesis which is a special application of a much more general problem – namely, how blood flow is matched to the metabolic needs of the organ which is perfused. The hypothesis implies that blood flow is actively regulated by metabolism and the simplest statement of this which was originally proposed by Roy & Brown (1879) and Gaskell (1880) is that the products of metabolism diffuse into interstitial fluid and affect vascular smooth muscle. This might occur under steady conditions where the rates of perfusion and metabolism were matched when the metabolites would 'set' the vascular 'tone' or under conditions where perfusion was inadequate or the level of metabolism raised, in which case the excessive outpouring of metabolites would cause vasodilatation, an increase in flow and a return to a normal perfusion/metabolism ratio.

This hypothesis is extremely plausible and is commonly invoked in textbooks of physiology to account for the increase in blood flow, for example which occurs in muscle during active exercise. In fact, each of the substances which might be thought to regulate flow during altered metabolism, oxygen, carbon dioxide, lactic acid, histamine, potassium, bradykinin, adenosine and phosphate have been carefully scrutinized and tested and it has been found

4. Haemodynamic considerations

that none of these variables can by itself account for the observed changes in blood flow (Barcroft, 1963; Berne, 1964; Haddy & Scott, 1968; Barcroft, Foley & McSwiney, 1970).

It is true that all these studies have been carried out in skeletal or cardiac muscle since, in these tissues, work and metabolic rate are more easily controlled than in brain. But unless it is proposed that vascular smooth muscle in cerebral vessels is radically different in its responses to the variables outlined above, the results of the various tests can be applied with some confidence to cerebral vessels. However, in the brain, the 'chemical' hypothesis has been stated in rather more definite terms than in other vascular beds (Betz & Heuser, 1967; Severinghaus, 1968) and the evidence although circumstantial suggests that most, if not all of the variables listed above may act first by altering $pH_{e.c.f.}$ and it is this which ultimately affects cerebral vascular smooth muscle. This evidence is considered fully in Chapter 8.

Passive factors

In an earlier section, it was mentioned that if the equilibrium described by $P_{t.m.} = T_c/r$ was disturbed by a rise in arterial pressure, one of the ways in which equilibrium could theoretically be restored would be a rise in tissue pressure. The mechanism of this might be as follows. A rise in arterial pressure causes a greater outward flow of fluid from the capillaries, a greater extravascular fluid volume and, if the organ is firmly or rigidly encased, a rise in tissue pressure. The rise in tissue pressure increases capillary and post-capillary resistance and hence a return of blood flow to control levels despite a sustained rise in arterial pressure.

Thus, the 'tissue pressure' hypothesis has been difficult to test in most vascular beds of the body because of the difficulty of measuring pressure within tissue (Hinshaw, Flaig, Logemann & Carlson, 1960). In the brain, however, c.s.f. pressure can be easily measured and, because of the relatively low compliance of the skull and duramater, can be regarded as an accurate index of tissue pressure. Even with the ease of this measurement, however, it is difficult to determine whether the characteristic relation between flow and pressure in cerebral vessels is caused by changes in tissue pressure or by active changes in arteriolar smooth muscle. To be certain, it would be necessary to distinguish between changes in resistance in pre- or post-capillary sections.

The concept of autoregulation

Then if the predominant change in resistance was in post-capillary vessels, changes in tissue pressure would be the most probable explanation. But this distinction cannot yet be made in cerebral vessels.

An alternative method is to consider the ways in which autoregulation of cerebral vessels can be impaired or abolished. Most, if not all, of these suggest damage to blood vessels which may be irreversible, e.g. following trauma (Fog, 1968), hypoxia (Häggendal, 1968; Freeman & Ingvar, 1968), pronounced intracranial hypertension (Zwetnow, Kjällquist & Siesjö, 1968). Insofar as this proposition is true, then autoregulation would be more satisfactorily explained by the myogenic rather than by the tissue pressure hypothesis.

The homestatic function of c.s.f.

If factors other than tissue or c.s.f. pressure are involved in the cerebral vascular response to changes in arterial pressure, there is every reason for believing that c.s.f. plays a crucial role in maintaining adequate blood flow to the brain when venous pressure is altered with straining or coughing, changes in posture or radial acceleration.

Static measurements of c.s.f. pressure in man in the lateral recumbent position showed frequency distribution ranging between 30 and 220 mm c.s.f. with a maximum frequency at 150 mm c.s.f. (Merritt & Fremont Smith, 1937), 148 mm c.s.f. (Masserman, 1934). Possibly because of differences in position and the effects of anaesthesia, rather lower average figures have been obtained in experimental animals, 100–130 mm saline (Hill, 1896), 90–150 mm saline (Bedford, 1935), 32–260 mm with a mean value of 105 mm (Goldensohn, Whitehead, Parry, Spencer, Grover & Draper, 1951). Measurement of the dynamic changes in c.s.f. pressure have confirmed that pressure rises immediately after cardiac systole and that there is a rise in c.s.f. pressure during expiration. The reported magnitude of these changes in pressure has depended on the method of measurement. Using undamped, open-ended manometers, Weed & McKibben (1919) found cardiac pulse changes in c.s.f. pressure of 1 mmHg, the respiratory changes being 4–5 mm saline. By using a wide bore needle and reducing the effects of damping, O'Connell (1943) found a cardiac pulse

4. Haemodynamic considerations

variation of 15 mm saline and a respiratory variation of 35 mm saline. Similar values were obtained by Goldensohn *et al.* (1951) using a wide bore needle and strain gauge and they also showed that when c.s.f. fluid pressure rose during hypercapnia, cardiac and respiratory fluctuations in pressure increased to some 70 mm saline.

These observations are of interest for they demonstrate first that although the arterial pulsations are damped in c.s.f., they are by no means negligible as suggested by Hill (1896). It is to be supposed that the arterial changes in c.s.f. pressure are possible because (*a*) the bony and membranous covering is not perfectly rigid and (*b*) the pressure changes are transmitted to the veins. Secondly, when c.s.f. pressure is raised (Bering, 1955; Wright, 1938; Ryder, Espey, Kimbell, Penka, Rosenauer, Podolsky & Evans, 1952), there is a corresponding increase in venous pressure, the veins cannot so easily collapse and the pulse pressure changes being undamped are more accurately reflected in c.s.f. pressure. Thirdly, the greater amplitude of the c.s.f. pressure fluctuation with respiration is further evidence of the dominant part played by changes in venous pressure in determining c.s.f. pressure.

In addition to changes with respiration, venous pressure is affected by changes in intrathoracic pressure brought about by coughing or straining. The high positive pressure thus induced even if transient would reduce perfusion pressure in cerebral vessels to levels sufficient to cause syncope. Further, since cerebral veins have no valves, this pressure would be transmitted to the small pial vessels which might easily burst. The rise in venous pressure however is transmitted proportionately to c.s.f. and the rise in c.s.f. pressure not only supports and protects small pial veins and capillaries but, by diminishing transmural pressure, effectively reduces cerebral vascular resistance and so allows a greater blood flow.

This protective and compensating mechanism appears to ensure that in the short term, i.e. for periods of a few seconds, perfusion pressure and flow are maintained at adequate levels. If intrathoracic pressure is held at high positive levels for longer periods, i.e. seconds to minutes as in the Valsalva manoeuvre, c.s.f. pressure starts to fall back toward normal levels (Bedford, 1936). The reason for this fall in pressure is not certain: it cannot be due to increased drainage into dural sinuses since pressure inside

Homeostatic function of c.s.f.

these is also high. It may be due to increased drainage into the veins surrounding the spinal cord. Whatever mechanism is involved, the progressive fall in c.s.f. pressure will cause a rise in transmural pressure and since venous pressure is unchanged, an adequate cerebral perfusion pressure can only be maintained if arterial pressure rises. As a result of vascular reflex activity, such a return of arterial pressure to normal levels or above occurs despite a fall in the volume of venous return and cardiac output. If this reflex activity is impaired by drugs (Finnerty, Guillaudeu & Fazekas, 1957) or disease (Johnson, Lee, Oppenheimer & Spalding, 1966), then the subject, faced with such a rise in intrathoracic pressure, becomes unconscious.

Effects of posture, tilting and acceleration

When a subject assumes or is tilted into an upright posture, c.s.f., jugular venous and carotid arterial pressures fall, Fig. 4.8. The opposite changes occur when the subject stands on his head. The changes in c.s.f. pressure are of the same magnitude as those in jugular venous pressure: the small decrease in pressures on standing upright and the large increase on being tilted head downwards is one among a number of pieces of evidence that suggest that the changes in venous pressure follow simply from regarding the contents of inferior and superior vena cavae as a hydrostatic column of blood rotating around the right auricle where the venous pressure varies only very slightly. The venous pressure thus appears to be the most important determinant of c.s.f. pressure; other possible causes have been considered in detail by Davson (1967). The relation between venous and c.s.f. pressure holds as positive or negative g is applied, that is over a range of -180 to $+80$ cm (Rushmer, Beckman & Lee, 1947), Fig. 4.9.

These changes in intracranial pressure have at least two important consequences for cerebral blood vessels and flow. First, if the difference between capillary and venous pressure remains relatively constant and venous pressure is in equilibrium with c.s.f. pressure, then the difference between capillary and c.s.f. pressure should remain constant irrespective of the position of the body. This implies that the pressure supported by capillary walls is similar at all points of the central nervous system and further

4. Haemodynamic considerations

that capillaries are protected against sudden changes in posture of the body or in intrathoracic pressure.

Secondly, the changes in intracranial pressure outlined above help to maintain adequate cerebral blood flow despite wide fluctuations of arterial pressure. The mechanisms involved are not yet completely clear but the more important features may be

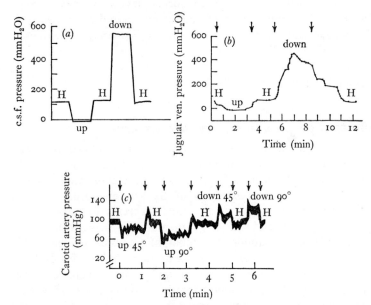

Fig. 4.8. (a) The changes in c.s.f. pressure (mm H_2O) which occurred when the subject was tilted as indicated by the arrows, from the horizontal (H) to the vertical head-up (up) position and to the vertical head-down (down) position. (b) Changes in jugular venous pressure (mm H_2O) during the same sequence of tilts. Symbols as in (a). (c) Changes in carotid artery pressure (mmHg) during tilting to the 45° and 90° head-up and head-down positions. Other symbols, as in (a). (Redrawn from Loman & Myerson (1935), *Am. J. Psychiat.* **92**, 792–813, Figs. 1, 2, 3.)

deduced from Fig. 4.8 and Fig. 4.10 which illustrates the response of a human subject to $+4.5\,g$ for fifteen seconds. The first obvious feature is the fall in arterial pressure which is roughly proportional to the magnitude of acceleration. In the example given in Fig. 4.10, mean arterial pressure fell briefly to less than 10 mmHg which, by itself would bring about cessation of cerebral blood flow and unconsciousness. However, venous pressure also

Effects of posture, tilting and acceleration

fell by 60 mmHg so that at no time was cerebral perfusion pressure less than 50 mmHg and we should predict that although cerebral flow fell, it was not severe. The subject experienced only transitory visual dimming. The fall in venous pressure therefore constitutes

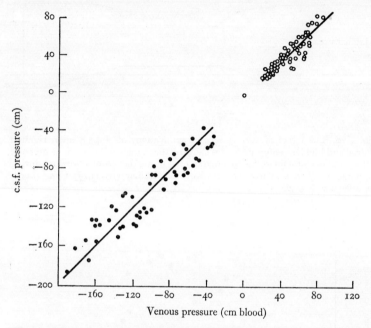

Fig. 4.9. The relation between simultaneous measurements of cerebrospinal fluid pressure and venous pressure during positive and negative radial acceleration, -6.5 to $+6.5g$. Open circles negative, closed circles positive acceleration. (From Rushmer, Beckman & Lee (1947), *Am. J. Physiol.* **151**, 355–65, Fig. 3.)

an important compensating mechanism whereby perfusion pressure is maintained at levels which ensure an adequate blood supply to the brain.

However, as is shown in Fig. 4.11, the changes in mean arterial pressure are always greater than those in venous pressure and there will thus always be a change in perfusion pressure and hence in blood flow. In the experiments of Henry, Gauer, Kety & Kramer (1951), cerebral venous oxygen saturation remained virtually constant with the application of up to $4.5\,g$ which suggests that cerebral blood flow was held within even narrower limits than

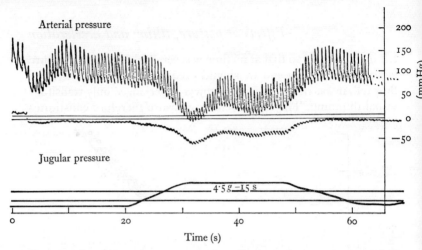

Fig. 4.10. The effect of two 15 second applications of +4.5 g upon cerebral arterial and jugular venous pressures showing that a high arterio-venous pressure differential is maintained due to a marked fall in jugular venous pressure during acceleration. (From Henry, Gauer, Kety & Kramer (1951), *J. Clin. Invest*, **30**, 292–300, Fig. 4.)

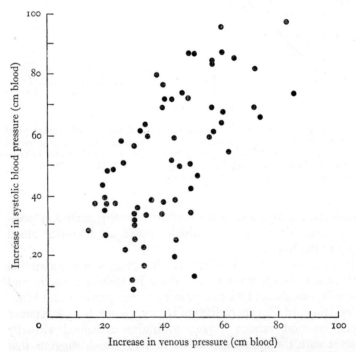

Fig. 4.11. The relation between the increase in systolic arterial pressure (cm blood) and increase in external jugular venous pressure (cm blood) during radial acceleration, −1.5 to −6.5 g. (From Rushmer, Beckman & Lee (1947), *Am. J. Physiol.* **151**, 355–65, Fig. 6.)

Effects of posture, tilting and acceleration

the calculated changes in perfusion pressure, Fig. 4.12. The results obtained by Scheinberg & Stead (1949) in patients in response to tilting were somewhat different since cerebral blood flow fell from an average of 61.5 to 51.6 or by 21 per cent of control. At the same time, mean arterial pressure fell by 34 per cent of control

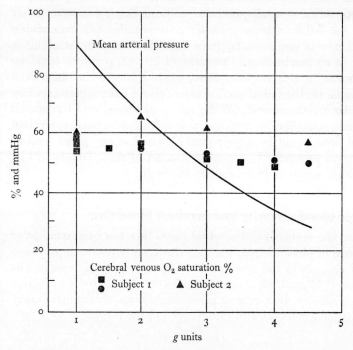

Fig. 4.12. The curve shows the relationship between mean arterial pressure measured at head level (mmHg) and units of g. The solid symbols show the cerebral venous oxygen saturation (per cent) for values of g units; ■ ● subject 1; ▲ subject 2. Cerebral oxygen venous saturation remains constant in spite of the progressive fall in arterial pressure with increasing positive acceleration. (From Henry, Gauer, Kety & Kramer (1951), *J. Clin. Invest.* **30**, 292–300, Fig. 1.)

and calculated vascular resistance by 16 per cent. The discrepancies between the results of Scheinberg & Stead (1949) and of Henry *et al.* (1951) may be explained in terms of the methods used. Scheinberg & Stead (1949) used the nitrous oxide method of determining cerebral blood flow which requires at least ten minutes steady state conditions. Under the conditions of tilting,

4. Haemodynamic considerations

these are hardly likely to be achieved since the changes in venous and c.s.f. pressure are always slower than in arterial. On the other hand, it is worth noting that in the paper of Henry et al. (1951), a fall of 10 per cent oxygen saturation is illustrated as coinciding with the maximum fall in arterial pressure: thus the results from the two series may be closer than at first appears.

Both series of results, however, indicate that some factor other than the fall in venous pressure is responsible for maintaining blood flow at levels not far from control whether the stimulus is a tilt or an acceleration. The changes in c.s.f. pressure could be important in this respect. Thus when a subject is tilted head upwards, (a) the fall in c.s.f. pressure could reduce post-capillary vascular resistance and, (b) the fall in perfusion and transmural pressure could lead to vasodilatation and, in consequence, cerebral blood flow would be higher than expected at any given level of perfusion pressure. The combined effect of these factors will be to ensure adequate blood flow to the brain.

Whole blood viscosity and cerebral blood flow

So far, the regulation of cerebral blood flow has been considered in terms of physical factors, e.g. perfusion or transmural pressures or of changes in the cross-sectional area of cerebral vessels. The Hagen–Poisseuille equation also shows that blood flow is a function of its viscosity. In a tube of given dimensions, this equation may be written for the flow of plasma (Q_p) in the form:

$$Q_p = G_p . P,$$

where G_p which is the conductance for plasma replaces the quantity $\pi r^4 / 8 \eta_p . \eta_w l$ (where η_p is the relative viscosity of plasma, η_w is the absolute viscosity of water, l is the length of the tube of radius, r). Similarly, for the flow of blood (Q_b) in the same tube, we may write:

$$Q_b = (G_p / \eta^*) . P,$$

where η^* is the apparent viscosity of the blood relative to plasma under the particular conditions of measurement. The value for η^* depends not only upon the volume fraction of red cells (the haematocrit value) but also upon the shearing stress applied, the radius of the tube and other factors.

Whole blood viscosity and cerebral blood flow

Viscosity and haematocrit

A number of equations relating relative viscosity to the volume fraction in more concentrated suspensions have been put forward, i.e. the relation between the relative viscosity of plasma and the protein concentration or the relation between the relative viscosity of blood and haematocrit. Few of these have taken account of the shear-dependency of the viscosity of blood. However the viscosity of blood has been experimentally related to haematocrit at defined shear rates by Bollinger, Luthy & Jenny (1967). Using a Wells–Brookfield viscometer (Wells, Denton & Merrill, 1961), they found a linear relation between whole blood viscosity and haematocrit over the range 35 and 52 per cent at three different shear rates. Over this range of haematocrit and at a shear rate of $11.5\ s^{-1}$, viscosity doubled, being 4.5 cP at a haematocrit of 35 per cent and 9.0 cP at 52 per cent. Similar values were found in the dog (Häggendall & Norbäck, 1966). These workers established that a linear relation existed between whole blood viscosity and haematocrit at different shear rates and they extended their observations to show that if red cells were diluted with dextran (mol. wt. 70000), the slope of the curve relating viscosity and haematocrit was not affected when haematocrit was reduced below 30 per cent. With high molecular dextran (mol. wt. 5×10^6) on the other hand, this relation was less obviously linear while the slope of the curve was, as would be expected, considerably steeper.

The relation between arterial haematocrit and cerebral blood flow

The effect of altered arterial haematocrit upon cerebral blood flow has been studied in patients with various types of chronic anaemia and polycythaemia. Kety (1950) reported that cerebral blood flow was reduced in patients with long-standing polycythaemia, i.e. with a haematocrit of > 65 per cent. In patients with pernicious anaemia and with a haemoglobin concentration of $< 7\ g\ 100\ ml^{-1}$, cerebral blood flow was found to average 96 ml $100\ g^{-1}\ min^{-1}$ (average control flow, $65\ ml\ 100\ g^{-1}\ min^{-1}$) (Scheinberg, 1951) while an approximately 50 per cent increase in cerebral blood flow was observed in patients with sickle cell anaemia (Heyman, Patterson & Duke, 1952).

4. Haemodynamic considerations

The effects of chronic anaemia or polycythaemia upon cerebral blood vessels are likely to be complex for, in addition to the undoubted changes in haematocrit and whole blood viscosity, there are also likely to be changes in cardiac output, possible changes due to the circulation of blood of reduced oxygen content and possible reflex changes. Precisely which factors are involved is uncertain. It may be of importance that in both the latter series in patients with anaemia, cerebral oxygen consumption was found to be markedly reduced. Cerebral oxygen consumption rose when the patients with anaemia (but not control subjects) inhaled high oxygen despite a small fall in cerebral blood flow and oxygen consumption returned towards normal levels (but not to them) when appropriate therapy was given. These observations suggest nutritional or other changes in the brain and blood vessels accompanying chronic anaemia and may well enhance the flow response to reduction in arterial haematocrit.

Further experiments have been carried out by Häggendal, Nillson & Norbäck (1966) in which cerebral blood flow was measured in the dog as haematocrit was altered by replacing volumes of blood with plasma or dextran of high and low molecular weight. The relation between cerebral blood flow and arterial haematocrit which they obtained is shown in Fig. 4.13, from which it can be seen that over the range 30–50 per cent haematocrit, cerebral blood flow is unaffected. Blood of high viscosity, i.e. with haematocrit 50–70 per cent has little effect upon cerebral blood flow, but as haematocrit is reduced below 30 per cent, blood flow steadily increases. In an accompanying paper (Häggendal & Norbäck, 1966) it was shown that if the viscosity of arterial blood of altered composition was also measured, a similar type of relation between blood flow and viscosity was obtained to that shown in Fig. 4.13. In this curve, the inflection occurred between 5 and 10 cP measured at a shear rate of 23 s^{-1}. Above this range, blood flow was independent of changes in blood viscosity, i.e. up to 35 cP; as viscosity was reduced below 5 cP, cerebral blood flow increased rapidly.

Blood viscosity and the size of blood vessels

The original observation of Fåhraens & Lundqvist (1931) that the apparent viscosity of blood diminishes in proportion to the

Blood viscosity and the size of blood vessels

radius of the tube has been amply confirmed (Bayliss, 1952; Prothero & Burton, 1962). The most important reason for this phenomenon is the presence of a marginal sheath of low viscosity due to the axial arrangement of red cells. The width of this sheath is 1 to 3 μm and its effect is most pronounced in small vessels: in tubes comparable to the size of arterioles, i.e. 20 μm

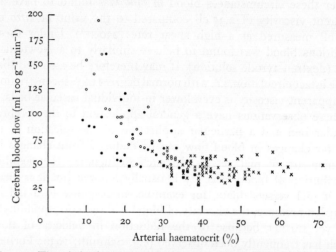

Fig. 4.13. The relation between cerebral blood flow in the dog (ml 100 g^{-1} min^{-1}) and arterial haematocrit (%). Haematocrit was altered by the replacement of 100–150 ml of blood by similar volumes of plasma, low molecular weight dextran (l.m.d., mean mol. wt. 40,000), high molecular weight dextran (h.m.d., mol. wt. 5×10^6) and concentrated red cells. Symbols as follows: × control; ○ homologous plasma; ● l.m.d.; ■ h.m.d. Mean arterial blood pressure > 100 mmHg; arterial oxygen saturation (S_{a, O_2}) > 85 per cent; arterial carbon dioxide tension (P_{a, CO_2}) = 20–35 mmHg. (From Häggendal, Nilsson & Norbäck (1966), *Acta Chir. Scand.* Suppl. **364**, 3–12, Fig. 3.)

diameter, the apparent viscosity has been calculated to be two thirds that in a larger tube, > 300 μm.

The viscosity of blood is also affected by the velocity of blood flow because of the anomalous non-Newtonian flow properties of blood. The apparent viscosity increases with reduced velocity of flow. Thus in the normal vascular bed, apparent viscosity falls steadily from the artery to the capillary and then rises again in the veins which have a low flow velocity in wide-bore vessels (Haynes, 1961; Whitmore, 1968). The apparent viscosity *in vivo* is therefore a complex function resulting from changes in viscosity

4. Haemodynamic considerations

in each segment of the vascular bed regarded as a series system. The measurement of apparent viscosity *in vivo* is difficult because of the necessity of excluding all active participation of the vessel walls; this has been accomplished by obtaining maximum dilatation of vessels following exercise and the local infusion of isoproterenol and acetylcholine (Djojosugito, Folkow, Öberg & White, 1970). Under these circumstances blood *in vivo* was found to have an apparent viscosity of 2.35 cP compared to the value *in vitro* of 4.88 cP measured at a high shear rate (230 s^{-1}). Under these conditions blood was found to behave similarly to a Newtonian fluid (dextran–tyrode solution). It may therefore be assumed that in the intact circulation, i.e. with normal degrees of vasoconstriction the apparent viscosity is even lower in individual vascular beds.

These observations have a general application to the cerebral vascular bed and a particular application to the question as to how far changes in blood flow can be deduced from changes in the diameter of pial vessels. In such deductions, the possible contribution of blood viscosity is usually ignored but it is clear that if pial vessels dilate, for example, in response to a fall in perfusion pressure, the viscosity of blood will rise correspondingly and yet further because of the reduction in velocity of flow which has frequently been observed (e.g. Schmidt, 1936; Forbes, Nason & Wortman, 1937). Under these circumstances any changes in vascular resistance caused by dilatation of vessels could be offset by changes in blood viscosity; and estimates of changes in blood flow could be quite erroneous.

5 THE CEREBRAL CIRCULATION: EXPERIMENTAL APPROACHES

Historical introduction

It seems clear that the significance of the blood supply to the brain was known to the Greeks for they named the great arteries of the neck carotid (Karotein = to stupefy) because they had observed that judicious pressure on these vessels caused anaesthesia sufficient for ritual circumcision to be carried out. But apart from a number of anecdotal types of observation which were made in succeeding centuries (for references, see Hill, 1896), no satisfactory experiments were possible until the anatomy of the blood supply to the brain was established. The first representation of the cerebral vessels which would be recognizable today was by Vesalius (1543). In his *De Humani Corporis Fabrica*, the arrangement of the dural sinuses and cerebral veins is well shown, Fig. 5.1 but the union of the carotid artery and jugular vein with the transverse sinus indicates a pre-Harveian grasp of the circulation.

The first tolerably accurate illustrations of the arterial arrangement at the base of the human brain were by Casserius (1645) and Veslingius (1651). The former omitted one of the posterior communicating arteries and the latter the anterior communicating arteries although, in a later illustration for the compendium of Thomas Bartholinus (1684), these arteries were inserted. The contribution of Willis' *Cerebri Anatome* (1664) has been discussed in Chapter 1.

A further anatomical treatise which exerted great influence over those who studied the cerebral circulation in the nineteenth and twentieth centuries was *'Observations on the Structure and Function of the Nervous System'* by Alexander Monro the younger (1783). Though more general in scope than *Cerebri Anatome* as its title implies, this book focused attention on the peculiar position of blood vessels of the brain which were encased in the rigid structure of the skull and membranes. The shortcomings of

5. *The cerebral circulation: experimental approaches*

Monro's observations have been discussed in the previous chapter, but the observations were to stimulate the study of two main problems in the nineteenth century: the origin of pulsations of the brain and the existence of vasomotor reactions of cerebral blood vessels.

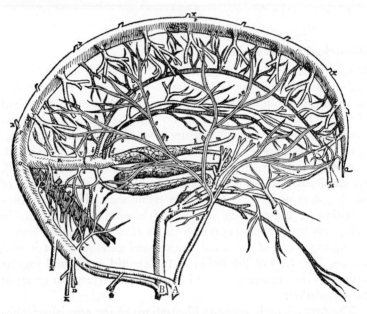

Fig. 5.1. The cerebral circulation of man showing the dural sinuses and cerebral veins together with the junction of the internal jugular vein (A) with the internal carotid artery (B). (From Vesalius (1543), *De Humani Corporis Fabrica*, p. 305.)

Ridley (1700) was probably the first to undertake a systematic study of pulsations of the brain and he concluded that they were arterial in origin. Haller (1755) confirmed this but also noted pulsations which were related in time to the respiratory cycle. Haller was the first to make use of trephine holes in the human skull for this purpose and this method was modified by Ravina (1811) who inserted a wooden cylinder into the trephine hole and sealed it with a watchglass. The method was further modified by Donders (1849, 1850) who screwed an air-tight window into the trephine hole and observed that when all the air had been evacuated, pulsations of the brain were no longer visible. He was

Historical introduction

thus the first to realize that the pulsations of the brain described by earlier workers were in fact an artifact because the trephine hole or other continuous rigidity defect allowed the brain to expand and c.s.f. to escape: when the defect was stopped by a rigid window, the ability of arteries to distend was reduced or abolished. Leiden (1866) introduced one further refinement which was to include a system for irrigating the surface of the brain beneath the 'window' and through which drugs could be introduced.

Thus, by the middle of the nineteenth century, techniques were available for the study of pial vessels *in situ* and it is of some interest that at this time, there was general interest among physiologists in vasomotor reactions and the reflex control of peripheral vessels. Weber (1831) is generally credited with having been the first to propose that the calibre of peripheral arteries was affected by nervous influences, basing this view upon the changes of skin colour which occurred with emotional upsets. Henle (1841) was the first to demonstrate the wealth of smooth muscle fibres in the media of small arteries and he gave the name 'vasomotor' to the nerves which supplied these fibres although, at the time, he did not know their true function.

The reflex control of blood vessels was studied by Claude Bernard (1851) who showed that excision of the superior cervical ganglion or the cervical sympathetic nerve caused a rise in temperature and flushing of the homolateral ear and face in the rabbit. He did not at first believe that the rise in temperature could be fully accounted for by the vasodilatation. However, following Brown-Sequard's demonstration (1853) that stimulation of the cervical sympathetic nerve caused vasoconstriction and a fall in skin temperature, Bernard carried out a further series of experiments (1853) which confirmed these findings, and he later published a review (1863) of the evidence at that time, which established clearly the reflex control of peripheral vessels and the important role of the sympathetic nerves. Similar conclusions had been reached independently by Waller (1853).

It was in this period that Donders carried out his first experiments and he showed that when the animal was made asphyxic, the pial vessels dilated (1849, 1850). In a later paper (1859), Donders confirmed the findings of Bernard and Brown-Sequard and showed that when the cervical sympathetic nerves were stimulated, the pial vessels on the same side constricted. Essentially

5. The cerebral circulation: experimental approaches

similar observations were made using the skull window technique by Kussmaul & Tenner (1857), Ackermann (1858), Schultz (1866), Reigal & Jolly (1871) and Lewin (1920).

However, a number of workers, using the same technique with some modifications were unable to confirm Donder's findings (for references, see Forbes & Wolff, 1928) and the question as to whether cerebral blood vessels had vasomotor properties which were independent of changes in arterial blood pressure remained open until the 1930s. It is not difficult to see why. Since few simultaneous measurements of pressure were made, it was impossible to say whether the observed changes in pial artery calibre represented vasomotor activity or the passive following of changes in arterial pressure. Secondly, the use of different species, different types and depths of anaesthesia and different techniques to reflect the dura mater and insert the window would be expected to give very varied results. Thirdly, it is possible that in many experiments the pial vessels were rendered unreactive by trauma or the presence of blood on the surface of the brain (Echlin, 1942) while the complete uncertainty as to the levels of oxygen and carbon dioxide partial pressures and pH of arterial blood would have made the interpretation of results difficult.

Some of these objections were partially overcome by Francois-Franck (1887) and Hürthle (1889) who measured the pressure from the cut end of the internal carotid artery while Gaertner & Wagner (1887) measured the volume of cerebral venous flow in an attempt to correlate changes in pial vessel calibre with blood flow. As will be shown later, this attempt was doomed to failure because of the large volume of contaminating blood from extra-cranial sources.

Measurements of intracranial pressure

Mosso (1881) reported on a number of experiments in which he had deduced changes in the volume of intracranial contents by observing or feeling movements of the fibrous membrane which covered defects in the cranial vault in human patients. These observations were of limited value since no satisfactory recordings could be made of the changes; but the principle commended itself to Roy & Sherrington who introduced a method of measuring changes in intracranial pressure using an oncometer (1890). These

Measurements of intracranial pressure

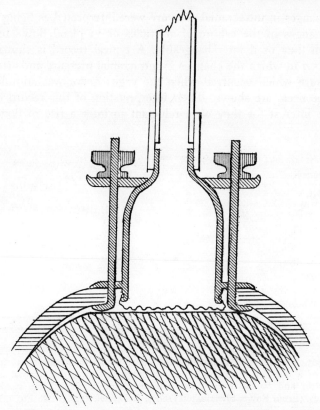

Fig. 5.2. The 'oncometer' used by Roy & Sherrington. The metal capsule was introduced into the trepan hole (about 22 mm in diameter in the dog) and clamped onto the bone by means of screws after the underlying dura mater had been removed. The cavity of the capsule was separated from brain by a thin membrane and was connected to a recording system which consisted of a piston connected to a lever. Protrusion of the brain and blocking of the trepan hole was prevented by a small (1 g) weight acting on the piston. (From Roy & Sherrington (1890), *J. Physiol.* **11**, 85–108, Fig. 1.)

workers, and later Bayliss & Hill (1895), introduced a rigid metal chamber into a trephine hole in the skull of a dog as shown in Fig. 5.2. The chamber cavity was separated from the brain and c.s.f. by a thin membrane and pressure changes in the oncometer caused by volume changes within the skull were transmitted to a recording lever. These workers simultaneously measured femoral arterial pressure and in some experiments, the jugular venous pressure, but none of the pressure measurements was calibrated.

5. *The cerebral circulation: experimental approaches*

Changes in intracranial pressure were interpreted as being due to changes in the calibre of arterioles or in blood flow – terms which they used interchangeably. A typical record is shown in Fig. 5.3 in which the changes in intracranial pressure and arterial pressure which occurred when the vagus nerve was stimulated in the neck, are shown. Their interpretation of this record is of some interest for they were reluctant to infer a rise in flow, as

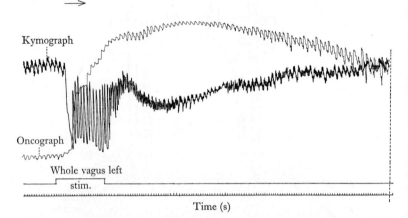

Fig. 5.3. The effect of stimulating the intact vago-sympathetic trunk in the neck of a dog for the periods indicated, upon: above, the femoral artery pressure; and below, intracranial pressure ('oncograph'). Time marks in seconds. (From Roy & Sherrington (1890), *J. Physiol.* **11**, 85–108, Fig. 9.)

they did for example when intracranial pressure rose in response to asphyxia. The reason for this reluctance was that the rise in intracranial pressure which occurred in the face of a marked fall in mean arterial pressure posed an awkward exception to their general conclusion that cerebral blood vessels did not possess vasomotor properties but passively followed changes in arterial pressure. Roy & Sherrington (1890) suggested that the 'cerebral congestion' shown in Fig. 5.3 was due to a rise in cerebral venous pressure though no records of such a rise in pressure were given.

Of course, it is impossible to infer any changes in cerebral blood flow from this type of record. Nor is it possible to be certain that the changes in intracranial vessels are due to changes in arteriolar calibre for they could equally well be due to changes in the volume of venous capacitance vessels the pressure in which,

Measurements of intracranial pressure

as is discussed in detail in Chapter 4, is more likely to affect the pressure of c.s.f. However, it is possible that by inserting an oncometer with a thin flexible membrane, these workers introduced an artifact which allowed changes in arterial pressure and distensibility to be transmitted more completely than normal. Certainly, the visible arterial pulsations in the 'oncograph' record in Fig. 5.3 would be consistent with a low inertia, moderately damped recording system.

Revival of the skull window technique

The technique of Roy & Sherrington (1890) has been criticized in some detail not only because of the assumptions which are inherently untenable but because the authoritative statements made by these workers, Bayliss & Hill (1895), Hill (1896) and Hill & McLeod (1900) on the basis of experiments using this technique effectively inhibited further experiments in this field for the next twenty years or more and gave rise to the erroneous opinions that cerebral vessels (*a*) did not possess vasomotor properties and (*b*) were not reflexly controlled.

The introduction of a satisfactory method for the measurement of cerebral blood flow had to wait until the late 1940s but an improved approach to the study of pial vessels and their response to various stimuli was adopted by Forbes in 1928. He revived the skull window technique of Donders and improved it by using a micrometer scale with which to quantify the changes in vessel calibre; photomicrography to record the changes and simultaneous measurements of arterial and, in some experiments, venous pressure. The method proved to be technically simple and capable of giving reproducible results and it formed the basis for a large number of experiments carried out in the next ten years in which the responses of pial vessels to a wide variety of stimuli were studied.

A typical photographed response is shown in Fig. 5.4 in which the pial artery and some veins are shown before and during faradic stimulation of the central end of the cut vagus nerve in the neck. The measured changes in pial artery diameter together with changes in femoral artery and c.s.f. pressure are shown in Fig. 5.5. This is essentially the same experiment as that shown in Fig. 5.3 and the same fall in arterial pressure and rise in c.s.f. pressure are shown. But in Fig. 5.4, there is an unequivocal

5. The cerebral circulation: experimental approaches

increase in pial arterial calibre and this is strongly suggestive of a vasomotor action which is independent of changes in systemic arterial pressure.

The results of these experiments are also of value because, being carried out by a small number of workers in two laboratories over a number of years, the techniques and form of anaesthesia became standardized so that it is possible to compare the results

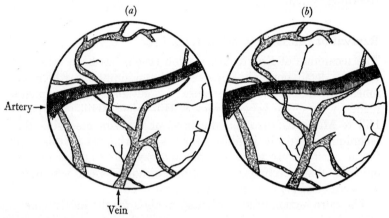

Fig. 5.4. The effect of stimulating the central cut end of the cut vagus nerve upon the calibre of pial vessels. Cat, anaesthetized with 1 per cent, isoamylethylbarbituric acid given intraperitoneally. (a) Control, 9 min before stimulation; arterial diameter 250 μm. (b) After 60 s of stimulation of left vagus nerve; arterial diameter 315 μm. The veins are unaffected. (Redrawn from Forbes & Wolff (1928), *Archs Neurol. Psychiat.*, Chicago **19**, 1057–86, Fig. 9. Copyright 1928, American Medical Association.)

of one series of experiments with another. At the same time, the technique has obvious limitations not all of which have been recognized. Thus, very few measurements of blood gas tensions or pH were made except in those experiments in which the effect of these variables was being measured. The results of such experiments indicated, and this has been abundantly confirmed since, that the magnitude of the responses to practically any stimulus is affected by the resting level of P_{a,O_2}, P_{a,CO_2} and pH_a. It is likely that these values varied between experiments and with time in individual experiments since it is usual to find that both P_{a,CO_2} and pH_a fall progressively in anaesthetized animals whether they are breathing spontaneously or being ventilated mechanically.

Secondly, the method yields only indirect information about

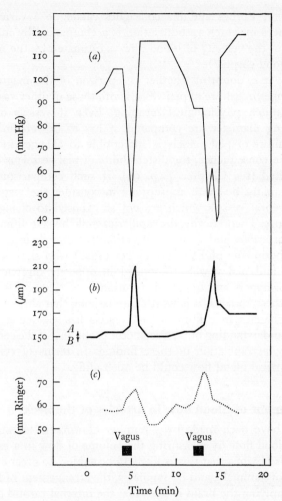

Fig. 5.5. The effect of vagal stimulation upon (a) femoral arterial pressure (mmHg), (b) pial artery diameter (μm) and (c) c.s.f. pressure (mm Ringer's solution). Cat ventilated mechanically. The change in arterial calibre, indicated A–B, is an estimate of the accuracy of measurement. (From Forbes & Wolff (1928), *Archs Neurol. Psychiat.*, Chicago **19**, 1057–86, Fig. 8. Copyright 1928, American Medical Association.)

the level of blood flow. There is no satisfactory formula for converting changes in the outside diameter of a single arteriole into changes of blood flow and which takes into account distensibility of the vessel, possible changes in blood viscosity, etc.

5. The cerebral circulation: experimental approaches

Furthermore, Forbes and his colleagues catalogue a variety of circumstances in which vasodilatation is accompanied by marked reduction in the velocity of blood flow as estimated by the movement of blood corpuscles.

Thirdly, it is doubtful whether comparison of the magnitude of the changes in calibre of pial vessels with those in other vascular beds has much physiological meaning. Even if vessels of the same resting diameter are compared, it has to be remembered that the calibre of pial vessels is susceptible to changes in c.s.f. pressure: in consequence, the distensibility of pial vessels is likely to be limited (see Chapter 4, p. 90). If such comparisons are sought, it might be more appropriate to consider the responses of intracerebral vessels which are not so exposed to changes in c.s.f. pressure. Theoretically, the capillaries could behave differently from pial arterioles and some confirmation that they do is afforded by the experiments of Kubie & Hetler (1928) who showed that capillaries and pial vessels responded in opposite directions to injection of hyper- and hypotonic solutions.

For all these reasons, it is worth emphasizing that although the study of changes in calibre of pial vessels has yielded a more complete understanding of the vasomotor properties of the cerebral circulation, extrapolation of these findings in terms of cerebral or even cortical blood flow could be misleading.

Measurement of blood flow in arteries of the neck

Attempts have been made by a number of workers to estimate cerebral blood flow by measuring the volume of flow in a carotid artery. It is obvious that this is likely to be subject to great errors. In laboratory animals, and in ungulates, the arrangement of blood vessels supplying the brain is such that the internal carotid artery supply is virtually completely replaced by the rete mirabile (Batson, 1944; Baldwin, 1964); in consequence, it is a matter of complete speculation as to what proportion of common carotid blood flow is destined for the brain. Under these circumstances, it is more appropriate to speak of 'cephalic flow' (Lucas, Kirschbaum & Assali, 1966).

In view of these uncertainties, attempts have been made to isolate cerebral vessels but, as would be expected, this involves such extensive and complicated ligation of vessels (Bouckaert &

Heymans, 1935; Green & Denison, 1956; Sagawa & Guyton, 1961) that there is grave risk to the cerebral vessels themselves. The fact that in the studies quoted, the volume of cerebral blood was low or the vessels unreactive indicates that this type of approach is unprofitable.

More complete separation of intra- and extracranial circulation exists in primates and in man and to a certain extent in rabbits (Schmidt & Hendrix, 1938). But it is clear that even in the monkey, the surgery necessary to implant flow devices in the internal carotid arteries can lead to interruption of reflex pathways and loss of reactivity of cerebral vessels (Dumke & Schmidt, 1943). Even with the use of more sophisticated and less damaging types of flowmeters for measuring flow in common or internal carotid arteries in monkeys (Yoshida, Sakamoto, Meyer & Handa, 1966), there is still the possibility of substantial extracranial blood flow being included in 'cerebral' blood flow measurements (Rapela, Green & Denison, 1967).

Measurement of venous outflow

Many of the objections made in the previous section also apply to the method of measuring the volume of cerebral venous outflow as an estimate of cerebral flow. In dogs and cats, the anastomoses between extracranial structures, cranial vault and intracranial veins and sinuses are legion and even if the more obvious anastomoses are interrupted, it is likely that a greater volume of blood would flow via other channels, e.g. orbital veins. Ingvar & Soderberg (1956) devised a method in the cat whereby sagittal sinus blood flow was measured after the cranial vault had been reconstituted with an artificial membrane. They used sagittal blood flow as an index of cortical blood flow but quite apart from the uncertainty as to whether other parts of the brain drain into the sagittal sinus, there is also the question as to whether the sagittal sinus drains facial structures. It is extremely difficult to test this latter possibility satisfactorily. For example, Purves & James (1969) showed that in the foetal sheep and newborn lamb dye injected into the muscular branches supplied by the external maxillary artery was not present in measurable amounts in sagittal venous blood. Although this gave some confidence that the sagittal sinus did not drain temporal or parietal extracranial areas,

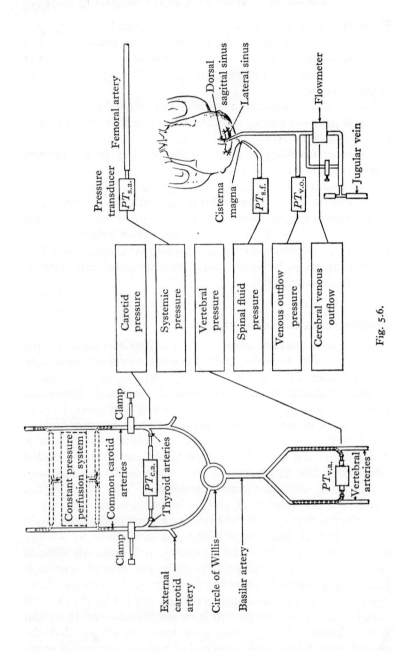

Fig. 5.6.

Measurement of venous outflow

it was clearly not possible to exclude that other areas, orbit and face were not so drained.

A more ambitious system was devised by Rapela & Green (1964) in which the outflow of straight and sagittal sinuses was measured by an electromagnetic flowmeter while the head was perfused either mechanically or naturally. The system is shown graphically in Fig. 5.6. Because of the numerous anastomoses, the lateral sinuses were blocked with bone wax to prevent contamination from extracranial sources. It was then necessary to compute the volume of brain being drained by the straight and sagittal sinus and an estimate of this was obtained by injecting a silicone rubber solution retrogradely. From the distribution of this injectate, the volume of brain drained was estimated as being between 50 and 70 per cent of the total. This method can only give a crude estimate of the volume of blood perfusing the brain and even with the extensive ligation cannot guarantee that the volume of blood measured is only coming from the brain. Possibly as a result of the surgery necessary for this preparation, very low rates of blood flow were obtained and the sensitivity of the blood vessels appeared to be impaired.

These results indicate that the cat and the dog are not satisfactory experimental animals for the measurement of cerebral blood flow when the method involves an estimation of carotid arterial or venous blood flow. A more satisfactory animal is the monkey in which there appears to be a much more complete separation of the intra- and extracranial circulations. Thus Reivich (1964)

Fig. 5.6. Procedure to measure cerebral venous flow and perfusion pressure in the cerebral vascular bed.

Right. Blood from the confluence of the sinuses was diverted through an electromagnetic flowmeter and returned to the jugular vein; outflow venous pressure ($PT_{v.o.}$) was measured upstream from the flowmeter. The spinal fluid pressure ($PT_{s.f.}$) was measured via the cisterna magna and the systemic artery pressure ($PT_{s.a.}$) by way of the femoral artery. Collateral communications between the intra- and extracranial venous circulations were effectively occluded by injecting bone wax into both lateral sinuses (marked with a cross).

Left. Perfusing pressure was measured in both common carotid arteries ($PT_{c.a.}$) downstream from the constricting clamps or from the constant pressure perfusion system and in the vertebral arteries ($PT_{v.a.}$) at the level of the second cervical vertebra at their entrance into the transverse canal. In the autoperfusion experiments, blood was shunted through a constant pressure perfusion system which consisted of a sigma-motor, peristaltic, finger type, pump and a pressure regulating system which included an electronically controlled shunt to a jugular vein. (From Rapela & Green (1964), *Circulation Res.* **14**, 1 205–11, Fig. 1.)

5. *The cerebral circulation: experimental approaches*

measured blood flow in the internal jugular veins using a type of thermistor flowmeter and was able to obtain rates of flow (ml 100 g brain^{-1} min^{-1}) which were similar to those found in most other species using alternative methods, e.g. nitrous oxide, Xenon-133, which did not involve surgery or cannulation of vessels. Further, he showed that the cerebral vessels were highly reactive to changes in P_{a,CO_2} and he was able to demonstrate that contamination from extracranial sources accounted for no more than 5 per cent of his measured internal jugular blood flow. Similar methods and results have been obtained by Meyer, Ishikawa & Lee (1964) also in the monkey.

Artificial perfusion of the brain

There are obvious advantages in being able to perfuse the isolated cerebral vascular bed with blood at either constant flow or pressure. It enables the investigator to distinguish between active and passive vascular responses and to study the effects of drugs, carbon dioxide, the effects of stimulating nerves, etc. without the changes in systemic pressure which so frequently obscure the response or which make interpretation impossible.

The first attempt to perfuse cerebral vessels was by Finesinger & Putnam (1933). They perfused one carotid artery with blood at constant pressure and recorded the inflow. This method was of little value in determining overall values for cerebral blood flow since other arteries were still intact; on the other hand, the system of perfusion had the merit that it could be combined with direct inspection of changes in calibre of pial vessels. It was thus possible for these workers to correlate changes in arterial inflow with those of pial vessel calibre and they showed (1) that stimulation of the cervical sympathetic nerve caused an average 15 per cent fall in inflow and marked pial artery constriction; (2) that stimulation of the vagus caused a rise of flow and pial artery dilatation, and (3) that the direct arterial infusion of adrenaline caused a fall in blood flow. Since all these changes occurred at constant perfusion pressure, it is reasonable to suppose that the changes in blood flow were caused by generalized alteration of vascular resistance of which the changes in pial artery calibre were a part.

A more satisfactory method of perfusing the isolated cerebral vascular bed of the cat was described by Geiger & Magnes (1947).

Their paper gives a comprehensive account of the venous drainage of the brain and of the ligation of veins which is necessary. In particular they stress the importance of ligating the sinuses which accompany the vertebral column and which if resistance is increased in cerebral or jugular veins can carry the entire venous drainage of the brain (Hill, 1896; Harris, 1941). The brain was perfused via one carotid artery after the vertebral arteries and other branches of the carotid artery had been tied. Venous blood was collected from the transverse sinuses and its volume measured with a flowmeter. Because of the large volumes of blood required for priming the perfusion apparatus (150 ml) and for analysis (250 ml), defibrinated ox blood was used for perfusion diluted to give an oxygen capacity of 14 volumes per cent.

The values obtained by Geiger & Magnes for blood flow were usually over 100 ml 100 g^{-1} min^{-1} and occasionally as high as 150 ml 100 g^{-1} min^{-1}. These very high values some two or three times as high as most subsequent measurements suggest that (*a*) the viscosity of the diluted blood was lower than that found in the cat; (*b*) despite all the care taken to exclude contamination from extracranial sources, this was in fact a significant factor, and (*c*) the vessels were dilated either by vasoactive substances or by a high carbon dioxide concentration in the perfusate due to inadequacy of the oxygenator. No measurements were made of blood gas tensions or pH and the sensitivity of the cerebral vessels to changes in carbon dioxide or [H^+] was not satisfactorily tested. Cerebral oxygen consumption was found to vary between 4.0 and 6.0 ml 100 g^{-1} min^{-1} and a close relation between oxygen consumption and the level of narcosis observed. Oxygen consumption increased with the injection of metrazol or other analeptics.

Despite the criticisms which can be levelled at these experiments, they constituted a remarkable pioneering effort and in many respects the results which were achieved were more physiological than any obtained since. Sagawa & Guyton (1961) attempted to perfuse the vascularly isolated brain of the dog. Like previous workers (Batson, 1940, 1942; Jewell, 1952) they found numerous anastomoses between intra- and extracranial circulations on both arterial and venous sides. From casts made of the arterial tree, they determined the number of arteries which would have to be tied or clamped in order to isolate the blood supply to the brain, Fig. 5.7. When the cerebral circulation had been isolated, it was

5. *The cerebral circulation: experimental approaches*

cross-perfused from a donor dog and the pressure–flow relationship measured, Fig. 5.8. This shows that over the physiological range of perfusion pressure, 120–140 mmHg, cerebral blood varied from approximately 40 to 120 ml 100 g^{-1} min^{-1} while the relation between pressure and flow was linear over the range tested. While the range of blood flow is similar to that obtained in the dog by

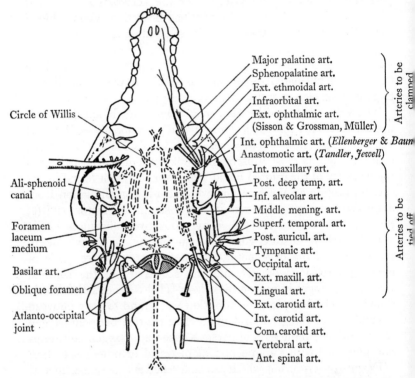

Fig. 5.7. Semi-schematic representation of the intra- and extracranial arteries, the vertebral arteries and the anastomoses between them on the right side. Ligatures and clamps employed in the study on the left side. Nomenclature is after Jewell, 1952. *J. Anat.* **86**, 83–94. (From Sagawa & Guyton (1961), *Am. J. Physiol.* **200**, 711–14, Fig. 1.)

other methods and in other species, the relation between pressure and flow is quite different (see Chapter 6). It is most probable that the very extensive surgery which was necessary to isolate the cerebral circulation caused damage either to the cerebral vessels or to their nerve supply, and, as a result, the outstanding

Artificial perfusion of the brain

feature of the relation between cerebral flow and perfusion pressure, namely, the independence of flow and pressure over the physiological range, was lost. It is difficult to see how the dog's brain can satisfactorily be perfused so that the normal reactivity of cerebral vessels is preserved and the volume of cerebral inflow or outflow accurately measured.

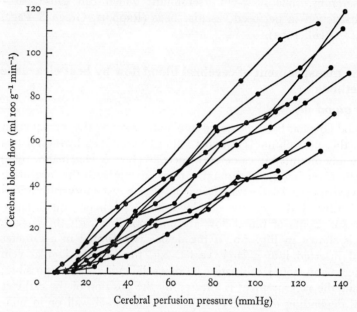

Fig. 5.8. Relation between cerebral blood flow (ml 100 g^{-1} min^{-1}) and cerebral perfusion pressure (mmHg). Results from ten dogs. Perfusion pressure was altered by varying the volume of perfusate. (From Sagawa & Guyton (1961), *Am. J. Physiol.* **200**, 711–14, Fig. 3.)

A further example of the difficulty encountered with the dog is the series of experiments carried out by Rapela, Green & Denison (1967). The preparation is shown in Fig. 5.6: the common carotid arteries are cannulated and the brain can be perfused either naturally or artificially by diverting carotid blood through a sigma-motor pump. Changes in perfusion pressure are achieved by clamping the carotid arteries but since the vertebral arteries are unaffected, pressure in them is measured and perfusion pressure is estimated as the mean carotid and vertebral pressures minus cerebral venous pressure.

5. The cerebral circulation: experimental approaches

As has been mentioned previously, this type of perfusion gives low values for cerebral blood flow and quite apart from the difficulties of ensuring that cerebral venous blood is not diverted elsewhere, e.g. via vertebral veins, and uncertainties of what volume of brain is being drained via the flowmeter, there is the unanswered question as to whether the perfusion pump liberates vasotonins, notably 5-HT (serotonin) which can cause vasoconstriction in perfused vascular beds (Rapport, Green & Page, 1948; Reid, 1952).

The measurement of cerebral blood flow by heat clearance methods

In blood vessels

Blood flow has been measured in blood vessels by the introduction of a thermoströhmur (Heymans & Bouckaert, 1931), by thermistors suitably mounted in catheters or needles (Betz & Herrmann, 1966) or by diverting blood through a rigid cuvette with the measuring device mounted in the wall (Reivich, 1964). Satisfactory calibrations in terms of the volume of blood flow and hence quantitative measurements of blood flow are only possible with the cuvette as is shown in Fig. 5.9. If the device is mounted in a catheter and inserted into a large vessel, e.g. the internal jugular vein (Herrmann, 1968), calibration becomes difficult, if not impossible; while the measurement is affected by the position of the catheter tip, depending on whether it is near the vessel wall or in midstream. This suggests that the thermistor is responding to the velocity rather than to the volume of blood flow. Furthermore, the catheter or needle is sensitive to and may indeed cause, turbulence. All these factors make this method of measurement unreliable.

In tissues

The measurement of regional or local cerebral blood flow is important because there is abundant evidence that blood flow varies between different parts of the brain. The changes in calibre of pial vessels give little indication of changes in blood flow while measurements of total cerebral blood flow whether by the collection of cerebral venous blood or by the nitrous oxide method (p. 123) may conceal local changes. The introduction of the heated thermo-

The measurement of blood flow by heat clearance

couple by Gibbs (1933b) seemed to offer a way of measuring such changes. The thermocouple was mounted in a needle which was introduced into tissue. The constantin needle was heated for part of its length and the difference in temperature between the heated and unheated parts was measured by a thermocouple.

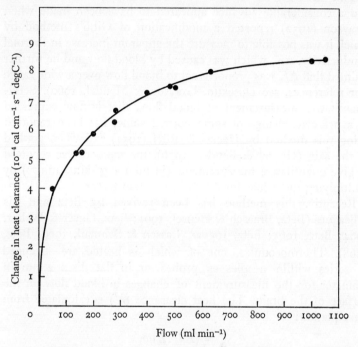

Fig. 5.9. Relation between flow (ml min^{-1}) and heat clearance in a rigid tube. The probe was placed in the marginal stream. (From Herrmann (1968), *Blood flow through organs and tissues*. Edinburgh: Livingstone, Fig. 2.)

Two procedures were used: (*a*) the reference junction was thermostatically controlled outside the animal's body and the temperature of the needle was repeatedly determined when heated (t_h) and when unheated (t_u). The value of ΔT in each situation was determined from the associated values of t_h and t_u. The heating obviously had to be intermittent. (*b*) Both the needle and the reference junction were introduced into the tissue, the reference junction being placed at a distance which was unaffected by heating the needle. Thus, ideally, changes in tissue tempera-

5. The cerebral circulation: experimental approaches

ture were automatically compensated for since both junctions were affected. The heating could then be constant and ΔT determined continuously. Gibbs emphasized that unless the probe was standardized in each experiment, the method could only be used for measuring qualitative changes in tissue blood flow.

This need for standardization was a very great disadvantage and a considerable advance appeared to have been made when Grayson (1952) reported a modification of Gibb's method by which it was possible to measure the apparent increase in thermal conductivity, ΔK, which was caused by blood flow and he further claimed that ΔK was proportional to blood flow over a wide range (for references, see Dosekun, Grayson & Mendel, 1960). Semi-quantitative measurement of blood flow, that is flow expressed as a per cent change of some normal value, has been reported using this method by Hensel & Ruef (1954) and when applied to the skin (Hensel & Bender, 1956), the method was reported to give quantitative measurements (in ml 100 g skin^{-1} min^{-1}) by multiplying the values for ΔK by a constant factor.

Recently, this method has been revived by Betz and his colleagues (Betz, Braasch & Hensel, 1961; Betz, Gayer & Weber, 1964; Betz, 1965; Betz, Ingvar, Lassen & Schmail, 1966; Betz, 1968). Thermocouples, one of which is heated are arranged in series within needles as probes or in flat discs 2 mm in diameter for the measurement of changes in blood flow on the surface of the brain. The heat clearance (λ') is calculated from the equation:

$$\lambda' = (KI^2 . R)/\delta,$$

where I is the heating current, R the electrical resistance of the heater, δ the temperature difference between the heated and unheated tip and K which is determined from measurements in substances of known thermal conductivity is the instrument constant. Blood flow is expressed in units of thermal conductivity cal cm^{-1} s^{-1} degC^{-1}.

Difficulties arose about the credibility and reliability of these methods when model experiments were carried out. Linzell (1953), using the Grayson type of thermocouple probe found that ΔK was only proportional to flow at low flow rates; at intermediate flow rates, ΔK was proportional to the square root of flow and at high rates of flow ΔK did not appreciably alter when flow was

The measurement of blood flow by heat clearance

changed by moderate amounts. This type of relation between ΔK and flow was confirmed by Graf, Golenhofen & Hensel (1957) in liver and in perfused bone marrow (Graf & Rosell, 1958). Further model experiments have been carried out by Bill (1962) who has shown from model experiments that there can be no standard relation between ΔK and flow when the needle is inserted blindly into tissue. The results will be least reliable when the probe is placed near large vessels and most reliable when placed near few and small vessels. If quantitative measurements are sought, the probe has to be standardized in each experiment and even then the standardizations apply only under certain conditions.

The model situations explored by Linzell, Bill and others were simple ones in which small tubes of different bore were placed in parallel in suitable material, e.g. gelatine and graphite. The situation in tissue is of course very much more complex, see for example Plate 2.1. Here the calibre of vessels varies considerably, the vessels are arranged irregularly and blood flow is likely to be non-homogeneous. It is therefore not surprising that when the probes or discs are used *in vivo* and the output compared to total blood flow of the organ, kidney (Betz *et al.* 1964), myocardium (Betz *et al.* 1961) or cerebral cortex (Betz *et al.* 1966), there is in general a non-linear relation and further that the shape and position of the curves vary considerably, Fig. 5.10.

The position may be summed up as follows. If perfusion of the tissue is homogeneous, then heat clearance values can be quantified; but such homogeneity can only be assumed after numerous measurements of heat clearance with time and may only occur under special circumstances *in vivo*, e.g. in an area of infarction. In the vast majority of other situations, homogeneous perfusion cannot be assumed and even if it were, there is no guarantee that other changes, e.g. local changes in the metabolic rate were not affecting the output of the measuring device. Because the output of these measuring devices is non-linear, because they can be calibrated only with great difficulty and because there is uncertainty as to the relation between heat clearance and blood flow over the range of flow encountered in capillary beds, it is doubtful whether quantitative measurements of blood flow can be made by this method. The experiments of Bill (1962) suggest that considerably more work with model systems of varying com-

5. *The cerebral circulation: experimental approaches*

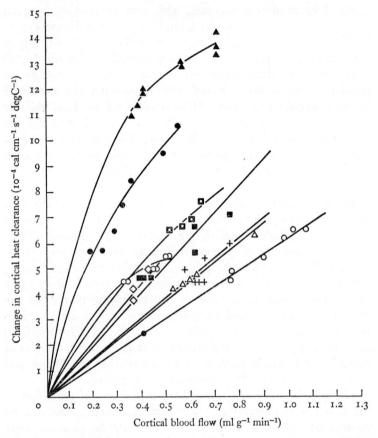

Fig. 5.10. The relation between cortical heat clearance (10^{-4} cal cm^{-1} s^{-1} degC^{-1}) and regional blood flow (ml g^{-1} min^{-1} obtained by clearance of ^{85}Kr). Nine experiments in two dogs and four cats anaesthetized with pentobarbitone: symbols refer to individual experiments. Linear correlation (five tests) was obtained in regions with predominantly small vessels (diameter < 100 μm). The other curves were obtained from areas with both smaller and larger vessels (diameter > 100 μm). (From Betz, Ingvar, Lassen & Schmail, 1966. *Acta Physiol. Scand.* **67**, 1–9, Fig. 6.)

plexity requires to be carried out before the correlation between heat clearance and blood flow can be validated. Meanwhile, the onus is upon individual authors to show by careful standardization controls that their measuring devices can give even qualitative estimates of changes in tissue blood flow.

Methods based on the use of inert, diffusible indicators
Nitrous oxide

The blood flow through an organ can be calculated according to the Fick principle

$$\text{Blood flow}(f) = Q.t/(A-V),$$

where Q is the quantity of a substance removed or added in time t divided by the difference between arterial and venous concentrations of the substance (A–V). Substances which are metabolized, e.g. oxygen or glucose can only be used if there is some independent way of determining the rate at which they are extracted from blood. The ideal substance is therefore inert and freely diffusible: these requirements are fulfilled by nitrous oxide which was chosen by Kety & Schmidt (1945) for the measurement of cerebral blood flow. Nitrous oxide has the additional advantage that it is easily administered by inhalation.

The principle of the method of measurement is simple. Nitrous oxide is inhaled until the brain tissue is saturated and this point is shown by equilibrium between arterial and cerebral venous nitrous oxide concentrations. The curves for arterial and venous nitrous oxide concentrations are constructed from values obtained during the process of saturation. The numerator of the Fick equation is the product of the concentration of nitrous oxide in brain (C), the weight of the brain (w) and λ, the blood–brain partition coefficient. The denominator is the integral of A–V nitrous oxide differences ($C_a - C_v$) measured from $t = 0$ until equilibrium, usually about ten minutes. Flow is therefore measured in absolute units or, more commonly, as ml $100\,\text{g}^{-1}\,\text{min}^{-1}$:

$$\text{Flow}(f) = \frac{C.\lambda}{\int_0^{10}(C_a - C_v)\,dt}. \tag{1}$$

The method has many advantages and was widely used in the 1950s to establish values for normal cerebral blood flow in man and blood flow in a wide variety of physiological and pathological conditions. However, there are a number of drawbacks. First, there is the vexed question of how accurately jugular venous blood samples reflect cerebral venous drainage only. This question has already been considered with respect to common laboratory

5. *The cerebral circulation: experimental approaches*

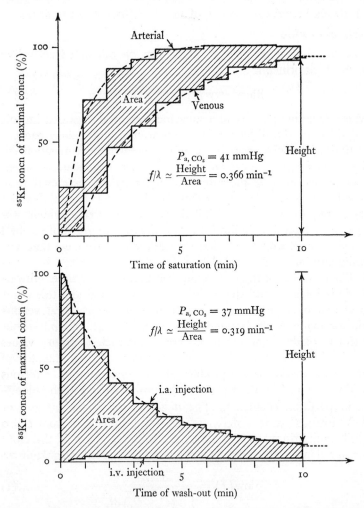

Fig. 5.11. Upper graph shows the results obtained by the outflow detection method of Kety & Schmidt during a 10 min period of saturation with ^{85}Kr. The shaded area was obtained as a sum of the arterio-venous differences. The lower graph shows the results obtained by the intra-arterial injection method using a digital ratemeter to record the ^{85}Kr gamma radioactivity remaining in the brain. In this subject the isotope recirculation was estimated experimentally by injecting the same dose of ^{85}Kr intravenously. Height is the actual amplitude at 10 min of desaturation. (From Lassen & Høedt-Rasmussen (1966), *Circulation Res.* **19**, 681–8, Fig. 1.)

The use of inert, diffusible indicators

animals and ungulates: in these animals, the method is of very limited usefulness unless most elaborate precautions are taken to separate the intra- and extracranial circulations. In man, also, there is doubt as to whether the composition of jugular venous blood is uniform since Gibbs & Gibbs (1934) showed that at the torcular, straight sinus flow was directed to the left lateral sinus and the greater part of the superior sagittal sinus flow entered the right lateral sinus.

Secondly, the method of estimating nitrous oxide in blood is tedious and in order to achieve an accuracy of ± 5 per cent, large blood samples are required.

Thirdly, experience showed that saturation of the brain was not always achieved in 10 min. This may have been due to non-homogeneous perfusion of the brain and/or low rates of blood flow. The calculated values would thus overestimate true blood flow.

Finally, failure to reach equilibrium by 10 min or irregularities in the arterial or venous curves during saturation could be due to fluctuations in alveolar concentrations of nitrous oxide: this was rarely monitored except in special circumstances, e.g. at altitude (Severinghaus et al. 1964).

For all these reasons, the original method of Kety & Schmidt was modified: (i) Since it was observed that after 8 min of nitrous oxide inhalation, the venous curve approached equilibrium values exponentially, the latter was calculated by extrapolation rather than by waiting for long periods for full saturation to occur (Lassen & Klee, 1965). (ii) The substitution of ^{85}Kr for nitrous oxide as indicator made analysis of arterial and venous blood samples both easier and more accurate (Lassen & Munck, 1955). (iii) A method which required saturation of the brain with nitrous oxide but did not require constant alveolar concentrations of this gas was devised by McHenry (1964) and for ^{85}Kr by Lassen & Klee (1965). The indicator gas was inhaled for 20–30 min, and a jugular venous sample taken to give the saturation concentration. The subject then inhaled air and arterial and venous samples were obtained for the construction of desaturation curves. Values for blood flow obtained by this method were the same as those obtained by Kety & Schmidt (1948a) while the curves were more even and therefore more accurately interpreted.

5. The cerebral circulation: experimental approaches

Radioisotopes

The modification suggested by McHenry (1964) was somewhat overshadowed by the introduction of an alternative method of measuring cerebral blood flow by Ingvar & Lassen (1962). The principle is similar, i.e. blood flow was computed from an indicator desaturation curve but the technique is quite different. ^{85}Kr dissolved in saline is injected as a bolus into the internal carotid artery and the clearance of gas from brain tissue is measured from the beta-emission by a suitably placed external scintillation counter. The penetration of beta particles is so slight that for adequate counting the surface of the brain has to be exposed. The method was subsequently improved by substituting the gamma emitter ^{133}Xe which enabled counting to be carried out with the skull intact and in human subjects (Glass & Harper, 1963; Lassen, Høedt-Rasmussen, Sorensen, Skinhøj, Cronquist, Bodforss & Ingvar, 1963). The technique has been described in detail (Høedt-Rasmussen, Sveinsdottir & Lassen, 1966).

Interpretation of desaturation curves

The assumption that desaturation is a function of perfusion only. The method assumes that the indicator is carried by arterial blood to the capillary bed where the gas, either ^{85}Kr or ^{133}Xe diffuses freely to the tissues and that the subsequent removal from tissue is a function of perfusion only. This assumption and the contrary assumption that radial diffusion is an insignificant factor derives from diffusion coefficients which were determined either for the solute in water or from steady state methods using excised tissues (Krogh, 1918; Kety, 1951; Roughton, 1952; Thompson, Cavert & Lifson, 1958). But there is evidence that extravascular tissue is not a homogeneous diffusion medium and that the diffusion coefficient when determined under truly transient conditions is several orders lower than that determined under steady state conditions (Dick, 1959; Fenishel & Horowitz, 1963; Perry, 1950; Hills, 1967). The evidence so far adduced had been derived either from situations which are far from physiological or from physiological situations but in which gross approximations have been made about the geometry of the cells and extracellular fluid which constitute the diffusion medium. Although, therefore, at the moment the case for blood–tissue exchange of non-polar

The use of inert, diffusible indicators

gases being limited by diffusion is far from complete, it is worth emphasizing that this assumption which is central to the techniques which have been described may be invalid.

Theoretical considerations. The analysis of the desaturation which is commonly used derives from the stochastic theory applied by Zierler (1965). He showed that the desaturation curve following a slug injection of indicator $c(t)$ equals:

$$c(t) = c(0) \cdot \int_0^t (1-H) \, dt, \qquad (2)$$

where $H = \int_0^t h \, dt$, h being the frequency function of transit times. Then

$$\bar{t} = \int_0^\infty (1-H) \, dt, \qquad (3)$$

where $\bar{t}$ is the mean transit time of the passage of the indicator through the tissue: and

$$\bar{t} = A/H \text{ min}, \qquad (4)$$

where A is the area under the curve measured from time zero to infinity without recirculation of tracer and H is the initial height assuming that the whole bolus has reached the counting area before clearance starts.

The relation between $\bar{t}$ and blood flow through the tissue is given by the equation discussed by Meier & Zierler (1954)

$$\bar{t} = V/F \text{ min}, \qquad (5)$$

where V is the equilibrium volume of distribution of the indicator and F, the total blood flow. If numerator and denominator are divided by W, the weight of the tissue,

$$t = (V/W)/(F/W). \qquad (6)$$

V/W is the volume of distribution of indicator per gram of tissue, i.e. it is the tissue–blood partition coefficient λ as defined by Kety (1951). F/W is the blood flow per gram of tissue, f. Thus

$$\bar{t} = \lambda/f \qquad (7)$$

and combining equations (4) and (7),

$$f(\text{ml g}^{-1} \text{ min}^{-1}) = \lambda \cdot H/A. \qquad (8)$$

Equation (8) is similar to that derived by Kety & Schmidt (1948a)

5. *The cerebral circulation: experimental approaches*

in order to derive f from the arterial and venous curves during saturation (equation 1).

A practical difficulty of this form of analysis is encountered in the measurement of the area A. Kety (1951) used an approximation by using values for H and A at 10 min (Fig. 5.11) and the same approximation may be made in the analysis of the tracer desaturation curve. Thus:

$$f \approx \lambda . H_{(10)}/A_{(10)}, \tag{9}$$

where $H_{(10)}$ is the difference between H_0 and H at 10 min and $A_{(10)}$ the area under the curve from $t = 0$ until $t = 10$ min. This approximation systematically overestimates blood flow and it cannot without modification be applied to other organs.

An alternative method of analysis of the desaturation curve can be used following saturation of the brain with the indicator. If no recirculation of indicator can be assumed after the subject inhales air and counting starts, the desaturation curve, $C(t)$ has been shown by Lassen (1968b) to be

$$C(t) = C(0) . \left[1 - 1/\bar{t} . \int_0^t (1-H) dt \right]. \tag{10}$$

If $H = 0$ for $t \to 0$, by differentiation,

$$\frac{\text{initial slope}}{\text{height}} = \left[\frac{-dC(t)dt}{C(0)} \right]_{t=0} = 1/\bar{t} \tag{11}$$

and since $1/\bar{t} = \bar{f}/\bar{\lambda}$, i.e. the ratio of mean blood flow to mean solubility coefficient, $\bar{f}$ can be calculated. The same primary information can thus be obtained by the bolus injection of indicator as by the saturation technique.

Difficulties about H_0. Although $\bar{f}$ can be calculated from the desaturation curve which follows either saturation of brain by indicator or after a bolus injection of indicator into the internal carotid artery, there are some theoretical and practical difficulties about the significance of H_0 and the use of H_0 for subsequent estimation of I_g and I_w (the intercepts at $t = 0$ for the desaturation curves for grey and white matter respectively). Difficulties also arise if for purposes of calculation, the bolus injection is assumed to be a delta function: clearly it is not.

Using a rapid manual form of injection, the initial plateau is usually delayed by some 3–4 s and lasts 3–4 s. The initial plateau has been described by Berg & Drift (1963) and by Fieschi, Agnoli & Galbo (1963) who used albumen ^{131}I as a non-diffusible indicator and for this reason and because the minimum carotid–jugular circulation time is approximately 4 s (Nilsson, 1965) it is probable that the plateau represents the complete arrival of the bolus of indicator to the counting area. With the bolus injection, therefore, the plateau represents something longer than a delta function and something less than complete saturation of the brain with indicator.

Two practical problems immediately arise: first, does the length of the arterial injection affect the calculated values for $\bar{f}$ and secondly, is there any evidence that a rapid impulse injection actually affects cerebral blood flow by reflex or other means?

The answer to the first question is clearly, yes: as is shown in Fig. 5.12. This shows that as the period of injection is lengthened, calculated $\bar{f}$ falls by up to 20 per cent of that when the injection is an 'impulse' one. The second question has not been formally studied in any detail but Betz et al. (1966) have shown that the injection of 0.5–1.0 ml of isotope solution rapidly into the internal carotid artery causes a disturbance of cerebral blood flow measured with a heat clearance device which lasts 1–4 min together with a fall in systemic pressure. This fall in pressure immediately suggests that the carotid sinus baroreceptors have been stimulated by a rise in sinus pressure as the injection is made and the changes in blood flow could also be brought about reflexly. This notion is confirmed by the observation that if the carotid sinus nerve on the side of the injection is cut, cerebral blood flow is unaffected by rapid injection of the indicator.

The dilemma is therefore posed as follows: a very rapid injection, i.e. approximating to a delta function, is essential to the theory which governs the calculation of $\bar{f}$ from the desaturation curve and yet such an injection can itself affect $\bar{f}$. On the other hand, if the period of injection is prolonged, $\bar{f}$ is also affected.

Before this particular problem is dealt with, another question must be considered, namely, the technique and justification of compartmental analysis of the desaturation curve.

Compartmental analysis of the desaturation curve. So far, the technique of deriving $\bar{f}$, the mean cerebral blood flow from

5. *The cerebral circulation: experimental approaches*

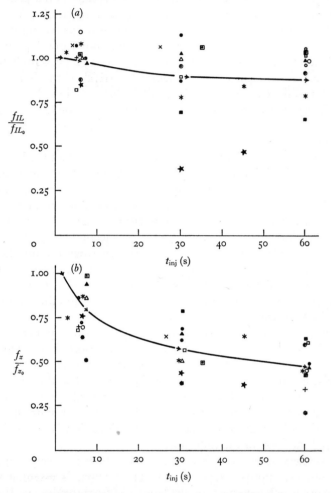

Fig. 5.12. (*a*) The relation between regional cerebral blood flow measured by intracarotid injection of ^{85}Kr and estimated from the initial slope of the desaturation curve and the period (s) over which the injection was made. Blood flow is expressed as the ratio of flow (any injection period, f_{IL}) to flow (injection period = 1 s; f_{IL_0}).

(*b*) The same relation as in (*a*) except that flow was estimated by the height-area method. Flow expressed as the ratio of flow (any injection period, f_z) to flow (injection period = 1 s; f_{z_0}). (From Hutten, Schwarz & Schultz (1969), *Cerebral blood flow*. Berlin: Springer, Fig. 1.)

The use of inert, diffusible indicators

the desaturation curve has been described. The desaturation curve however may also be considered as a composite curve representing clearance of indicators from different areas of brain each of which may have a different value for λ and each of which is perfused

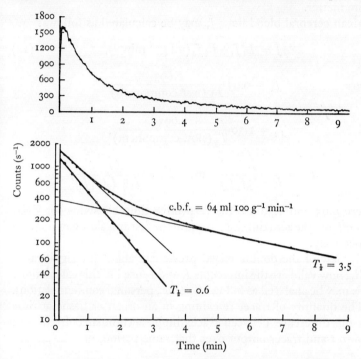

Fig. 5.13. Upper diagram shows the clearance curve of ^{85}Kr obtained over the right temporal region. The isotope was dissolved in about 10 ml of saline and injected into the right internal carotid artery.
Lower diagram shows the semi-logarithmic plot of the same clearance curve. Note the final rectilinear part of the curve indicating a slow perfusion phase ($T_{\frac{1}{2}} = 3.5$ min). Graphical analysis yielded a fast component with $T_{\frac{1}{2}}$ of 0.6 min. The average flow ($\bar{f} = 64$ ml 100 g^{-1} min^{-1}) was obtained by calculating a weighted average of the two flow rates using the method described in the text. (From Capon, Cleempoel, Lenaers & Martin (1968), *Cerebral circulation*. Amsterdam: Elsevier, Fig. 3.)

at different rates. The evidence for the non-homogeneity of flow and its bimodal distribution will be considered in a later section. At the moment we will consider a simple model in which two principal compartments are perfused in parallel. Simple exponential stripping of the desaturation curve yields two components,

5. *The cerebral circulation: experimental approaches*

Fig. 5.13, a fast component having a half-time of 0.6 min and a slow component having a half-time of 3.5 min. For the moment, the fast component may be considered to represent principally blood flow through grey matter, the slow component blood flow through white matter.

Mean cerebral blood flow, $\bar{f}$, may be calculated as follows:

$$\bar{f} = L_1 f_1 + L_2 f_2 \text{ (ml g}^{-1}\text{ min}^{-1}\text{)}, \tag{12}$$

where
$$f_1 = 0.95 \frac{0.693}{T_{\frac{1}{2}} \text{ (fast component)}}$$

and
$$f_2 = 1.30 \frac{0.693}{T_{\frac{1}{2}} \text{ (slow component)}},$$

$$L_1 = \frac{I_1}{I_1 + I_2(f_1 f_2)} \quad \text{and} \quad L_2 = \frac{I_2(f_1 f_2)}{I_1 + I_2(f_1 f_2)},$$

where 0.95 and 1.30 are the respective time constants for ^{85}Kr, I_1 and I_2 are the zero time intercepts of the fast and slow components respectively.

In view of the doubts raised previously about the significance of H_0, how valid are the intercepts I_1 and I_2 used in this calculation? This may be analysed as follows (Zierler, personal communication).

The quantity of tracer remaining in an organ or tissue at time t is the difference between tracer input accumulated from time zero to t and tracer output over that same period, or

$$q(t) = \int_0^t [i(\tau) - i(\tau) * h(\tau)] \, dt$$

$$= I(t) - I(t) * h(t), \tag{13}$$

where $i(\tau)\,dt$ is the quantity of tracer entering the system between t and $t + dt$, $h(t)$ is the probability density function of transit times through the tissue, and $i(t) * h(t)\,dt$ is tracer output over the interval between t and $t \pm dt$, where $i(t) * h(t) = \int_0^t i(t - \tau) h(\tau)\, d\tau$.

If the input $i(t)$ is defined as a constant injection over a time interval from zero to a, and zero injection thereafter or,

$$i(t) = m[U(t) - U(t - a)], \tag{14}$$

The use of inert, diffusible indicators

where $U(t)$ is the unit impulse. Then equation (13) becomes

$$q(t) = mt - m\int_0^t H(\tau)\,d\tau - U(t-a)\left[m(t-a) - m\int_a^t H(\tau-a)\,d\tau\right]$$

$$= mt[1 - H(t)] + m\int_0^t \tau h(\tau)\,d\tau - U(t-a)$$

$$\times \left[m(t-a) - m\int_a^t H(\tau-a)\,d\tau\right]. \quad (15)$$

If a is sufficiently long, $H(a) = 1$ and $\int_0^a th(t)\,dt = \bar{t}$, the mean transit time. In that case, for $a \leqslant t$, and for $t' = t-a$

$$q(t') = m\bar{t} - m\left(t' - \int_0^t H(\tau)\,d\tau\right)$$

which is the equation for washout of tracer after loading to steady state.

At the other extreme, a can be so brief that $i(t)$ becomes a delta function, $q_0\delta(t)$. In this case,

$$q(t) = q_0[1 - H(t)] \quad \text{for} \quad i(t) = q_0\delta(t), \quad (16)$$

which is the equation for external monitoring following instantaneous injection.

All cases in which a falls between these extremes are troublesome and do not readily lead to calculation of blood flow or of mean transit time, which measures specific (or relative) blood flow (= blood flow per unit weight or volume).

Consider further equation (15). If $h(t)$ is assumed to have a particular form, then it is possible to treat this further. If, for the sake of argument, we deal with two homogeneous compartments in parallel, then

$$h(t) = \alpha k_1 e^{-k_1 t} + (1-\alpha)k_2 e^{-k_2 t} \quad (0 < \alpha < 1) \quad (17)$$

then, $\quad H(t) = 1 - \alpha e^{-k_1 t} - (1-\alpha)e^{-k_2 t}$

and $\quad \int_0^t H(\tau)\,d\tau = t - \dfrac{\alpha}{k_1}(1 - e^{-k_1 t}) - \dfrac{1-\alpha}{k_2}(1 - e^{-k_2 t})$

$$= t - \bar{t} + \frac{\alpha}{k_1}e^{-k_1 t} + \frac{1-\alpha}{k_2}e^{-k_2 t}. \quad (18)$$

From equations (15) and (18),

$$q(t') = \frac{m\alpha}{k_1}e^{-k_1 t'}(1 - e^{-k_1 a}) + \frac{m(1-\alpha)}{k_2}e^{-k_2 t'}(1 - e^{-k_2 a}) \quad (19)$$

5. *The cerebral circulation: experimental approaches*

and the derivative is

$$\frac{dq(t')}{dt'} = m\alpha\, e^{-k_1 t'}(1-e^{-k_1 a}) - m(1-\alpha)\, e^{-k_2 t'}(1-e^{-k_2 a}). \quad (20)$$

After a sufficiently long time, if $k_1 \gg k_2$, $e^{-k_1 t'} \to 0$ while $e^{-k_2 t'}$ is still detectably greater than zero. Then,

$$q(t') \approx \frac{m(1-\alpha)}{k_2}(1-e^{-k_2 a})\, e^{-k_2 t'} \quad \text{for} \quad e^{-k_1 t'} \approx 0$$

and
$$\frac{dq(t')}{dt} \approx -m(1-\alpha)(1-e^{-k_2 a})\, e^{-k_2 t'} \quad \text{for} \quad e^{-k_2 t'} \approx 0$$

and
$$\frac{d\ln q(t')}{dt} = \frac{dq(t')}{q(t')dt'} = -k_2 \quad \text{for} \quad e^{-k_2 t'} \approx 0. \quad (21)$$

Therefore, after a sufficiently long time, even though a is too long for $i(t)$ to be a delta function and too short for loading to have been to a steady state, a plot of the logarithm of $q(t)$ against time falls linearly, with a slope $-k_2$.

It is further possible to conclude that even if a is too long for $i(t) = q_0(t)$, and too short for $H(a) = 1$, if $h(t)$ is the sum of two exponentials representing two homogeneous compartments in parallel, $q(t)$ is still the sum of two exponentials. The slopes, k_1 and k_2, are indeed the desired slopes, but the intercepts obtained by curve peeling from this $q(t)$ are not the desired intercepts, m/k_1 and $m(1-\alpha)/k_2$, which, for $\lambda = 1$, represent the quantity of tracer in each component at equilibrium. The intercepts differ from the desired intercepts by the factors $(1-e^{-k_1 a})$ and $(1-e^{-k_2 a})$, respectively.

Thus, it is possible to obtain the slopes k_1 and k_2 by this method; corrections have to be applied to the observed intercepts and it is not possible to estimate a.

The brain–blood partition coefficient, λ. If the intercepts I_g and I_w are themselves dependent upon the input function which is somewhere between an impulse injection and an infusion sufficient to cause saturation, they become correspondingly arbitrary. The brain–blood partition coefficient, λ, therefore has to be measured directly and the technical details for this have been given by Veall & Mallett (1965a) together with values for λ and fractional weights of grey and white matter in man, cat and dog. Similar measurements have been made in the baboon (James,

Millar & Purves, 1969) and in the foetal sheep and newborn lamb (Purves & James, 1969).

With a knowledge of λ for homogenized brain, grey and white matter, blood flow may be calculated

$$\text{Blood flow (ml 100 g}^{-1}\text{ min}^{-1}) = 100\lambda_{(t)}\frac{0.683}{T_{\frac{1}{2}(t)}}, \qquad (22)$$

where (t) is the tissue for which λ has been measured (Ingvar, Cronqvist, Ekberg, Riseberg & Høedt-Rasmussen, 1965). Two precautions are required. First, as is shown in Table 5.1, ^{133}Xe (and other non-polar gases) is approximately twice as soluble in red cells as in plasma: λ is therefore dependent upon blood haematocrit. The relation between λ_T and haematocrit may be expressed (Veall & Mallett, 1965a)

$$\lambda_T = \frac{S_T}{S_E H + S_P(100-H)} \qquad (23)$$

where the subscript T refers to the tissue in question, S_E is the solubility of the indicator in red blood cells, S_P the solubility in plasma and H the haematocrit. Using the values in Table 5.1, the nomogram shown in Fig. 5.14 may be constructed and appropriate corrections applied if λ is to be used for calculation of blood flow under conditions where H is abnormal, particularly if it is abnormally low.

The second precaution which has to be considered is the use of values for λ which are derived from homogenized brain, grey or white matter for the purposes of calculating regional cerebral blood flow from an arbitrary area of cortex and from an uncertain

Table 5.1. *Relative solubility of* ^{133}*Xe in human tissues relative to air at* $37°C$ (from Veall & Mallett, 1965a. *Phys. Med. Biol.* **10**, 375–80)

Fluid phase	Number of observations	^{133}Xe concentration relative to air (= 1000) $\bar{x} \pm$ S.E.
Water	9	90.3 ± 0.5
Plasma	12	102.5 ± 0.9
Red cells	12	210.0 ± 4.3
Grey matter	14	119.6 ± 3.2
White matter	14	225.3 ± 4.5
Brain homogenate	12	161.6 ± 3.5

5. The cerebral circulation: experimental approaches

volume of brain. The proportion of grey and white matter in brain was originally determined by dissection (Kappers, 1926) but the results were so varied as to be of only limited value. If the solubilities of ^{133}Xe or ^{85}Kr in homogenized brain, grey and white

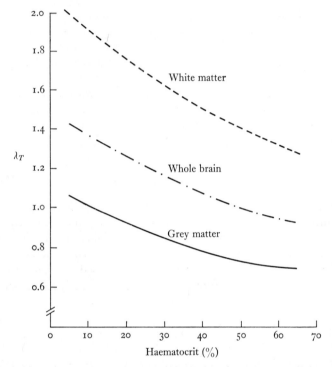

Fig. 5.14. The relation between the brain–blood partition coefficient λ and arterial haematocrit in man, for white matter, whole brain and grey matter. (From Veall & Mallett (1965a), *Physics. Med. Biol.* **10**, 375–80, Fig. 1.)

matter S_h, S_g and S_w respectively, are known, the fractional proportion of grey matter, F_g, can be calculated:

$$F_g = \frac{S_w - S_h}{S_w - S_g}. \tag{24}$$

In man, this has been found to be 0.63 (Veall & Mallett, 1965a) and, in the cat, 0.73 (Ingvar & Lassen, 1962).

By the use of sixteen external counters, Wilkinson, Bull, Boulay, Marshall, Ross Russell & Symon (1969) have been able to show that this fractional proportion of grey matter varies considerably over the cortex of the hemisphere. Over the region of the insula and basal ganglia, F_g was 0.5–0.6: in intervening regions, over the corona radiata and corpus collosum, F_g was much less, 0.3–0.4. These estimates agreed well with the distribution of grey matter found by dissection after death.

Problems arising from non-homogeneous blood flow

Thus far, analysis of the isotope decay curve whether following brain saturation or after an impulse intra-arterial injection has been considered primarily in terms of deriving values for mean cerebral flow or $\bar{f}$. Mean cerebral blood flow, however, is of strictly limited physiological interest. With the present techniques available, it is certainly necessary to determine mean cerebral blood flow in order to calculate cerebral oxygen consumption and to measure the gross effects of various manoeuvres, e.g. tilting, administration of carbon dioxide. But there is every indication that perfusion of the brain as well as the rate of oxygen consumption is non-homogeneous: and the effort, as a first step, to break down mean cerebral flow into grey and white matter flow represents a move to characterize the rates of perfusion of different parts of the brain separately.

In this context, three questions have to be asked: first, what evidence is there for non-homogeneous flow in the brain, secondly, what evidence is there that compartmental analysis of the tracer decay curve corresponds to this type of non-homogeneous flow and thirdly, how far does the compartmental type of analysis compare with those types of analysis already considered.

Evidence for non-homogeneous blood flow in the brain.

In Chapter 2, evidence was discussed which showed that there was a wide spectrum of capillary density in the brain, the density being greatest in the various layers of the cortex and least in sub-cortical white matter. This suggested that there might be parallel differences in the rate of perfusion and this was confirmed quantitatively by Landau, Freygang, Rowland, Sokoloff & Kety (1955). They injected the radioisotope $^{131}ICF_3$ and, by autoradiography, mapped out the distribution of the isotope in cats brain, Table 5.2. This confirms a very wide difference in the rate of perfusion of cerebral cortex and white matter with intermediate values in the cerebral nuclei and other structures.

5. The cerebral circulation: experimental approaches

Table 5.2. *Local cerebral blood flow in the unanaesthetized cat. Mean values from ten experiments using* $^{131}ICF_3$. (From Landau et al. 1955. *Trans. Am. neurol. Assoc.* **80**, 125–9)

	Blood flow (ml g^{-1} min^{-1} ± S.E.)
Superficial cerebral structures:	
Cortex:	
Sensory motor	1.38 ± 0.12
Auditory	1.30 ± 0.05
Visual	1.25 ± 0.06
Miscellaneous–association	0.88 ± 0.04
Olfactory	0.77 ± 0.06
White matter	0.23 ± 0.02
Deep cerebral structures:	
Medial geniculate ganglion	1.22 ± 0.04
Lateral geniculate ganglion	1.21 ± 0.08
Caudate nucleus	1.10 ± 0.08
Thalamus	1.03 ± 0.05
Hypothalamus	0.84 ± 0.05
Basal ganglia and amygdala	0.75 ± 0.03
Hippocampus	0.61 ± 0.03
Optic tract	0.27 ± 0.02
Cerebellum, medulla and spinal cord:	
Cerebellum:	
Nuclei	0.79 ± 0.05
Cortex	0.69 ± 0.04
White matter	0.24 ± 0.01
Medulla:	
Vestibular nuclei	0.91 ± 0.04
Cochlear nuclei	0.87 ± 0.07
Pyramids	0.26 ± 0.02
Spinal cord:	
Grey matter	0.63 ± 0.04
White matter	0.14 ± 0.02

However, it is likely that even this study does not completely summarize small differences in flow which occur in cerebral cortex or individual nuclei. Evidence has already been given that the rate of cortical perfusion varies considerably over the hemisphere (Wilkinson *et al.* 1969). Ingvar & Lassen (1962) also showed that there were at least two tissue compartments which differ as to blood flow in small volumes of cerebral cortex. Further evidence on this point has been given by Fieschi, Bozzao, Agnoli & Kety (1964) who showed the clearance of hydrogen gas from the

The use of inert, diffusible indicators

caudate, lateral geniculate and thalamic nuclei fitted a double exponential function and therefore proposed that sub-cortical grey nuclei are not homogeneous with respect to blood flow.

On the other hand, Espagno & Lazorthes (1965), Häggendal, Nilsson & Norbäck (1965) using local microinjections of ^{133}Xe or ^{85}Kr in the cortex did not find any evidence of heterogeneity. According to them, the two main components of tracer desaturation curves represent blood flow through cortex and subjacent white matter respectively; the cerebral cortex itself being homogeneously perfused.

Some experiments which showed that this was not so were carried out by Fieschi, Isaacs & Kety (1968). Using an autoradiographic technique, they calculated regional blood flow twice in each animal by injecting antipyrine tagged with ^{14}C or ^{131}I and were able to minimize experimental errors.

Local blood flow per unit weight of tissue is calculated from the equation
$$C_{1,(T)} = K_1 \int_0^T C_a e^{-k_1(T-t)} dt, \qquad (25)$$

where $C_{1,(T)}$ is the mean concentration of tracer in individual structures at the end of the experiment, T = time of infusion before circulation is arrested and $C_a(t)$ is the arterial concentration of the tracer. This method assumes that venous blood emerging from a tissue is continuously in equilibrium with the mean tissue concentration of the tracer: this is only true if the tissue is homogeneously perfused and in this case K_1 is independent of the time of infusion.

But if the tissue is not homogeneously perfused, a disparity occurs between the average concentrations of tracer in tissue and venous blood since the average concentration in the tissue is determined by the weights of the component concentrations while that for venous blood is determined by the blood flows. The clearance of a mixed tissue equilibrated with an inert gas would describe the following curve:
$$C_{(T)} = C_{(0)}(W_1 e^{-K_1 t} + W_2 e^{-K_2 t} \ldots + W_n e^{-K_n t}) \qquad (26)$$

where Ks relate to the perfusion rates and Ws to the weightings of individual components. With the autoradiographic technique, estimation of mean tissue concentration is made at a single time (T). In equation (26), the derivative of $C_{(T)}/C_{(0)}$ approaches: $W_1 K_1 + W_2 K_2 \ldots W_n K_n$ (which is the true average flow rate)

5. The cerebral circulation: experimental approaches

only as T approaches zero (Ingvar & Lassen, 1962). Therefore for Ts of increasing duration, the slope becomes progressively weighted in favour of the components with smaller Ks and mean blood flow calculated from equation (25) becomes lower in heterogeneous tissue.

The effect of altering T upon the calculated flow rate is shown in Table 5.3 (Fieschi, Isaacs & Kety, 1968). The first series of columns (controls) indicates the reproducibility of measuring blood flow using the different isotopes: the second and third series of columns indicate the effect of lengthening T with either ^{131}I or ^{14}C labelled antipyrene and the fourth group of columns shows the per cent disparity in calculated flow as T is lengthened from 0.75 to 1.75 min. This is therefore an index of heterogeneity of blood flow: it is most obvious in the cortex and sub-cortical nuclei and least obvious in sub-cortical white matter.

Computation of mean cerebral blood flow. It is probable that if the methods outlined in the previous section are pursued, a continuous distribution of rates of blood flow will be found between and within various structures of the brain and spinal cord. The question then arises, can a composite clearance curve be determined and if it can, is it possible to fit the curve in terms of only two exponentials? Further, is it possible to fit the coefficients and exponents of the two terms to the actual averages obtained from the distribution?

In preliminary studies, Kety (1965) reported a bimodal distribution from values obtained by Landau *et al.* (1955) (Table 5.2) with one peak for white matter having K values between 0.20 and 0.30 and representing 37 per cent of the brain, $(K = F/\lambda W)$. A second peak representing 42 per cent of the brain for grey matter structures largely from cerebral cortex had values of between 0.80 and 0.90. Analysis of a clearance curve obtained from such a distribution yields two components with coefficients and exponents almost identical with the average weights and blood flow values of white and grey matter.

A further step has been taken by Reivich (1969) and by Reivich, Slater & Sano (1969). In the first of these papers, a mathematical model is described for the analysis of cerebral clearance curves consisting of a bimodal Gaussian distribution of exponentials. In the second paper, data are given from ten experiments in cats

Table 5.3. *Local cerebral blood flow (ml g^{-1} min^{-1}) in cats with ^{14}C and ^{131}I antipyrine infused for different time intervals.* (From Fieschi, Isaacs & Kety, 1968. In *Blood flow through organs and tissues*, ed. Bain, W. H. & Harper, A. Edinburgh: Livingstone.)

Structure	Controls			^{131}I short			^{14}C short			Mean of means		
	^{131}I (1 min)	^{14}C (1 min)	$\frac{^{131}I}{^{14}C}$	^{131}I (0.75 min)	^{14}C (1.75 min)	$\frac{^{131}I}{^{14}C}$	^{131}I (1.75 min)	^{14}C (0.75 min)	$\frac{^{131}I}{^{14}C}$	Short $\frac{^{131}I}{^{14}C}$	Long $\frac{^{131}I}{^{14}C}$	% diff from short
1. Lateral geniculate	0.43	0.40	1.08	0.75	0.63	1.20	0.75	0.66	1.13	0.71	0.69	− 3.0
2. Medical geniculate	0.49	0.39	1.27	0.77	0.66	1.17	0.81	0.76	1.07	0.77	0.74	− 4.1
3. Caudate	0.43	0.35	1.20	0.64	0.60	1.06	0.57	0.59	0.97	0.61	0.58	− 4.6
4. Thalamus	0.36	0.35	1.04	0.68	0.48	1.41	0.63	0.52	1.21	0.60	0.55	− 7.9
5. Motor-sensory cortex	0.41	0.32	1.30	0.61	0.57	1.08	0.59	0.60	0.98	0.61	0.58	− 5.1
6. Auditory cortex	0.39	0.34	1.13	0.62	0.49	1.26	0.56	0.51	1.09	0.56	0.52	− 7.2
7. Visual cortex	0.38	0.33	1.14	0.67	0.53	1.25	0.49	0.50	0.98	0.56	0.48	−12.9
8. Association cortex	0.34	0.30	1.15	0.53	0.45	1.18	0.40	0.44	0.90	0.49	0.42	−12.9
9. Hippocampus	0.30	0.26	1.16	0.41	0.37	1.13	0.38	0.39	0.97	0.40	0.37	− 7.1
10. Superior colliculus	0.41	0.32	1.28	0.72	0.58	1.24	0.57	0.62	0.93	0.67	0.58	−13.9
11. Sub-cortical white	0.125	0.125	0.186	0.186	0.173	1.08	0.155	0.145	1.07	0.166	0.164	− 0.9
1–4. Mean sub-cortical nuclei	0.43	0.37	0.71	0.71	0.59	1.20	0.69	0.63	1.09	0.67	0.64	− 4.9
5–8. Mean cortex	0.38	0.32	0.61	0.61	0.51	1.19	0.51	0.51	0.99	0.55	0.50	− 9.5

5. *The cerebral circulation: experimental approaches*

in which cerebral blood flow was measured in twenty-five regions using ^{14}C antipyrine autoradiography. From the actual flow and weight data, a composite clearance curve from the whole brain in response to a step function change in inert gas concentration in arterial blood was obtained and this curve was analysed by means of both the bimodal Gaussian and the two exponential models. The results were then compared with the original values

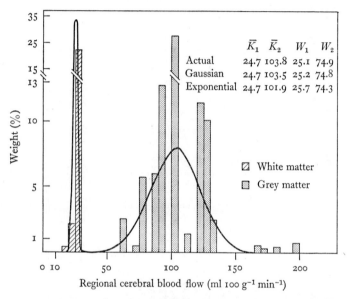

Fig. 5.15. The histogram represents the actual distribution of blood flow in the brain of the awake cat while the curve represents the distribution predicted by the bimodal Gaussian model. The values for the mean white ($\bar{K}_1$) and grey ($\bar{K}_2$) matter flows and their respective weights (W_1 and W_2) are shown together with values predicted by the Gaussian and exponential models. (From Reivich, Slater & Sano (1969), *Cerebral blood flow*. Berlin: Springer, Fig. 1.)

for flow and weight. The results are shown in Fig. 5.15 and from this it was concluded that the two exponential model is a very good description of the cerebral clearance curves with a maximum error of about 8 per cent and that the bimodal Gaussian curve predicts grey matter flow with significantly less error. In addition, it produces a significantly better fit to the data and provides information regarding the standard deviation of grey and white matter blood flow.

Consistence of the various methods of calculating and measuring blood flow. In the previous section, it was shown that analysis of the distribution of local blood flow in the brain using the Gaussian method gives essentially the same answer as the simple two-phase exponential method and that the results of both these methods are remarkably close to the actual situation. There is further evidence of consistency between results obtained by the nitrous oxide method which is based upon measurement of diffusible tracer in venous blood and the xenon clearance method which is based upon the measurement of tissue concentration of tracer. Although the assumptions on which each technique is based are different, a comparison of the results obtained by each method, Table 5.4, shows that these are very similar (Lassen & Høedt-Rasmussen, 1966). Similar results were obtained by Ingvar *et al.* (1965). By the outflow detection method (calculated to 10 min of ^{85}Kr saturation), cerebral blood flow was 50.4 ml 100 g^{-1} min^{-1}, S.D. = 4.9 ($n = 11$). By residue detection method (after intra-arterial injection of ^{133}Xe) the corresponding values were 49.8 ml 100 g^{-1} min^{-1}, S.D. = 4.1 ($n = 7$).

It should be emphasized that in determining the clearance of ^{133}Xe, in these experiments, either one counter which was collimated to 'see' the whole of one hemisphere or multiple external counters were used; in this way, the danger of extrapolating from a clearance curve obtained from one atypical area was avoided.

In summary therefore, there now seems to be good evidence from a variety of sources that despite the considerable variation of blood flow in various parts of the brain, mean cerebral flow can be computed from the relatively simple two-phase exponential treatment and that the results agree closely with those obtained using the more formal Gaussian bimodal treatment and with those obtained by the outflow detection method or one of its modifications. How far the fast and slow components should be labelled as representing blood flow through grey and white matter respectively has not yet been settled satisfactorily and the use of these terms is a convention rather than actuality.

Inhalation techniques

An obvious disadvantage of the outflow detection (e.g. nitrous oxide) or residue detection (e.g. ^{85}Kr or ^{133}Xe) techniques described above, is the necessity for puncturing the jugular vein and/or

Table 5.4. Cerebral blood flow determined by ^{85}Kr with the outflow detection method (Kety–Schmidt) and the residue detection method (intra-arterial injection and external counting). (From Lassen & Hoedt-Rasmussen, 1966. Circulation Res. **19**, 681–8)

			Outflow detection method (Kety–Schmidt)			Residue detection method (injection method)			
Case	Sex age	Side	P_{a,CO_2} (mmHg)	f/λ Measured (min^{-1})	f/λ† Corrected (min^{-1})	P_{a,CO_2} (mmHg)	f/λ Measured (min^{-1})	f/λ* Corrected (min^{-1})	Diagnosis
1	M 73	R	44	0.177	0.177	44	0.199	0.199	Cerebral thrombosis
2	F 48	L	30	0.282	0.339	41	0.385	0.330	Presenile dementia
3	M 32	L	51	0.386	0.378	48	0.366	0.383	Organic dementia
4	M 52	L	34	0.412	0.425	36	0.452	0.438	Cerebral arteriosclerosis
5	M 37	R	38	0.424	0.418	37	0.350	0.356	Organic dementia
6	M 65	R	37	0.388	0.388	37	0.327	0.327	Tumour of right hemisphere
7	F 54	R	48	0.575	0.540	44	0.477	0.511	Epilepsy
8	F 54	L	41	0.366	0.345	37	0.319	0.338	Presenile dementia
9	F 60	L	44	0.419	0.384	38	0.378	0.405	Organic dementia

* Side of carotid and jugular puncture.
† f/λ corrected is calculated from the observed values using the equation f/λ corrected = f/λ measured $1/(1 + \Delta 0.003)$ where Δ = mean P_{a,CO_2} – measured, i.e. the correction is 3 per cent mmHg of deviation from the mean P_{a,CO_2} value.

The use of inert, diffusible indicators

internal carotid artery. The ideal method would involve the inhalation of a radioactive tracer and the measurement of a saturation or desaturation curve by means of external counters. Unfortunately, the method raises two problems, (a) the input is not a delta-function and thus there must be correction for recirculation, and (b) the tracer desaturation curve will be affected by contamination from extracerebral tissues.

The first attempt to measure blood flow without blood sampling was by Conn (1955) but a systematic study of the problem was undertaken by Mallett & Veall (1963, 1965). They performed a two-compartment analysis of ^{133}Xe desaturation curves with a correction for recirculation (Veall & Mallett, 1966) and the curves were typically elevated at the tail and gave longer half-times for both slow and fast components. Further studies by Jensen, Høedt-Rasmussen, Sveinsdottir, Stewart & Lassen (1966), Ueda, Hatano, Molde & Gondoaira (1965), Ibister, Schofield & Torrance (1966), Häggendal et al. (1965), Johnson & Gollan (1965) have shown that the long half-times derived from inhalation clearance curves and which yielded correspondingly low values for regional blood flow were due to extracerebral contamination.

An attempt was made by Obrist, Thompson, King & Wang (1967) to overcome this difficulty by undertaking a three-compartment analysis of which the slowest component was taken to represent homogeneous blood flow through extracerebral tissues. By convoluting the curves with computer assistance, Obrist et al. (1967) were able to obtain values for grey and white matter and mean cerebral flow which were similar to those obtained by Ingvar et al. (1965), see Table 5.5. On the other hand using a two-compartment analysis they obtained similar values to those of Veall & Mallett (1966) which were substantially lower.

A further attempt to simplify the method was introduced by Agnoli, Prencipe, Priori, Bozzao & Fieschi (1969) who counted activity over the skull after a bolus intravenous injection of ^{133}Xe in saline. By measuring the arterial levels of tracer, they were able to correct for recirculation and they then analysed the residual curve which approximated to that following a delta-function input, in terms of three components, assuming for extracerebral tissue a mono-exponential decay. This analysis was, again, only possible using a digital computer. Veall (1969) has more recently suggested the possibilities of using analogue computer

5. The cerebral circulation: experimental approaches

Table 5.5. *Comparison of normal values for cerebral blood flow obtained by different methods* (from Obrist et al. 1967. Circulation Res. **20**, 124–35)

Series	Method	Mean c.b.f. $\pm$ s.d. (ml 100 g^{-1} min^{-1})		
		f_1	f_2	$\bar{f}$
Ingvar et al. (1965)	^{133}Xe internal carotid injection	79.7 $\pm$ 10.7	20.9 $\pm$ 2.6	49.8 $\pm$ 4.6*
Obrist et al. (1967)	^{133}Xe inhalation, 3-compartment analysis	74.5 $\pm$ 9.9	24.8 $\pm$ 3.5	54.7 $\pm$ 6.1
Veall & Mallett (1966)	^{133}Xe inhalation, 2-compartment analysis†	52.9 $\pm$ 8.0	9.9 $\pm$ 2.0	31.5 $\pm$ 4.9
Obrist et al. (1967)	^{133}Xe inhalation, 2-compartment analysis	51.2 $\pm$ 6.2	9.1 $\pm$ 1.7	30.2 $\pm$ 4.1

* Based on the compartmental blood flows and tissue weights reported for these subjects ($n = 7$).

† Blood flows were calculated from the published half-times, corrected for recirculation ($n = 6$), utilizing $\lambda_1 = 0.80$ and $\lambda_2 = 1.51$ and mean tissue weights obtained by Obrist et al. (1967).

facilities and a programme which takes account of a carefully conducted study which must be undertaken to determine how extracerebral tissues affect the cerebral clearance curve should make this approach a practicable proposition.

Molecular hydrogen as an indicator

Hydrogen is inert and diffusible and could theoretically be used as an indicator in the same way as other gases, e.g. nitrous oxide, ^{85}Kr, ^{133}Xe to measure cerebral blood flow provided (*a*) that its concentration in venous blood is in equilibrium with that in tissue and (*b*) that the arterial concentration is lowered to zero. A further advantage of its use is that its concentration or partial pressure in tissue or in plasma can be easily and rapidly measured using a conventional Clarke electrode with a polarizing voltage of $+200$ to 300 mV. Over this range of voltage, the output of the electrode is specific for hydrogen and is proportional to the concentration of hydrogen molecules 'seen' at the tip of the electrode (Hyman, 1961; Aukland, Bower & Berliner 1964).

Its possibility as an indicator has been explored in a number of ways. Clark & Bargerson (1959) used hydrogen electrodes mounted in catheters to detect right-to-left shunts in patients

The use of inert, diffusible indicators

with septal defects. Aukland (1965), Aukland, Bower & Berliner (1964) have shown that hydrogen electrodes can be used to measure the volume of blood flow in kidney skeletal and cardiac muscle. Fieschi, Bozzao & Agnoli (1965) and Gotoh, Meyer & Tomita (1966) made preliminary studies which showed that the hydrogen electrode could be used to determine regional cerebral blood flow while more recently Shinohara, Meyer, Kitamura, Toyoda & Ryu (1969) have measured flow in one cerebral hemisphere by sampling blood continuously from the ipsilateral lateral sinus using a hydrogen electrode in a through-flow cuvette. They calculated blood flow from the desaturation curve using either the Stewart–Hamilton principle or by compartmental analysis making use of the fact that the brain:blood partition coefficient for hydrogen (λ), is 1.0 (Fieschi, Bozzao & Agnoli, 1965). Mean hemispheric blood flow in the monkey was 43.3 ml 100 g^{-1} min^{-1}, S.D. $\pm$ 4.3 and this value differed little whether hydrogen was administered as a bolus in saline by rapid intra-arterial injection, by intra-arterial infusion lasting four to ten minutes or following inhalation of hydrogen. By compartmental analysis, the fast component represented an average blood flow of 116 ml 100 g^{-1} min^{-1}, the slow component an average blood flow of 32 ml 100 g^{-1} min^{-1}. All these values, for mean cerebral blood flow and for fast and slow components agree closely with those obtained using ^{85}Kr as indicator (Lassen & Ingvar, 1961; Ingvar & Lassen, 1962; Lassen *et al.* 1963). This study also provided further evidence that (*a*) intracarotid injection of indicator resulted in its distribution principally to the ipsilateral hemisphere; (*b*) blood sampled from the lateral sinus was not significantly contaminated from extracerebral sources and (*c*) that multi-exponential clearance curves could be obtained from hydrogen electrodes placed on the surface of the cerebral cortex, indicating that perfusion of the cortex is not homogeneous.

Hydrogen clearance as a measure of local blood flow

Although the clearance of radioactive gases can be used for the measurement of blood flow in circumscribed areas of brain (Rosendorff & Cranston, 1969), the method requires local injection of the tracer with lack of certainty as to how far it diffuses and, in consequence, uncertainty as to the specificity of the measurement. With stereotactically placed hydrogen electrodes, however, the

5. *The cerebral circulation: experimental approaches*

point at which measurements are made becomes much clearer. Clearance curves obtained from the caudate nucleus are illustrated in Fig. 5.16 (Bozzao, Fieschi, Agnoli & Nardini, 1968). The curves indicate (*a*) that in the caudate nucleus, blood flow is not homogeneous and (*b*) that the mean value for blood flow is similar to that obtained by Landau *et al.* (1955) using autoradiography, and (*c*) that in this part of the brain, as in the cerebral vascular bed as a whole, blood flow is independent of mean arterial pressure, certainly over the range 55 to 150 mmHg, Fig. 5.17.

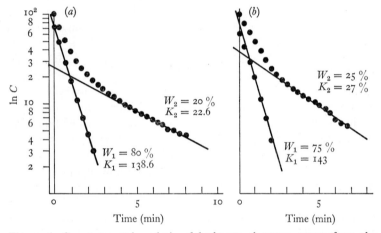

Fig. 5.16. Compartmental analysis of hydrogen clearance curves from the caudate nucleus of the cat. Mean arterial pressure, flow rate and P_{a,CO_2}: (*a*) 115 mmHg, 114.5 ml 100 g min^{-1}, 37.5 mmHg; (*b*) 55 mmHg, 114 ml 100 g^{-1} min^{-1}, 37 mmHg. (From Bozzao, Fieschi, Agnoli, & Nardini (1968), *Blood flow through organs and tissues*. Edinburgh: Livingstone, Fig. 3.)

Hydrogen electrodes can be refined yet further and used for the measurement of blood flow at the capillary level. Using a small electrode suitable for measuring hydrogen clearance on the surface of the cerebral cortex (Lübbers & Baumgärtl, 1967), measurements of the distribution of cortical capillary blood flow have been reported (Lübbers, Kessler, Knaust, McDowall & Wodick, 1966; Lübbers, 1968*a*). By the use of microelectrodes analogous to those described by Silver (1965), measurement of flow in individual capillaries has been attempted (Stosseck & Acker, 1969; Stosseck, 1970).

A further recent development suggests that microelectrodes capable of measuring the clearance of hydrogen from discrete

The use of inert, diffusible indicators

areas of tissue and which, by changing polarity, can also measure tissue oxygen tension, may be of considerable value in determining the mechanisms of oxygen diffusion at the capillary level in brain or other tissue (Heidenreich, Erdmann, Metzger & Thews, 1970).

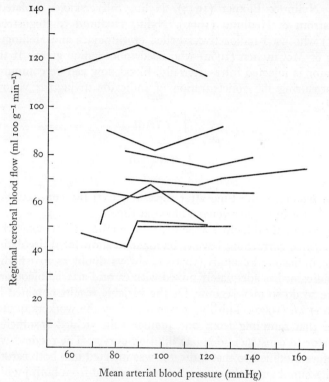

Fig. 5.17. The relation between mean blood flow in the caudate nucleus of the cat obtained from hydrogen clearance curves (see Fig. 5.16) and mean arterial blood pressure. (From Bozzoa, Fieschi, Agnoli & Nardini (1968), *Blood flow through organs and tissues*. Edinburgh: Livingstone, Fig. 1.)

Non-diffusible indicators
Use of the Stewart–Hamilton principle

The indicator dilution principle is based on the relation

$$\bar{t} = V/F,$$

where $\bar{t}$ denotes the mean transit time of the indicator, V the volume of distribution of the indicator in tissue and F the blood

5. The cerebral circulation: experimental approaches

flow (Stewart, 1921; Moore, Kinsman, Hamilton & Spurling, 1929). This method which is used routinely for the measurement of cardiac output has been adapted for the measurement of cerebral blood flow by Gibbs, Maxwell & Gibbs (1947) who used Evans blue; Nylin & Blömer (1955), Nylin, Silfverskiold, Lofstedt, Regnström & Hedlund (1960), Nylin, Hedlund & Regnström (1961) who used radioactive labelled erythrocytes and Hellinger, Bloor & McCutchen (1962) who used indocyanine green. If the indicator is injected intra-arterially, blood flow can be computed by measuring the concentration of indicator in jugular venous blood:

$$F = \frac{\int_0^\infty C(t)\,dt}{\int_0^\infty t\,.\,C(t)\,dt},$$

where t denotes the time after injection, $C(t)$ the concentration of indicator in jugular venous blood at time t.

A number of difficulties arise with this method. First, somewhat complicated corrections have to be made to allow for recirculation of the indicator. Secondly, there is always doubt as to whether the indicator has adequately mixed with carotid arterial blood and failure to do so may account for the variable results obtained by Nylin et al. (1960). Thirdly, it was found by the workers quoted above that sampling from one jugular vein yielded insufficient indicator to construct adequate dilution curves. The method was therefore modified so that indicator was injected into both carotid arteries simultaneously and indicator recovered from both jugular veins. Fourthly, as with other methods in which indicator is injected into the carotid artery, there is doubt as to how far jugular blood is contaminated from extracerebral sources.

For all these reasons, the method is unlikely to yield accurate results while the complexity of the injection and sampling methods together with the calculations required are unlikely to commend themselves to investigators.

Measurement of the cerebral mode transit time

The circulation time for non-diffusible indicators, e.g. ^{131}I albumen (Fedoruk & Feindel, 1960), ^{131}I hippuran (Oldendorf, 1962; Oldendorf & Kitano, 1965; Taylor & Bell, 1966; Kak & Taylor,

Non-diffusible indicators

1967) has been used as an index of cerebral blood flow particularly in a variety of pathological conditions in man. The indicator is injected intravenously and an activity curve is obtained from gamma emission with a scintillation counter placed externally over the occiput. The curve is then differentiated and the mode circulation time measured.

The relation between the mode transit time and cerebral blood flow is not a simple one since it depends, among other factors, upon cerebral blood volume. Fieschi, Agnoli, Battistini & Bozzao (1966) demonstrated an approximately linear relation between the reciprocal of the mode transit time and regional cerebral blood flow measured independently and over a relatively restricted range of blood flow, i.e. 15 to 65 ml 100 g^{-1} min^{-1}. The scatter of results however was considerable so that, for example, for the same mode transit time of 5 s, blood flow could be, within 95 per cent limits of probability, anywhere between 24 and 46 ml 100 g^{-1} min^{-1}. When 5 per cent carbon dioxide was inhaled, however, the relation between mode transit time and blood flow was lost.

The dependence of this method on a constant cerebral blood volume was demonstrated by Harper, Rowan & Jennett (1968). If cerebral blood flow in the monkey was altered by the administration of carbon dioxide or by hyperventilation, the relation between the mode transmit time and blood flow was linear at low levels of blood flow, Fig. 5.16: but as blood flow increased and as c.s.f. pressure which was used as an index of cerebral blood volume, also increased, the relation was no longer linear. It would seem therefore that this method, used as an index of cerebral blood flow, is inadequate and could be misleading unless cerebral blood volume is measured independently and appropriate corrections made. There is some evidence which will be discussed on p. 313 that under certain circumstances, there is a close relation between changes in cerebral blood volume and blood flow (Ingvar & Risberg, 1967; Risberg & Ingvar, 1968).

Cerebral blood flow measured by indicator fractionation

The principle of measuring the distribution of radioisotope indicators in various organs was introduced by Sapirstein (1958). The method consisted of injecting a known quantity of isotope (^{42}KCl, $^{86}RbCl$) intravenously and measuring the concentration

5. *The cerebral circulation: experimental approaches*

(activity/gram) in various organs after death. If cardiac output was also measured, the distribution of isotope could be expressed as blood flow, i.e. as a fraction of cardiac output. Alternatively, the activity in a tissue could be monitored with an external counter from the moment of injection and if a plateau of activity was achieved, this could be extrapolated to $t = 0$ and the activity expressed as a fraction of the injected activity.

The principle is based on the assumption that the indicator is removed from the circulation during one passage and although this was the case with these indicators in peripheral organs, e.g. skeletal muscle, no satisfactory plateau of activity developed in brain and no extrapolation to $t = 0$ was possible. Sapirstein (1962) subsequently introduced a modification whereby extracerebral flow could be distinguished from cerebral flow by the use of ^{131}I antipyrine and ^{42}KCl. Variable fractions of cardiac output were obtained: but confidence in the method was somewhat shaken by the finding in four human subjects of 'negative' cerebral flow. In fact, even with the more complex method of separating cerebral from 'cephalic' blood flow, it was probable that the fundamental assumption that the indicator should be extracted was not being fulfilled.

More recently, the introduction of nuclide-labelled microspheres has made it possible to measure regional or organ blood flow according to Sapirstein's principle. One of the most complete studies using microspheres to determine the distribution of cardiac output has been in the sheep foetus (Rudolph & Heyman, 1967) and this paper includes most of the precautions and controls which require to be observed. The method involves the injection of labelled microspheres whose range of diameter must be known and selected for the animal. The extraction of microspheres is very nearly complete in one circulation: the animal is killed and activity/gram of organ determined. The method has been adapted to brain by Meyer, Tschetter, Klassen & Resch (1969) and in preliminary experiments, these workers obtained values for blood flow in cerebral cortex, cerebellum, sub-cortical white matter, caudate nucleus and thalamus using carbon ^{169}Yb microspheres which agreed with those of Landau *et al.* (1955). Although the method suffers from the disadvantage that only one measurement can be made and that it is confined to experimental animals, it offers some vindication for Sapirstein's pioneering efforts and

Non-diffusible indicators

may prove to be a valuable addition to the methods available for measuring regional blood flow in the brain under particular conditions.

Other methods
Rheoencephalography

This term was introduced by Jenkner (1957) to describe the technique whereby the electrical impedance of the head is measured. Typically, a voltage of 2–3 V at 30 MHz and a current of 2 mA are used with two electrodes placed on the scalp in the parietal region and a reference electrode placed on the forehead. The technique and its applications have been fully documented by Jenkner (1962): other workers who have used the technique but who have been undecided as to what exactly is being measured have described it as 'impedance cephalography' (Waltz & Ray, 1965), 'impedance plethysmography' (Nyboer, 1959) 'electroplethysmography' (Moskalenko, Cooper, Crow & Walter, 1964). It is likely that the electrical impedance can be affected by a range of variables both inside and outside the skull. Attempts have been made to relate the recorded changes in impedance with changes in cerebral blood flow, blood volume, c.s.f. composition and pulsatile flow of blood (Lifshitz, 1963). No clear correlation has emerged. Specifically, there appears to be no correlation between electrical impedance changes and cerebral blood flow (Jelsma & McQueen, 1968).

Cerebral angiography

The intra-arterial injection of opaque dyes and X-ray angiography has been of very considerable value in determining various features of the cerebral arterial system and the size of venous drainage vessels (see Chapter 1): and it is widely used in clinical practice for the diagnosis of intracranial lesions. Attempts have been made to use rapid serial angiography as a measure of cerebral blood flow (Tönnis & Schiefer, 1959; Dilenge & David, 1964) but as with other non-diffusible indicators, the variable that is measured is the cerebral circulation time: and, since no measure of cerebral blood volume is usually obtained at the same time, it is unlikely that there is any simple relation between the circulation time and cerebral blood flow.

5. *The cerebral circulation: experimental approaches*

Summary

In this chapter, the methods which have been, or are commonly, used for the measurement of cerebral blood flow have been considered in some detail together with the assumptions which underlie these methods. As in other vascular beds, the measurement of blood flow is a difficult matter and there is the additional difficulty that in the brain the direction of flow is so complex and the anastomoses so profuse that it is virtually impossible to compare one method of measurement with another and in particular with the simple direct measurement of volume per unit time, a comparison that physiologists have rightly insisted on before any particular method can be validated.

For all these reasons, it is not surprising that there are a variety of methods now available for the measurement of cerebral blood flow and that each year shows further developments. From this array, the investigator is faced with two problems: which animal and which method of measurement is most satisfactory for the particular problem to be studied. It is unfortunate that for experimental studies common laboratory animals, the dog and the cat, are of limited usefulness and it would seem that the animals of choice, particularly if the results are to be remotely relevant to man, are primates. Here it is possible always to relate measurements of flow to the volume of venous drainage, a relation which has been shown to be tolerably consistent. The elucidation of physiological principles from experiments in man for obvious reasons is more difficult and requires, as in other fields of clinical investigation, a high degree of ingenuity.

Of the various methods available, the most comprehensively tested is that introduced by Lassen & Ingvar (1961). It has two advantages. First, it can be modified to give information about cerebral blood flow in most species and to give information about flow in discrete structures of the brain. Secondly, the method stands on a sound theoretical basis and is directly comparable with other methods, e.g. the Fick method using nitrous oxide. Whether the investigator uses molecular hydrogen as indicator rather than ^{85}Kr or ^{133}Xe, whether he uses a simple method of interpreting the desaturation curves, e.g. the initial slope or the *H/A* method rather than with external counters and computer assisted convolution of curves, will depend the question asked,

Summary

the species studied, the accuracy required and the money available.

A review of the literature reveals, however, that confusion has arisen most commonly not because faulty methods have been used but because important variables have not been adequately controlled. This applies most particularly to pharmacological studies of the cerebral circulation (Chapter 12) and in the intervening chapters an account is given of the way in which these variables may operate and how, if uncontrolled, they may confound results obtained by the most theoretically sound and immaculately executed technique for measuring cerebral blood flow.

6 CEREBRAL BLOOD FLOW AND ARTERIAL PRESSURE

Observations on the responses of pial vessels to changes in arterial pressure

The most important proposition of the Monro–Kellie doctrine was that cerebral blood flow varied with arterial pressure. An account has been given in the previous chapter of how persuasive this proposition was for those workers who studied the cerebral circulation in the second half of the nineteenth century and the first thirty years of the twentieth century. The papers of Roy & Sherrington (1890) and of Bayliss & Hill (1895) were particularly influential in confirming the idea that blood flow passively followed changes in pressure and as late as 1928, Kubie & Hetler were concerned to show that their results were consistent with the Monro–Kellie doctrine in every particular.

Some discrepancies had been observed in the meantime. For example, Biedl & Reiner (1900) made long continued measurements of systemic arterial pressure and the pressure in the circle of Willis. They observed on numerous occasions spontaneous fluctuations in these variables which were opposite in direction and concluded that cerebral vessels had an independent vasomotor function which enabled them to respond to a rise in arterial pressure with vasoconstriction. Heymans & Bouckaert (1929) used an isolated dog's head preparation perfused with blood from a donor animal and connected to the body only by the spinal cord and vagosympathetic trunk. They obtained evidence that cerebral vessels dilated in response to an increase in aortic pressure and concluded that this dilatation was reflex in origin. They did not test the effect of altering the perfusion pressure so they obtained no information about the direct effects of altered pressure upon cerebral vessels.

The decisive experiments were carried out by Fog who reported them first in his doctoral thesis (1934) and in a subsequent series of papers (1937, 1938, 1939b). Using the cranial window technique

Responses of pial vessels to changes in arterial pressure

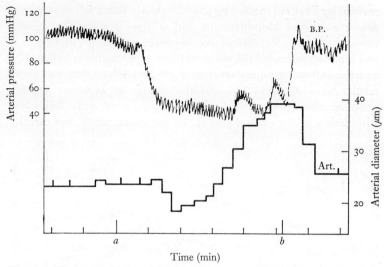

Fig. 6.1. The reaction of a pial artery (below) expressed as a change in the external diameter (μm) to a fall in arterial pressure (above) of long duration. The interval between a and b represents the period of stimulation of the carotid sinus nerve. Time in minutes. (From Fog (1937), *Archs Neurol. Psychiat., Chicago* **37**, 351–64, Fig. 3A. Copyright 1937, American Medical Association.)

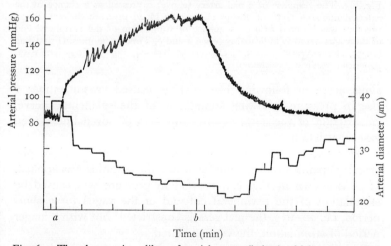

Fig. 6.2. The changes in calibre of a pial artery (below) which accompanied an increase in arterial pressure (above) when adrenaline was injected intravenously between a and b. Time in minutes. (From Fog (1939b), *Archs. Neurol. Psychiat., Chicago* **41**, 260–8, Fig. 3B. Copyright 1939, American Medical Association.)

6. Cerebral blood flow and arterial pressure

revived by Forbes (1928), Fog showed clearly that a fall in systemic pressure caused vasodilatation and a rise in pressure caused vasoconstriction. Representative responses are shown in Figs. 6.1 and 6.2. Fog extended the scope of these experiments considerably and the following points emerge. First he showed that the same results were obtained however the change in systemic arterial pressure was accomplished. Thus a fall in pressure following stimulation of the central end of the cut vagus nerve, the carotid

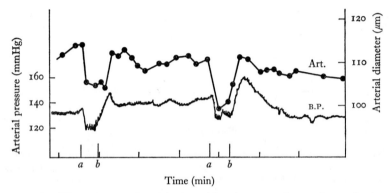

Fig. 6.3. The response of a pial artery (above) expressed as a change in the external diameter (μm) of abrupt changes in blood pressure (below). Blood pressure was lowered as a consequence of stimulation of the peripheral end of the vagus nerve as indicated between a and b. Time in minutes. (From Fog (1937), *Archs Neurol. Psychiat., Chicago* **37**, 351–64, Fig. 2. Copyright 1937, American Medical Association.

sinus nerve or following haemorrhage caused vasodilatation; a rise in pressure following stimulation of the splanchnic nerve, constriction of the aorta or after injection of adrenaline caused vasoconstriction.

Secondly, Fog observed that the changes in calibre of the pial vessels depended on the time for which the stimulus was applied. If as is shown in Fig. 6.3, the fall in pressure was caused by stimulation of the peripheral cut end of the vagus for a short period, i.e. 20–30 s, the pial vessels constricted; but with a longer period of stimulation, the vessels dilated.

Thirdly, Fog was able to show that these changes in pial artery diameter occurred with changes in systemic pressure whether the vagus, aortic depressor, cervical sympathetic or carotid sinus nerves were intact or not. Fourthly, he observed that the effect of a change in pressure, notably a fall, depended on the control level of arterial pressure. If the pressure was already low, i.e. < 70 mmHg, the effect was marked: if high, the changes upon lowering pressure were negligible. Finally, Fog observed that if systemic pressure was very low, i.e. < 40 mmHg, the ability of the pial vessels to respond to changes in pressure was lost, and they then followed pressure changes passively.

A contemporary paper by Forbes, Nason & Wortman (1937) contained essentially the same results and they showed that if the change in blood pressure could be avoided by using a compensator, no change in pial artery diameter was observed when, for example the vagus or carotid sinus nerves were stimulated. In a later paper (Forbes, Nason, Cobb & Wortman, 1937), these workers showed that the one exception to this general rule was the VIIth cranial nerve for stimulation of the geniculate ganglion caused a pial vasodilatation without changes in arterial pressure.

This group of experiments was decisive for it showed unequivocally that cerebral vessels possessed an independent vasoactive mechanism which effectively opposed changes in systemic pressure. The reign of the Monro–Kellie doctrine was over and subsequently physiologists have turned their attention to elucidating how the calibre of vessels (and in later experiments, blood flow) is adjusted to changes in pressure.

As far as they go, the experiments of Fog and of Forbes suggest that the property of adjusting to changes in pressure is an intrinsic one, that is, it is a property of vascular smooth muscle or of local changes in the chemical environment brought about by changes in perfusion. The demonstration that the changes in calibre or vessels could occur whether the nerves were intact or not and after cocaine had been applied locally (Fog, 1938) would suggest this. However, the recent observation that the vasodilatation in response to lowered perfusion pressure could be abolished by injection of atropine or other cholinergic inhibitors (Mchedlishvili & Nikolaishvili, 1970) suggests that a nervous cholinergic mechanism is involved under conditions of diminished blood supply to the brain. In view of these discrepant findings, the question of mechanisms involved in pial arterial dilatation or constriction must remain open.

There are, however, some other aspects of the earlier experiments in which the calibre of pial vessels was used as the index of

6. Cerebral blood flow and arterial pressure

response to a variety of stimuli which require to be considered since they must modify the interpretations put upon the results obtained.

The calibre of pial vessels as a measure of vascular responses

It is very simple to translate pial artery diameter into blood flow (F) by using the relation: $F/\triangle P = Kr^4$ derived from Poiseuille's law. For a variety of reasons, such a relation is probably not valid; and even if it were, there would of course be no justification for extrapolating cortical blood flow from the response of a single vessel.

The relation derived from Poiseuille's equation has not satisfactorily been validated for small vessels; blood flow is pulsatile, it is uncertain whether blood under these circumstances behaves as a Newtonian fluid, the vessel wall is distensible and the apparent viscosity of blood is likely to be critically affected by the cross-section of the vessel. Apart from these theoretical considerations, the changes in perfusion pressure and the changes in pial vessel calibre which accompany them will cause complex changes in both the volume and velocity of flow. For example Forbes, Nason & Wortman (1937) observed that changes in the velocity of movement of red cells within the pial vessels did not necessarily follow changes in vessel diameter. Thus when systemic pressure fell and the vessels dilated, blood flow was retarded. Forbes, Nason & Wortman (1937) use this observation to reconcile their results and those of Schmidt (1936) who, using a thermocouple on the surface of the parietal cortex, noted a reduction in estimated flow when the vagus was stimulated.

If the possible relations between pial vessel calibre and flow are complex and unlikely to be expressed in a simple mathematical equation, the computation of flow from pial vessel responses becomes even more complicated when capillary responses are taken into account. It has been held by many reviewers that intracerebral vessels are unlikely to change in calibre since after the leptomeningeal reflection, the vessels are tightly invested in glia. This notion, which is reminiscent of the Monro–Kellie doctrine itself, cannot be true for as was shown in Chapter 2, there is clear evidence of adaptation of capillary cross-section to such stimuli as hypoxia and it was presumably an increase in the

Pial vessel calibre as a measure of blood flow

blood volume of brain which was observed by early workers who used the volume of brain as an index of blood flow (Roy & Sherrington, 1890).

For all these reasons, the measurements made by Fog (1937) and others using the cranial window technique, however accurately, cannot be considered to give quantitative measurements of blood flow to the cortex. A further difficulty arises in interpreting the results of this group of workers. The results in many cases are given in pooled form standardized as a percentage change of control diameter and it is obvious that the values are obtained from vessels of different sizes, different animals and at different times of the same experiment. Although these factors may not obscure a qualitative change, they could markedly affect quantitative assessments, for example, the amount by which the vessel calibre changed in response to a change in pressure before and after section of a vasomotor nerve. The possible variations in response which are included in pooled data could also include changes in blood gas tensions, pH and the level of anaesthesia.

These comments are not intended to detract from Fog's achievement. They merely emphasize that it is difficult, if not impossible, to extrapolate from observations on the changes of calibre of pial vessels to changes in blood flow to the brain and further that the observations cannot be regarded as conclusive concerning the mechanisms which may be involved in maintaining a stable level of flow within the brain.

Blood pressure and flow

The introduction by Kety & Schmidt (1945) of the nitrous oxide method made it possible to make quantitative measurements of the relation between perfusion pressure and blood flow. This method is suitable for use in man and a series of studies over the subsequent ten years confirmed the remarkable stability of cerebral blood flow in the face of large changes in perfusion pressure. A typical pattern of response is shown in Fig. 6.4 which confirms that when all the nerves are intact and blood gas tensions carefully controlled, blood flow remains constant over the range 60–160 mmHg; as perfusion pressure falls below this range, blood flow falls to very low levels.

This type of relation between perfusion pressure and blood

6. *Cerebral blood flow and arterial pressure*

flow has been found in man (for references, see Kety, 1950; Lassen, 1959). More recent studies have been carried out by Greenfield & Tindall (1965) and by Wullenweber (1965). In a typical study (Finnerty, Witkin & Fazekas, 1954) cerebral blood flow remained constant until perfusion pressure was reduced to

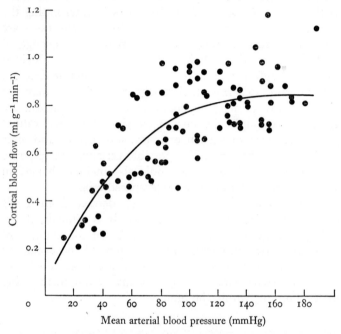

Fig. 6.4. The relation between cortical blood flow (ml g^{-1} min^{-1}) and mean arterial pressure (mmHg). Values from eight dogs, anaesthetized with pentobarbitone, paralysed with suxamethonium and ventilated mechanically. Arterial pressure was altered by haemorrhage and replacement of blood, P_{a,CO_2} maintained at control levels throughout. (From Harper (1966), *J. Neurol. Neurosurg. Psychiat.* **29**, 396–403, Fig. 1.)

50 per cent of physiological levels. If pressure was further reduced, blood flow fell precipitately and there were then clear signs of cerebral hypoxia.

The changes in blood flow were the same whether hypotension was induced by tilting (Finnerty, Witkin & Fazekas, 1954), by spinal anaesthesia (Kety, King, Horvath, Jeffers & Hafkenshiel, 1950; Kleinerman, Sancetta & Hackel, 1958) or by ganglion blocking drugs (Bessman, Alman & Fazekas, 1952). Further,

Häggendal, Löfgren, Nilsson & Zwetnow (1966) have shown in dogs that a similar relation holds if perfusion pressure is reduced as a result of increasing c.s.f. pressure. Up to a critical c.s.f. pressure of approximately 100 mmHg, blood flow was unaffected. Above this pressure, as perfusion pressure was reduced < 60 mmHg, blood flow progressively fell.

Hypertension has been induced by infusion of vasopressin (Sokoloff, 1959), by L-noradrenaline (King, Sokoloff & Wechsler, 1952; Moyer & Morris, 1954; Sensenbach, Madison & Ochs, 1953) and in these subjects and in those with essential hypertension (Hafkenshiel, Crumpton & Friedland, 1954), blood flow was found to be within physiological limits. The proposition however that the constancy of blood flow is maintained above physiological limits should be treated with caution since the values from experiments quoted above were expressed as pooled data and were obtained from one or two observations in each subject.

The relation between pressure and flow found in man has also been confirmed in a number of animal species, using a variety of methods; heat clearance (Carlyle & Grayson, 1955), electromagnetic probes (Rapela & Green, 1964; Yoshida, Meyer, Sakamoto & Handa, 1966), inert gas clearance in dogs (Harper, 1965; Häggendal, 1965), in the newborn lamb (Purves & James, 1969) and in the baboon (James, Millar & Purves, 1969).

The mechanisms whereby blood flow is maintained at constant levels despite large changes in perfusion pressure are not clear and the evidence will be reviewed. First, however, the conditions under which the normal relation of flow and pressure are disturbed will be considered.

Alterations in the relation between flow and pressure

It has been shown that the relation between cerebral flow and pressure illustrated in Fig. 6.4 can be altered in certain circumstances. Sagawa & Guyton (1961) developed a system of perfusing the cerebral vascular bed of the dog which because of the numerous anastomoses between intra- and extracerebral circulations involved considerable surgery, ligations, etc. They found that when perfusion pressure was altered, blood flow varied linearly. This relation between flow and pressure was also observed by Rapela & Green (1964) who arranged a system whereby the dog's brain

6. Cerebral blood flow and arterial pressure

could be perfused either naturally or by a pump. With natural perfusion, the normal relation between flow and pressure was observed: with artificial perfusion, constancy of flow was not usually maintained. A number of reasons have been put forward to explain these results. In Sagawa & Guyton's preparation, the extensive surgery may have caused periods of ischaemia or damage to blood vessels and there is evidence from other sources (Echlin, 1942) that cerebral vessels are very sensitive to trauma and characteristically lose their usual responses, e.g. to a change in P_{a,CO_2}. However, in the preparation of Rapela & Green, the extent of surgery was common to both types of perfusion. In this case, the abnormal relation of flow and pressure may have been due to the release of vasoactive substances in the pump circuit.

If P_{a,CO_2} is raised and maintained some 7–11 mmHg above control levels, the relation between flow and pressure is maintained (James, Millar & Purves, unpublished observations); that is although blood flow is substantially above control over the physiological range, as pressure is reduced to < 60 mmHg, flow falls to very low levels. However, if P_{a,CO_2} is raised to higher levels, > 60 mmHg, the normal relation between flow and pressure is lost and this is illustrated in Fig. 6.5 (Harper, 1966). The reason for this type of response is not known but it has been suggested that since under these conditions, the cerebral blood vessels are maximally dilated, no further dilatation can occur in response to a reduction in perfusion pressure. It is possible that such a mechanism may explain the findings of Sagawa & Guyton (1961) since on the one hand, tissue carbon dioxide will have risen if the brain was under-perfused: on the other hand, the P_{CO_2} of the perfusing blood may have been abnormal, but this must remain conjectural for no measurements of P_{CO_2} were made.

Further evidence that the relation between flow and pressure may be disturbed by abnormal blood gas tensions has been provided by Rapela & Green (1968). They confirmed the findings of Harper (1966) and showed that the disturbance was related to the hypercapnia rather than to the marked fall in venous pH. Häggendal (1968) showed that severe arterial hypoxia also caused the loss of the characteristic relation between blood flow and pressure. On the other hand, this relation was preserved during hypocapnia, severe enough to produce evidence of tissue hypoxia.

Alterations in the relation between flow and pressure

Similar results were obtained by Freeman & Ingvar (1968) who caused short-lived arterial hypoxia severe enough to cause EEG changes by administering 100 per cent nitrous oxide, stopping the respirator or by bleeding. These manoeuvres caused intense post-hypoxia hyperaemia lasting $1\frac{1}{2}$ to 2 hours and when this had subsided, the normal relation between flow and pressure was abolished.

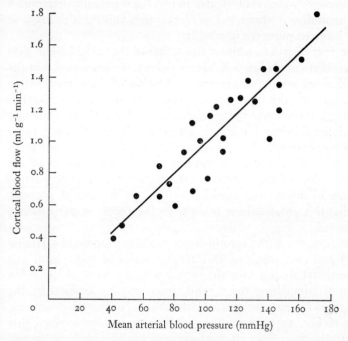

Fig. 6.5. The relation between cortical blood flow and mean arterial pressure. Values from four dogs prepared as in the experiment illustrated in Fig. 6.4. P_{a,CO_2} raised and maintained at 68–86 mmHg. (From Harper (1966), *J. Neurol. Neurosurg. Psychiat.* **29**, 398–403, Fig. 2.)

It is possible that all these experiments have a common factor, namely a departure of the chemical environment of vascular smooth muscle from physiological limits and which is extravagant enough to damage the normal contractile processes. If this is so, then it would suggest that an important part of the vascular pattern of response to changes in perfusion pressure is indeed some property of the smooth muscle itself.

6. Cerebral blood flow and arterial pressure

The neural contribution to the pressure–flow relationship

The evidence which might settle the general question of whether cerebral blood vessels are neurally regulated is considered in Chapter 10. In the present section, two more restricted questions are asked: first, is there any evidence that the acute changes of cerebral vessels to changes in pressure or the steady-state pressure–flow response which is illustrated in Fig. 6.4 is neurally determined and, secondly, is there any evidence that the carotid sinus or aortic baroreceptors are involved?

The first person to connect the action of the carotid sinus with the cerebral circulation was Meyer (1875). He suggested that the carotid sinus was an aneurysmal dilatation of the carotid artery which affected the blood flow to the brain and caused insanity: the bigger the aneurysm, the madder were the patients. Although this relationship was soon shown not to be universal (Schafer, 1877; Binswanger, 1879), Meyer's observations focused attention upon the carotid sinus and Sicilliano (1900) although ignorant of the true function of the carotid sinus, proposed that some part of the carotid artery was sensitive to changes in arterial pressure and that this protected the brain from the effects of anaemia and hyperaemia.

The function of the carotid sinus was first proposed by Hering (1925) and as a result of an intensive series of histological and experimental studies over the next few years, most of the reflex effects of stimulating the carotid sinus were understood by the early 1930s (Hering, 1927; De Castro, 1928; Heymans & Bouckaert, 1929; Keller, 1930; Rein, 1931). All these authors stressed that the carotid sinus from its strategic position was likely to protect the brain in some way. The only dissenter from this piece of teleology was Koch (1931) who argued that the carotid sinus was where it was for purely developmental reasons.

At this time, there were two main theories as to how the carotid sinus might affect cerebral blood flow. It could either initiate reflex vasomotor activity to cerebral blood vessels or it could affect primarily peripheral vascular beds and, by causing a redistribution of cardiac output, a stable systemic pressure and blood flow to the brain could be ensured. The experiments which were carried out to try and resolve this question have only been partially successful. Rein (1929, 1931) showed that if the common

Neural contribution to the $\Delta P/F$ relation

carotid artery on one side was clamped and if the carotid sinus nerve was intact, blood flow approximately doubled in the opposite carotid artery. This increase in blood flow was abolished if the carotid sinus nerve was cut and it was enhanced if the vagi and aortic depressor nerves had been cut. Rein's conclusion was that reflex vasodilatation of cerebral vessels took place in response to the reduction of pressure in the carotid sinus and pointed out that this response was opposite to that in all other vascular beds except the coronary (for references, see Hering, 1932). Rein was not, of course, able completely to counter the criticism that the changes he observed were due to the reflex increase in systemic arterial pressure, beyond pointing out that the changes in pressure were small in comparison to the changes in blood flow.

Heymans & Bouckaert (1929) tackled the problem by means of cross-circulation experiment in dogs. The head of the recipient dog was isolated from its body except for the spinal cord and vago-sympathetic trunks and was perfused through one carotid artery from the donor dog. Pressure was continuously measured in the recipient dog in the remaining carotid artery and in the femoral artery. When the carotid sinus nerve in the recipient dog was stimulated, arterial pressure fell at both points and Heymans & Bouckaert interpreted this finding as indicating that the cerebral vessels had relaxed in the same way as vessels in other vascular beds had. They thus agreed with Rein that cerebral blood vessels were reflexly affected by the carotid sinus but differed in concluding that vascular beds responded in parallel: Rein said that the responses were opposite. The most serious objection to the experiments of Heymans & Bouckaert was that nowhere do they state whether the external carotid artery of the recipient dog was clamped and to this could be added that whether the artery was clamped or not is irrelevant since the anastomoses between intra- and extracranial circuits is so great in the dog as to render it virtually useless for this type of experiment.

In subsequent experiments, Heymans & Bouckaert (1932) repeated Rein's study, measuring carotid arterial flow with an electric or mechanical ströhmur while pressure was altered in the opposite carotid sinus *ad modum* Moissejeff (1927). Their results did not confirm those of Rein since they suggested that cerebral blood flow followed changes in arterial pressure passively and the idea of a vasomotor reflex from the carotid sinus which regulated

6. *Cerebral blood flow and arterial pressure*

the tone of cerebral vessels was abandoned. Keller (1930) had come, rather more tentatively, to the same conclusions.

Ask-Upmark (1935) studied the responses of pial vessels to stimulation of the carotid sinus nerve or to alteration of pressure within the carotid sinus and obtained the same somewhat confusing results as those of Gollwitzer-Meier & Schulte (1932) who studied retinal vessel responses. In both series of experiments, variable or biphasic responses were obtained which the authors interpreted as being due either to passive responses to the change in systemic arterial pressure or to reflex vasomotor activity.

The results of this group of experiments, then, although variable tend to support the view that cerebral vessels respond to changes in arterial pressure passively: Ask-Upmark (1935) made the *caveat* that although vasomotor responses might occur, they were overwhelmed by the passive response.

It should be recalled that these experiments were carried out before Fog (1934) and Forbes, Nason & Wortman (1937) had shown that the responses of pial arteries to changes in arterial pressure were active but probably not neurally controlled. In view of these findings, only the results of Rein (1929, 1931) are explicable and no further studies to clarify these discrepancies have been made until recently (Rapela, Green & Denison, 1967). These workers again used the dog, but used a complex system of ligation and cannulation from which they could calculate the effective perfusion pressure and measure the volume of venous blood which drained somewhere between 50 and 70 per cent of the brain. They measured the proportionate change in blood flow and perfusion pressure which occurred with graded occlusion of the common carotid artery and expressed this as a change in vascular conductance (flow/Δ pressure). They found a significant fall in conductance after denervation of the carotid bifurcations and a rise in conductance when the carotid sinus nerves were stimulated but no significant change in the steady-state relation between pressure and flow. They argued that since section of the sinus nerves and a low pressure in the carotid sinus, i.e. conditions in which baroreceptor reflex effects would be reduced or abolished, caused opposite vascular effects it was unlikely that baroreceptors were involved.

Two questions arise concerning these results. First, the very low values for cerebral venous flow and for the change in flow with comparatively large changes in perfusion pressure raise doubts as to how reactive the vessels were after the extensive preliminary surgery. As a consequence the values for change in vascular conductance would be meaningless. Secondly, changes in carotid sinus pressure brought about by clamping the common carotid artery are also likely to affect carotid body chemoreceptor afferent activity: it is not therefore possible to interpret the results purely in terms of baroreceptor activity. For these reasons and those previously expressed, the question of baroreceptor participation in the cerebral blood flow response to changes in pressure must remain open; and it will continue to remain so until the effects of baroreceptor stimulation by physiological means is tested upon the vascularly isolated brain which is perfused at constant pressure or flow.

By contrast, there is now mounting evidence that the cerebral pressure–flow relation can be affected by division or stimulation of the vagi or cervical sympathetic nerves. Yoshida, Meyer, Sakamoto & Handa (1966) measured the changes in cerebral blood flow which occurred immediately following an abrupt rise in pressure caused by clamping the thoracic aorta in monkeys. They found that there was an immediate rise in calculated vascular resistance, i.e. perfusion pressure increased proportionately more than did blood flow, and that this increase in resistance was largely abolished after cervical sympathectomy. This would suggest that sympathetic vasoconstrictor activity was involved in the vascular response to acute hypertension although this does not exclude an intrinsic mechanism which also opposes an increase in pressure. These results have been severely criticized by Rapela *et al.* (1967) on the grounds that a proportion of 'internal carotid arterial flow' measured by Yoshida *et al.* (1966) was in fact perfusing extracranial tissue. This possibility cannot be excluded in any species though it is less likely in the monkey in which the anastomoses between intra- and extracranial circuits is limited. But Yoshida *et al.* (1966) showed that the responses of blood vessels served by the external carotid artery to changes in arterial pressure were exactly opposite, namely that there was the expected fall in calculated vascular resistance and this would suggest that their description of the internal carotid arterial response was probably valid.

A further attempt to assess the importance of the vagus and sympathetic nerves in the relation between pressure and flow was

6. Cerebral blood flow and arterial pressure

made by James, Millar & Purves (1969) in the baboon. These workers first confirmed that the steady-state relation between pressure and flow when all nerves were intact is that shown in Fig. 6.4 but that when the ipsilateral cervical sympathetic nerves

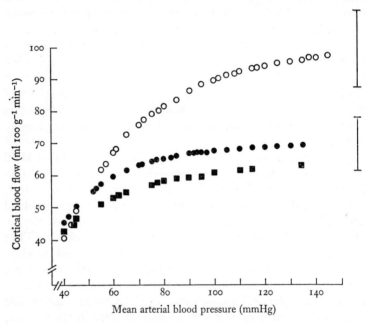

Fig. 6.6. The relation between cortical blood flow (ml $100 \text{ g}^{-1} \text{ min}^{-1}$) and mean arterial pressure (mmHg). Values from thirteen baboons, anaesthetized with pentobarbitone, paralysed with gallamine and ventilated mechanically. Blood gas tensions and pH maintained within physiological limits throughout. Response with nerves intact, solid circles; following sympathectomy, open circles; and during sympathetic stimulation at between 5 and 10 s^{-1}, 15 V and 100 μs duration, solid squares. The vertical lines at the right represent the 95 per cent confidence limits for values of asymptotes towards which the upper curves tend. (From James, Millar & Purves (1969). *Circulation Res.* **25**, 77–93, Fig. 7.)

had been cut, the relation was that shown in Fig. 6.6. Not only was blood flow some 50 per cent higher than control over the physiological range of perfusion pressure, but blood flow was no longer independent of pressure. When the sympathetic nerve was stimulated at constant frequency and intensity, blood flow fell below control levels.

Neural contribution to the $\Delta P/F$ relation

These results are of interest for several reasons. First, they show that the action of the sympathetic nerve is opposite to that found in other vascular beds, namely it is most obvious at high pressure, and it must be confessed that this is difficult to explain in terms of what is known of the baroreceptor–sympathetic pathways. Secondly, these results furnish yet one more example of the way in which the normal pressure–flow curves can be modified, and confirm the suggestion by Yoshida et al. (1966) that sympathetic vasoconstrictor activity is involved in the vascular response to a rise in pressure. Thirdly, the results suggest that the effect of stimulation of the sympathetic nerve is dependent upon the initial (that is, following sympathectomy) level of blood flow. A similar relation between sympathetic stimulation and the changes in blood flow has been observed when the latter are caused by changes in P_{a,CO_2}.

The effect of vagotomy upon the cerebral vascular response to changes in arterial pressure was also studied by James, Millar & Purves (1969) but in less detail. They found that following division of either the vagi or aortic depressor nerves, a relatively modest reduction of perfusion pressure to 80–90 mmHg caused a rapid and marked fall in blood flow which caused death of the animal or irregularities of the heart beat, and which was avoided only by rapid replacement of blood. These results indicate that the activity mediated by these nerves is involved in maintaining dilatation of blood vessels in the face of a reduction in pressure and this is consistent with the observation by Mchedlishvili & Nikolaishvili (1970) that the normal dilatation of pial vessels in response to hypotension is abolished after the injection of atropine or other cholinergic inhibitors.

Summary

The evidence which has been considered in this chapter indicates unequivocally that in all species so far examined, cerebral blood flow is independent of perfusion pressure, certainly over the physiological range and over a variable range below it. There is therefore now no reason to believe as suggested by earlier workers, that blood flow follows changes in pressure passively. The mechanisms whereby blood flow is stabilized are not yet clear and the evidence is conflicting. The earlier workers suggested

6. *Cerebral blood flow and arterial pressure*

that the changes in vessel calibre in response to changes in pressure were brought about by intrinsic mechanisms, notably the property of vascular smooth muscle to oppose changes in tension. More recent evidence has provided strong support for the view that part, at least of the vascular response to pressure is regulated by activity in vagus, aortic depressor, cervical sympathetic and VIIth cranial nerves; the sympathetic nerves causing constriction of vessels at high pressure and activity in the other nerves causing vasodilatation at low pressures. It may be that the reflexly initiated changes in flow ensure a rapid response, for the myogenic response develops more slowly, i.e. over 20–30 s. Such reflex activity is presumably initiated by the aortic and carotid sinus groups of baroreceptors but so far no experiments have shown unequivocally whether or not they are involved or how their activity regulates the vasomotor nerves.

Modest changes in blood gas tensions appear to have little effect upon the relation between cerebral blood flow and pressure. However, severe hypercapnia, hypoxia or hypotension abolishes this normal relationship and there is evidence that they may cause long-lasting or irreversible damage to cerebral blood vessels.

It is suggested that none of these questions is likely to be answered until the effects of changing the perfusion pressure are measured in the vascularly isolated but fully innervated cerebral vascular bed.

7 REGULATION OF CEREBRAL VESSELS BY CARBON DIOXIDE

Historical introduction

That an increase in P_{a,CO_2} causes a dilatation of cerebral vessels and an increase in cerebral blood flow is the one point on which all, or nearly all, observers are agreed. Changes in the calibre of pial vessels viewed through the skull window were noted by many of the workers who used this technique in the late nineteenth century when the animals were made asphyxic (e.g. Donders, 1850, 1859; Reigal & Jolly, 1871). They also noted that the dilatation occurred without a change in arterial pressure which could account for it: indeed, asphyxia was invariably accompanied by a rise in intracranial pressure and in view of this reduction in perfusion pressure, the dilatation became more significant.

Roy & Sherrington (1890) used changes in brain volume as an index of cerebral blood flow and observed that asphyxia caused an increase in brain volume which was dissociated in time from the accompanying changes in arterial pressure. In interpreting their results, they were clearly in a quandary because if they suggested that carbon dioxide had an independent vasomotor activity, this would have invalidated the Monro–Kellie doctrine. They therefore concluded that the rise in brain volume was due to the rise in systemic venous pressure which also occurred. This interpretation might be valid but for the subsequent demonstration that (a) the inhalation of carbon dioxide was accompanied by only small changes in jugular venous pressure and (b) that an increase in venous pressure caused either by tourniquet obstruction (Jacobson, Harper & McDowall, 1963a) or by raising the intracranial pressure (Kety, Shenkin & Schmidt, 1948b) was without effect on cerebral blood flow, certainly up to 400 mmH$_2$O.

Roy & Sherrington (1890) also stated 'the chemical products of cerebral metabolism contained in the lymph which bathes the walls of the arterioles of the brain can cause variations of the calibre of the cerebral vessels... in this re-action the brain possesses

7. Regulation of cerebral vessels by carbon dioxide

an intrinsic mechanism by which its vascular supply can be varied locally in correspondence with local variations of functional activity.' This was a perceptive remark and evidence is discussed in more detail in Chapter 11 which might show whether in fact there is such a correspondence between regional or local blood flow and metabolism in the brain. But the experimental findings which form the basis for this conclusion require some examination. Roy & Sherrington found that the intravenous injection of a variety of acids caused a large increase in brain volume and intracranial pressure without much change in arterial pressure. They recalled that Langendorff & Seelig (1886) had shown that as a result of anaerobic metabolism of the brain, acid (? lactic) was produced, and they therefore prepared a filtered extract of dogs brain which had died from haemorrhage and had been kept for some four hours under anaerobic conditions at 37 °C. Intravenous injection of this filtrate (red to litmus paper) caused the same rise in brain volume as with acid and although they were unwilling to concede that carbon dioxide might affect cerebral vessels independently, they were prepared to allow this action for the products of cellular metabolism.

It is a matter for wonder that Roy & Sherrington obtained the results which they did, for subsequent workers have found the injection of acids to be either without effect (Harper & Bell, 1963) or to cause a fall in cerebral blood flow (Schieve & Wilson, 1953a). Furthermore, the injection of acid causes brisk hyperpnoea and hypocapnia in the spontaneously breathing animal, a combination which would be expected either to cause a fall in cerebral blood flow or to offset any rise in flow. The distinct possibility therefore arises that the increase in brain volume observed by Roy & Sherrington was not due to the injection of acid produced by glycolysis of the brain but some other vasoactive substance possibly liberated by contact with glass used in the preparation of the filtrate.

Certainly Bayliss & Hill (1895) were unable to confirm their results and because of their authoritative statement to this effect and the even more vigorous assertions by Hill (1896) that carbon dioxide had no independent action, the matter was not pursued until 1928 when Forbes revived the cranial window technique for studying the calibre of pial vessels. Wolff & Lennox (1930) made the first systematic study of the effects of hypoxia and

Historical introduction

hypercapnia induced separately. Their results were consistent and clearcut. A rise in P_{a,CO_2} caused dilatation of pial arteries and a fall in P_{a,CO_2} caused by hyperventilation caused a less marked constriction. The former response is illustrated in Fig. 7.1. The changes with oxygen were less marked but Wolff & Lennox were able to show that high oxygen levels caused vasoconstriction and

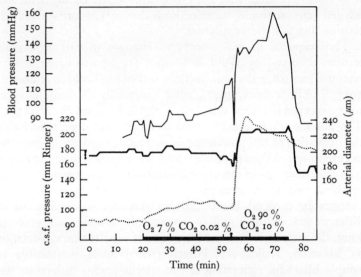

Fig. 7.1. The effect of reducing the oxygen and of changing the carbon dioxide content of the blood upon, from above down, femoral artery pressure (mmHg), diameter of the pial artery (μm) and pressure of the cerebrospinal fluid (mm Ringer). For the period indicated, the cat breathed air with one third of its normal oxygen concentration and the resulting hyperpnoea caused a reduction of 17 per cent of the carbon dioxide of the blood. Then the cat breathed a mixture of 90 per cent oxygen and 10 per cent carbon dioxide. The diameter of the artery increased 15 per cent. (From Wolff & Lennox (1930), *Archs Neurol. Psychiat., Chicago* **23**, 1097–120, Fig. 7. Copyright 1930, American Medical Association.)

the inhalation of low oxygen concentrations caused vasodilatation. Although arterial pressure was not controlled, the changes in vessel calibre were in the opposite direction to those predicted by a change in pressure (Fog, 1937, 1938) and were therefore more significant. Three points are of interest. First, it was clear that the vessels were more sensitive to changes in carbon dioxide than in oxygen since the latter had to be reduced substantially before dilatation was observed. In this respect, the cerebral vessels

7. Regulation of cerebral vessels by carbon dioxide

behave in a similar fashion to ventilation (Dripps & Comroe, 1947). Secondly, the experiments give some indication of the speed of response of the vessels to change in blood gas tensions and allowing for the relative infrequency of measurements, it seemed that the lag before vessels changed in calibre was some 30 s from the time of administration of the gas mixtures. Thirdly, it is worth recalling that the changes in vessel calibre with low oxygen tension would include those due to low carbon dioxide tension due to hyperventilation.

The response of pial vessels to changes in carbon dioxide tension observed by Wolff & Lennox (1930) was confirmed by later workers using the same technique (Sohler, Lothrop & Forbes, 1941; Lubin & Price, 1942), while essentially the same changes were observed in retinal vessels by Cobb & Fremont-Smith (1931). The latter workers also observed as others had done before that the inhalation of carbon dioxide caused a large and abrupt increase in cerebrospinal fluid pressure which was assumed to be the consequence of dilatation of blood vessels and an increase in brain volume.

About this time, attempts were being made to measure the changes in cerebral blood flow in response to inhalations of different gas mixtures and two principal methods were evolved. First, the introduction of the technique of jugular vein cannulation by Myerson, Halloran & Hirsch (1927) made it possible to sample blood for estimation of the arterio-venous difference and so obtain an estimate of flow. Secondly, local blood flow could be estimated using the heated thermocouple technique introduced by Gibbs (1933b) The former method depended on the assumption that cerebral metabolic rate did not alter with changes in P_{a,CO_2}, an assumption which was to be validated by Kety & Schmidt (1948b). By both methods, it was confirmed that cerebral blood flow varied with carbon dioxide tension (Lennox & Gibbs, 1932; Irving & Welch, 1935; Gibbs, Gibbs & Lennox, 1935b; Irving, 1938; Gibbs, Gibbs, Lennox & Nims, 1942; Nims, Gibbs & Lennox, 1942). In some of these studies, flow in the femoral vein was also measured and was shown to fall with high carbon dioxide and rise with low levels (Lennox & Gibbs, 1932; Irving, 1938) thus confirming the earlier observations of Bronk & Gesell (1927); and from these studies arose the generalization that during the inhalation of carbon dioxide or asphyxia, 'cardiac' output was redistributed so that the brain was preferentially perfused.

Historical introduction

Patterson et al. (1955) claimed that a threshold for response to an increase in carbon dioxide level existed, i.e. that P_{a,CO_2} had to rise by some 4 mmHg before any increase in flow was detectable. The use of the term threshold in this sense is misleading, for it implies that the probability of a change in flow below that level of P_{a,CO_2} is zero. Consideration of other data obtained by the same group, that a fall in flow was detected when P_{a,CO_2} had fallen by some 2 mmHg from control (Wassermann & Patterson, 1961) and the vascular response curves to changes in carbon dioxide level caused by inhalation of the gas and hyperventilation (Kety & Schmidt, 1946; Kety & Schmidt, 1948b, Reivich, 1964; Harper & Glass, 1965; James, Millar & Purves, 1969) shows that such a threshold for response to carbon dioxide is improbable. The results of Patterson et al. (1955) should therefore be interpreted literally, i.e. their method of measuring blood flow by the determination of the A–V oxygen difference was insufficiently sensitive to detect changes in flow of less than approximately 10 per cent of control values. No indication of the sensitivity of their method was given by Patterson et al. (1955): clearly it can vary in different hands for using the same method, Noell & Schneider (1944) were able to detect a change in flow when P_{a,CO_2} had risen by only 2 mmHg.

Quantitative measurements

The introduction of the nitrous oxide technique by Kety & Schmidt (1948a) meant that blood flow could be measured quantitatively and with only trivial surgical manoeuvres in man and experimental animals, and a series of studies followed which confirmed the earlier qualitative results, namely, that cerebral blood flow varied with P_{a,CO_2} (Kety & Schmidt, 1948a, b; Lewis, Sokoloff & Kety, 1955; Patterson, Heymann, Battey & Ferguson, 1955). Estimation of the carbon dioxide response was from isolated points in any given individual but from pooled data, it was possible to say that the inhalation of 5 per cent carbon dioxide caused a 50 per cent increase in flow; the inhalation of 7 per cent carbon dioxide caused flow approximately to double while either active or passive hyperventilation caused flow to be reduced progressively until a limit was reached at about 40 per cent of control when signs of disordered sensation were apparent. The response curve so obtained was

7. Regulation of cerebral vessels by carbon dioxide

similar to an hyperbola with evidence of diminished sensitivity (Δ flow/$\Delta P_{a,CO_2}$) at low P_{a,CO_2}, < 40 mmHg.

The shape of the vascular response curve to changes in P_{a,CO_2} has been established for a limited range in the dog (Harper & Glass, 1965) in the baboon (James, Millar & Purves, 1969) and over a much wider range in the monkey (Reivich, 1964). Comparable, though more restricted, curves are available for man

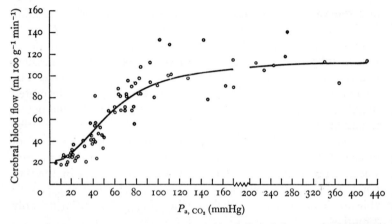

Fig. 7.2. The relation between cerebral blood flow (ml 100 g^{-1} min^{-1}) and P_{a,CO_2} (mmHg) obtained from a series of rhesus monkeys, anaesthetized with pentobarbitone and paralysed with gallamine. The regression equation,

$$y = 20.9 + \frac{92.8}{1 + 10570 e^{-5.251 \log x}},$$

describes the line of best fit. (From Rievich (1964), *Am. J. Physiol.* **206**, 25–35, Fig. 3.)

(Kety & Schmidt, 1948 a, b): a study by Alexander, Wollman, Cohen, Chase & Behar (1964) on human subjects under halothane anaesthesia, suggested that the relationship between blood flow and P_{a,CO_2} was linear but this is inconsistent with other data and follows from the fact that no more than two readings were obtained in any individual subject.

Over the range 15 to 150 mmHg P_{a,CO_2}, the relation between cerebral blood flow and P_{a,CO_2} is sigmoid (Fig. 7.2, Reivich, 1964) the range of greatest change in flow being 40–80 mmHg while above 100 mmHg and below 20 mmHg, changes in flow are considerably reduced. Over the range of maximum sensitivity, the increase in blood flow with P_{a,CO_2} is of the order 1.3 ml

100 g^{-1} min^{-1} mmHg^{-1} P_{a,CO_2} and this is similar to that obtained in primates by other workers (Kety & Schmidt, 1948b; Patterson et al. 1955; Wasserman & Patterson, 1961; James, Millar & Purves, 1969).

Reivich (1964) has also used his data and changes in blood pressure to derive an equation which describes the changes in vascular resistance. Some of the theoretical difficulties of this type of treatment have been discussed in Chapter 4, while further difficulties arise if it is assumed that blood flow is independent of pressure above the physiological range, i.e. within the range of pressure encountered as P_{a,CO_2} is raised. The pressure–flow relation in this range is uncertain since it has been determined under non-physiological conditions, e.g. following administration of noradrenaline. For these reasons, quantification of vascular resistance is of doubtful physiological value. On the other hand, the demonstration that the changes in flow greatly exceed changes in pressure indicate a substantial reduction in vascular resistance as P_{a,CO_2} is increased. Further, the data of Reivich permit the changes in cerebral blood flow with P_{a,CO_2} to be predicted and a figure of 1.11 ml 100 g^{-1} min^{-1} mmHg^{-1} P_{a,CO_2} has been derived as a correction factor to be applied to values for blood flow to standardize them for P_{a,CO_2}, 40 mmHg.

Regional changes in blood flow with carbon dioxide

It cannot be assumed that blood vessels in individual regions of the brain respond to increases in P_{a,CO_2} equally or even in the same direction, for it is known that there are marked differences in resting flow (Landau, Freygang, Rowland, Sokoloff & Kety, 1955) and that there are local differences in response to other pharmacological agents (Freygang & Sokoloff, 1958). Using the autoradiographic technique devised by Landau et al. (1955), Hansen, Sultzer, Freygang & Sokoloff (1957) found that changes in grey matter blood flow in response to the inhalation of 5 per cent carbon dioxide were proportionately greater than white, +67 per cent and +54 per cent of control respectively, while James, Millar & Purves (1969) showed that over the limited range of P_{a,CO_2} 35–65 mmHg, grey matter flow increased at the rate of approximately 1.4 ml 100 g^{-1} min^{-1} mmHg^{-1} P_{a,CO_2} while white matter flow increased by only an average of 0.46 ml 100 g^{-1} min^{-1}

7. Regulation of cerebral vessels by carbon dioxide

mmHg^{-1} P_{a,CO_2}. Some further evidence on this point has been given by Cross & Silver (1962) who used changes of tissue P_{O_2} as an index of changes in blood flow and who observed considerable variation in the increases when the animals inhaled high carbon dioxide–air mixtures.

Vascular responses to carbon dioxide and age

There is now clear evidence that an increase in blood flow which is at least as great as in the adult occurs in both the sheep foetus and newborn lamb as P_{a,CO_2} is increased over the range 30 to 65 mmHg (Purves & James, 1969). The changes, particularly in grey matter blood flow, show the same alterations in rate, being greatest in the range 40–50 mmHg P_{a,CO_2} and progressively less as P_{a,CO_2} is reduced below this. There is also evidence that at this age, the vascular response to carbon dioxide is reduced or abolished with hypotension, or after vagotomy and enhanced after sympathectomy: these features are discussed more fully in later sections.

For reasons of ethics and practical consideration, no comparable data are available in the human and no satisfactory studies have been carried out in the young animal at or about the time of sexual maturity to determine whether there are changes in vascular sensitivity to carbon dioxide which accompany the large increases in cerebral blood flow and oxygen consumption which occur at this time. There is some evidence however that the vascular response to carbon dioxide is affected by age even when such factors as hypertension, arteriosclerotic and other degenerative changes which occur are taken into account. Fazekas, Alman & Bessman (1952) found a small reduction in the response to carbon dioxide in a heterogeneous group of elderly subjects when compared to normal controls. Schieve & Wilson (1953b) found that the average control level of blood flow fell with age being 62, 57 and 55 ml 100 g^{-1} min^{-1} at 21–35, 35–45 and 50–76 years respectively, while the increase in flow above control when 5 per cent carbon dioxide was inhaled was 69, 50 and 40 per cent respectively. Other workers have shown that in the human, the response is complex and in three groups of elderly subjects, (i) normotensive but with evidence of arteriosclerosis, (ii) hypertensive, (iii) hypertensive with evidence of arteriosclerosis, they found that the latter two groups responded to the inhalation of carbon dioxide with an increase in flow similar to control while in the first there was neither a significant increase in arterial pressure or in flow (Novack, Shenkin, Bortin, Goluboff & Soffe, 1953).

It is probable that in addition to the structural changes in cerebral vessels which occur with age, there are other differences which may be generally classified as the 'level of excitability' and which may affect both the resting and experimental values for blood flow. These may be difficult to control but may be important in explaining the discrepant results obtained. An approach which circumvents some of these difficulties has been to measure not only the acute vascular response to hypercapnia (or hypoxia) but the rate and completeness of the subsequent adaptation if the stimulus is maintained for some hours (Haining, Turner & Pantall, 1970).

These workers measured local blood flow in the frontal cortex and cerebellum in rats six and twenty-four months old, using a hydrogen clearance method (Auckland, 1965) and found (a) no age dependent resting levels of flow or changes at 5 min in response to the inhalation of either high carbon dioxide or low oxygen gas mixtures. If hypoxia or hypercapnia was maintained however, the compensating changes in flow which consisted of a progressive fall in blood flow were completed rapidly in the young rats and only very slowly in old rats. These results would suggest that the acute response to carbon dioxide alters only slightly with age but that the adaptive powers are diminished.

Other factors which affect the vascular response to carbon dioxide
Spasm of cerebral vessels

Spasm of cerebral vessels can be seen directly on the surface of the brain or by angiography. It occurs in response to manipulation of the vessels particularly during surgery in which blood enters the sub-arachnoid space or following spontaneous sub-arachnoid haemorrhage in man (Crompton, 1964; Echlin, 1965; Ecker & Riemenschneider, 1951; James, 1968). In addition, there is evidence that haemorrhage is accompanied by a fall in blood flow and a significant fall in P_{a,CO_2}, presumably as a result of hyperventilation (Froman & Crampton-Smith, 1967; James, 1968). Echlin (1965) has further shown that with the vasoconstriction induced by blood and compounded by the fall in P_{a,CO_2}, the vessels are

7. Regulation of cerebral vessels by carbon dioxide

capable of only very slight dilatation in response to the inhalation of carbon dioxide. The nature of the vasoactive substance in blood has been sought (Buckell, 1964) but not identified. The possibility that cerebral vessels may be affected in this way should always

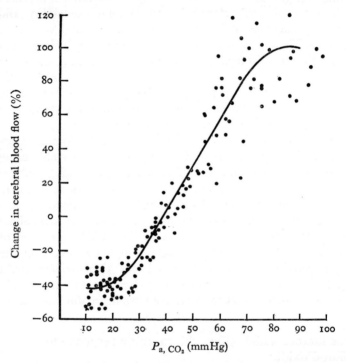

Fig. 7.3. The relation between cerebral blood flow, expressed as a per cent change from flow at $P_{a, CO_2} = 40$ mmHg and P_{a, CO_2}. Dogs anaesthetized with thiopentone and paralysed with suxamethonium chloride. Values from ten dogs with normal blood pressure as carbon dioxide was added to the inhaled gas mixture and during hyperventilation. (From Harper & Glass (1965), *J. Neurol. Neurosurg. Psychiat.* **28**, 449–52, Fig. 1.)

therefore be considered in any experiments in which blood vessels are manipulated or damaged. Such a type of spasm may well explain, for example, the evident lack of reactivity of vessels in the experiments of Dumke & Schmidt (1943) in which the placing of bubble flowmeters in the internal carotid arteries of monkeys was attended by considerable surgery.

Systemic arterial pressure

The normal cerebrovascular response to carbon dioxide can be markedly affected by the level of systemic pressure. This relation has been studied by two series of workers, Harper & Glass (1965) and Häggendal & Johansson (1965). The effect of reducing perfusion pressure upon the vascular response to carbon dioxide is illustrated in Figs. 7.3–5 which make it clear that at approximately

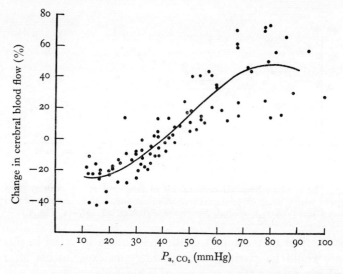

Fig. 7.4. The relation as in Fig. 7.3 between cerebral blood flow and P_{a,CO_2}. Values from seven dogs in which mean arterial pressure had been lowered by haemorrhage and maintained at 100 mmHg. (From Harper & Glass (1965), *J. Neurol. Neurosurg. Psychiat.* **28**, 449–52, Fig. 2.)

50 mmHg mean arterial pressure, the vascular response to carbon dioxide is abolished. Similar lack of response to carbon dioxide was observed in foetal sheep or in newborn lambs in which there was evidence of circulatory failure, a spontaneously falling arterial pressure and P_{a,O_2}, and a rise in P_{a,CO_2} and arterial [H$^+$] (Purves & James, 1969). The mechanism for this relation between the vascular response to carbon dioxide and perfusion pressure is not known. A possible factor could be the reduction in P_{CO_2} as the circulating volume is reduced, as cerebral flow falls and as the animal hyperventilates, but this should be corrected by the administration of

7. Regulation of cerebral vessels by carbon dioxide

carbon dioxide and would not explain the unresponsiveness of vessels in the newborn. A fall in pressure would tend to cause a rise in sympathetic activity and insofar as this affects the vascular response to carbon dioxide (see later section, p. 197) it would tend to oppose vasodilatation. Finally, a serious reduction in cardiac output and cerebral blood flow may well cause reversible metabolic alterations to vascular smooth muscle which render them less than normally reactive.

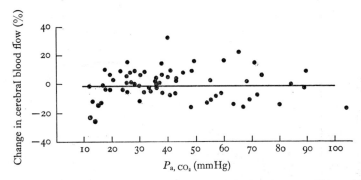

Fig. 7.5. The relation between cerebral blood flow and P_{a,CO_2}. Values from five dogs in which mean arterial pressure was maintained at 50 mmHg. (From Harper & Glass (1965), *J. Neurol. Neurosurg. Psychiat.* **28**, 449–52, Fig. 3.)

It is possible that one part of the mechanism involved in this response could be clarified by repeating the experiments after sympathetic nerve section or α-receptor blockade.

Interactions with oxygen

Quantitative measurements of the vascular response to hypoxia and hypercapnia with sufficient precision to enable us to say whether hypoxia potentiates the response to carbon dioxide have not yet been carried out. The study by Shapiro, Wasserman & Patterson (1966) suggested from isolated measurements in individual subjects that the effects of hypoxia and hypercapnia upon cerebral blood flow were additive. It has also been shown by James, Millar & Purves (1969) that the effect of increasing carbon dioxide upon the vascular response to hypoxia is additive, and by analogy with the comparable studies on the regulation of ventilation (Neilsen & Smith, 1951; Lloyd, Jukes & Cunningham, 1958) it may be inferred as unlikely that hypoxia potentiates the

Other factors affecting the vascular response

vascular response to carbon dioxide. This may also be considered as some evidence that the carotid body chemoreceptors are not involved.

There is some evidence that at low P_{a,CO_2}, < 20 mmHg, the low level of cerebral blood flow induces tissue hypoxia and that this in turn causes vasodilatation so that flow is in fact higher than it would otherwise be and the tissues thereby protected. This possibility has been tested by comparing values for blood flow at a constant level of P_{a,CO_2} (19 mmHg) when the subjects breathed air, 100 per cent oxygen at 1 atmosphere or 100 per cent oxygen at 2 or 3.5 atmospheres. Flow was found to be lower at high oxygen concentrations but the results cannot be considered conclusive since the high oxygen level might have acted by eliminating tissue hypoxia and therefore a supposed stimulus to dilate vessels: alternatively, it might have acted directly by constricting the blood vessels. This question therefore remains unresolved.

The speed of vascular response to carbon dioxide

The early experiments of Wolff & Lennox (1930) which have already been quoted gave some indication of the rate at which pial vessels responded to changes in carbon dioxide. The latent period after the gases were given and before which changes in vessel diameter were observed was certainly longer than that for ventilation to change, and allowing for the infrequency of measurements, appeared to be about 20–30 s of which 4–5 s may have represented the lung-to-brain circulation time. From two of their diagrams, it would seem that the time constant for the response was 3–4 min.

The nitrous oxide method for determining blood flow is not suitable for measuring the speed of response because of the time necessary to make a single measurement. However, the modification introduced by Lewis, Sokoloff & Kety (1955) enabled a more rapid resolution to be made, and this suggested that the latent period was not greater than 1 min while blood flow was still increasing in response to carbon dioxide after 4 min. Similar estimates of the speed of response were obtained by Shapiro, Wasserman & Patterson (1965) who measured the A–V oxygen difference intermittently.

7. *Regulation of cerebral vessels by carbon dioxide*

More recently Severinghaus & Lassen (1967) used a more rapid sampling technique and showed the latent period is barely greater than the lung-to-brain circulation time, and the response has a time constant of some 20 s. This is illustrated in Figs. 7.6 and

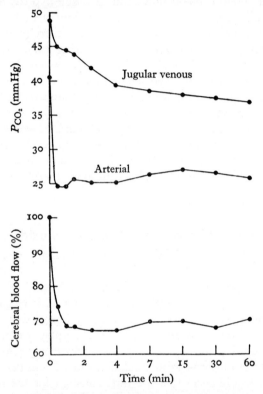

Fig. 7.6. The change in P_{CO_2} of jugular venous blood and arterial blood in an unanaesthetized human subject following a step reduction in alveolar P_{CO_2}. Below, changes in cerebral blood flow calculated from the arterio-venous oxygen difference and expressed as per cent of control. Cerebral blood flow followed the arterial P_{CO_2} rather than the jugular venous P_{CO_2} washout curve. (From Severinghaus & Lassen (1967), *Circ. Res.* **20**, 272–8, Fig. 4.)

7.7. This result is of wider interest for it suggests that the changes in vascular smooth muscle are so rapid that they must be more closely related to changes in arterial P_{CO_2} than to changes in tissue or mixed venous P_{CO_2}. Alternatively, they could be reflex in origin: this point could be directly tested in further experiments.

The speed of vascular response to carbon dioxide

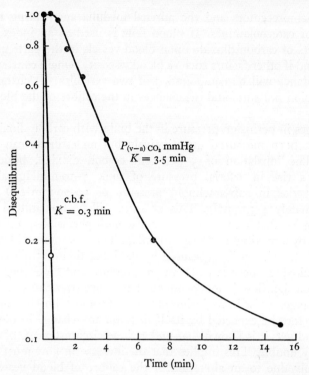

Fig. 7.7. Time course of approach to equilibrium of cerebral blood flow (c.b.f.) and jugular venous P_{CO_2} following a step reduction in alveolar P_{CO_2}. C.b.f. was computed as $1 - [(F_t - F_{60})/(F_0 - F_{60})]$, where F is flow estimated as percentage of control ($F_0 = 100$) at the time in minutes. The jugular venous P_{CO_2} transient was computed as $1 - [(\Delta t - \Delta 60)/(\Delta 0.5 - \Delta 60)]$, where Δ is P_{v, CO_2} minus P_{a, CO_2} at the time indicated in minutes. This type of presentation confirms that the changes in cerebral blood flow have little relation to the changes in P_{v, CO_2}. (From Severinghaus & Lassen (1967), *Circ. Res.* **20**, 272–8, Fig. 3.)

Mode of action of carbon dioxide upon cerebral vascular smooth muscle

The results of experiments cited in the last section raise the important question of how carbon dioxide acts upon vascular smooth muscle in general, that of the cerebral vessels in particular. In the intact animal the vascular effects of carbon dioxide are likely to be extremely complex since apart from local effects, a rise in P_{a, CO_2} brought about by inhalation of carbon dioxide or asphyxia affects the cardiac output, blood pressure, the aortic and carotid

7. Regulation of cerebral vessels by carbon dioxide

sinus chemoreceptors and the adrenal medulla augmenting the release of catecholamines. If blood flow is used as an index of the effects of carbon dioxide upon blood vessels in the particular organ, and if other factors such as blood viscosity remain constant, the resistance will be compounded of two variables: (i) changes in perfusion pressure, and (ii) changes in the calibre of the blood vessels.

Changes in perfusion pressure in the brain with carbon dioxide have not been measured quantitatively but an estimate can be made. The inhalation of 5 per cent carbon dioxide, besides causing a rise in arterial pressure of some 7-10 mmHg, also causes a rise in sub-arachnoid pressure of 50-70 mmH$_2$O or approximately 4-5 mmHg. This pressure will be transmitted to the veins so that under these circumstances perfusion pressure will rise by a maximum of only 5 mmHg. There is some evidence that with the rise in P_{a,CO_2} caused by the inhalation of 5 per cent carbon dioxide, the relation between pressure and blood flow is unaffected, i.e. flow is independent of pressure over the physiological range. A rise in perfusion pressure of some 5 mmHg would not therefore be expected by itself to cause any change in blood flow. And yet, it is known that under these circumstances, blood flow may double. This implies that the increase in flow must be principally due to an alteration of the calibre of blood vessels. The question may therefore be restated: what factors cause the alterations in cerebral vessel calibre in response to changes in carbon dioxide level?

If this question were asked of any peripheral vascular bed, the evidence would indicate the factors outlined above, namely the direct action of carbon dioxide, reflex and hormonal factors. With respect to the brain, however, the idea that reflex control of blood vessels is negligible is so firmly entrenched that Sokoloff (1960) could state the position as: 'The fact that cerebral vessels respond to CO_2 like other vascular beds do only after denervation is further evidence that the CO_2 effect upon cerebral vessels is not mediated by nervous mechanisms but is a direct action upon the smooth muscle walls.' Examination of this statement forms a convenient basis for the following sections.

Carbon dioxide and cerebral vascular smooth muscle

Evidence for the direct action of carbon dioxide upon cerebral vessels

Direct evidence for an action of carbon dioxide upon cerebral vessels is virtually non-existent. There is however a wealth of speculation, see for example Severinghaus, 1968: and there is more general evidence of the action of carbon dioxide upon smooth muscle which may or may not be applicable to cerebral vascular smooth muscle. This evidence derives first from the observation made by Casteels (1970) that changes in pH can affect the fluxes of potassium and chloride ions in taenia coli smooth muscle preparations, Figs. 7.8 and 7.9. A rise in pH from 7.4 to 8.9 increases the efflux rate of ^{36}Cl by a factor of 2 whereas a fall in pH to 5.6 has no such effect and it should be noted that this effect can only be obtained in a potassium-free solution. Fig. 7.9 demonstrates the fact that a pH of 5.6 reduces ^{42}K efflux (a) and a pH of 8.9 slightly increases it (b): but if chloride is replaced by a non-permanent ion, the effect of the pH of 8.9 is abolished and that of the pH of 5.6 is reduced, (c) and (d). A pH of 5.6 slows down the efflux of sodium by a factor of about 1.5 and the effect of pH could thus be to inhibit the membrane mechanism for sodium extrusion and potassium uptake. Since the membrane takes on positive changes with acidification, it would be predicted that the permeability to anions should increase; and this has been found in red cell membranes (Passow, 1961). But smooth muscle cells do not appear to conform to this model since they became more permeable to chloride and potassium ions in alkaline medium. Although it is too early to propose a mechanism which would explain the action of hydrogen ions, the evidence suggests that they affect chloride ion fluxes principally.

The effect of changes in carbon dioxide level and pH upon taenia coli smooth muscle has been studied in a second way by Wienbeck, Golenhofen & Lammel (1968). They showed that if the P_{CO_2} of the perfusing solution was increased at constant pH, i.e. by increasing bicarbonate, the spontaneous electrical activity and tension developed was inhibited; if it was reduced at constant pH, the activity increased. If the carbon dioxide level was held constant and pH reduced, activity increased and raising the pH had the opposite effect. These observations are of interest but

7. Regulation of cerebral vessels by carbon dioxide

scarcely form the basis for a proposal of the mechanism whereby changes in $[H^+]_{e.c.f.}$ affect smooth muscle. Conceivably, the ultimate effects could be intracellular or mediated by alterations in calcium fluxes since $[Ca^{2+}]$ also affects smooth muscle activity.

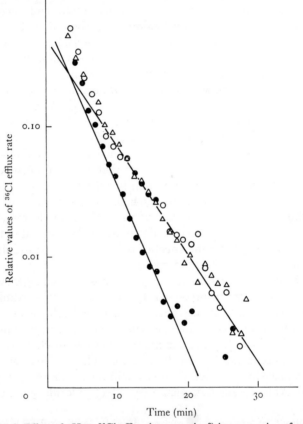

Fig. 7.8. Effect of pH on ^{36}Cl efflux (counts min^{-2}) in a potassium-free solution, pH 7.4, (open circles). A pH of 8.9 units (solid circles) increases the efflux rate by a factor of 2. A pH of 5.6 units (open triangles) has no effect on the chloride efflux. (From Casteels (1970), *Smooth muscle*, ed. Bülbring, Brading, Jones & Tomita. London: Arnold, Fig. 15.)

It would therefore seem to be important that further experiments be carried out on vascular smooth muscle to determine whether or how changes in carbon dioxide or $[H^+]_{e.c.f.}$ are effective. The only evidence on this point, and for obvious reasons widely

Carbon dioxide and cerebral vascular smooth muscle

quoted, derives from experiments carried out in 1911 by Cow. He examined the effect of carbon dioxide in perfusate upon tension developed in rings of carotid artery. The origin of the arterial ring, intra- or extracranial, is not specified. The perfusion fluid was oxygenated but not carbonated and the change in tension which developed when carbon dioxide was bubbled into the

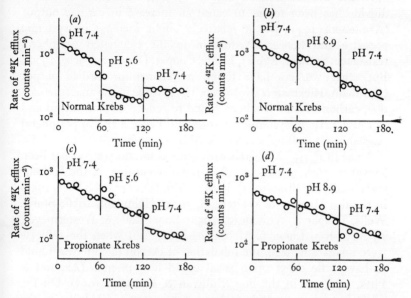

Fig. 7.9. (a), (b) The effect of pH on ^{42}K efflux (counts min^{-2}) in normal Krebs solution at 35 °C. An acid pH decreases the efflux and alkalinization increases it. Semi-log scale. (c), (d) If chloride has been replaced by a non-permanent ion, the effect of pH 5.6 and 8.9 on the ^{42}K-efflux has disappeared. (From Casteels (1970), *Smooth muscle*, ed. Bülbring, Brading, Jones & Tomita. London: Arnold, Fig. 17.)

solution presumably reflected a large fall in pH. The effect on the carotid artery is not illustrated but, according to the author, was similar to that seen in gastric artery and was smaller than the effects of adrenaline, alcohol and changes in temperature.

Comparison of the cerebral vascular response to carbon dioxide with that in other vascular beds

There is clear evidence that the action of carbon dioxide upon the cardio-vascular system as a whole and on individual vascular

7. Regulation of cerebral vessels by carbon dioxide

beds is complex, and this is compounded by the conditions of the experiment, notably the type of anaesthesia, surgical manipulation, maintenance of hydration, plasma volume and e.c.f. volume. For these reasons, studies require to be both systematic and well controlled and very few such have been reported. Consequently, the literature gives a confused picture.

In the unanaesthetized human subject, inhalation of carbon dioxide has been found to cause an increase in cardiac output (Asmussen, 1943) but a relatively slight increase in systemic pressure and hence a reduction in the calculated peripheral resistance (Sechzer, Egbert, Linde, Cooper, Dripps & Prices, 1960). But other workers have found no change in cardiac output (Burnham, Hickman & McIntosh, 1954) while in the anaesthetized dog, carbon dioxide has been found to reduce cardiac output and to cause a rise in systemic pressure and hence a rise in peripheral resistance (Wendling, Ekstein & Abbond, 1967).

A rise in P_{a,CO_2} associated with acute respiratory arrest has been shown to cause a marked fall in renal blood flow in the harbour seal (Bradley & Bing, 1942), in the rabbit (Forster & Nyboer, 1955) in the dog (Draper & Whitehead, 1944; Holmdahl, 1956; Stone, Wells, Draper & Whitehead, 1958; Bohr, Ralls & Westermeyer, 1958; Nahas, Ligon & Mehlman, 1960). But when the subjects were unanaesthetized, the inhalation of carbon dioxide was found to have little effect upon renal blood flow in man (Axelrod & Pitts, 1952) or in the dog (Sullivan & Dorman, 1955). On the other hand in vascularly isolated but innervated kidneys in the dog, inhalation of 20 per cent carbon dioxide caused vasodilatation and this was enhanced by section of the nerves. These discrepant results clearly cannot be accounted for by arguing that the local action of carbon dioxide is to dilate renal vessels while a superimposed reflex action mediated by the sympathetic nerves and perhaps augmented by the release of catechol amines opposes such an effect. This may be so but the evidence is incomplete and will remain so until technically more perfect methods of perfusing the kidney are devised.

Similar confusion exists over the responses of mesenteric vessels. Mohammed & Bean (1951) found that a rise in P_{a,CO_2} of blood perfusing vascularly isolated segments of the mesenteric bed caused vasoconstriction while Brickner, Dowds, Willitts & Selkurt (1956) found that blood flow in the intact mesenteric bed varied

Carbon dioxide and cerebral vascular smooth muscle

with carbon dioxide levels. But in these experiments high carbon dioxide was always given last and the possibility arises that the dilatation represented deterioration at the end of lengthy studies. The precise response therefore remains undecided.

Vasodilatation of the skin with inhalation of carbon dioxide is a common observation in class experiments and has been confirmed by Diji & Greenfield (1957) who found an increase in heat loss when carbon dioxide was injected subcutaneously or when the hand (Diji, 1959) or whole body (Liljestrand & Magnus, 1922) was immersed in water with a high carbon dioxide content.

The injection of acid into the femoral arteries of dogs was reported to cause vasodilatation of vessels in the hind limb (Deal & Green, 1954) but clearly this did not distinguish between the responses of skin and muscle. The most clear cut results were obtained by McArdle, Roddie, Shepherd & Whelan (1957) who found that the inhalation of carbon dioxide in unanaesthetized human subjects caused intense vasoconstriction in the forearm muscles and that this was only slightly affected by nerve block. Under anaesthesia, a mild and transient vasodilatation was observed (McArdle & Roddie, 1958). These results confirm the earlier and less direct observations of Bronk & Gesell (1927), Irving & Welch (1935).

The problem has been analysed some way further by Fleischman, Scott & Haddy (1957) who calculated the resistance offered by segments of the vascular bed in the forelimb of anaesthetized dogs in series. Their results are summarized in Fig. 7.10 and demonstrate that the effect of carbon dioxide varies considerably, dilating post-arteriolar vessels more completely than it does arteries or arterioles. If this action applies generally, then it implies that an important action of carbon dioxide is to regulate the capacitance vessels and hence the venous return to the heart. In general this would favour peripheral pooling of blood and reduced venous return.

The action of carbon dioxide upon the heart has been measured in isolated perfused preparations by McElroy, Gerdes & Brown (1958), the rabbit (Vaughan Williams, 1955), and the dog (Price & Helrich, 1955). The effect is a reduction in the frequency and force of contraction and in the rate of conduction, while in the perfused guinea pig heart, coronary flow varied with the hydrogen ion concentration (McElroy *et al.* 1958).

7. Regulation of cerebral vessels by carbon dioxide

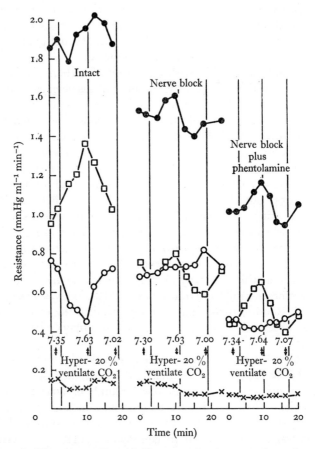

Fig. 7.10. The average effects of pH upon total and segmental vascular resistances in the dog forelimb under control conditions with nerves intact, with nerves blocked and after administration of phentolamine. Solid circles, total vascular resistance; open circles, vascular resistance in arteries; open squares, vascular resistance in small vessels, and crosses, in veins. (From Fleischman, Scott & Haddy (1957), *Circ. Res.* **5**, 602–6, Fig. 2.)

A number of points of interest emerge from this brief survey. First, there does not seem to be any generalization as, for example, that suggested by Sokoloff (1960) which satisfactorily describes the response of peripheral vascular beds to a rise in P_{a,CO_2}. This is partly because the evidence simply is not good enough; but even where it is satisfactory, it is clear that the direct action of carbon dioxide upon vascular smooth muscle causes diametrically different results in different beds. And when the vessels are innervated, the responses are even more complex, for there is abundant evidence that the responses of peripheral vascular beds to carbon dioxide is largely a function of a generalized sympathetic nervous system discharge (Levy, Dogeest & Zieske, 1966; Manley, Nash & Woodbury, 1964; Tenney, 1956a; Richardson, Wasserman & Patterson, 1961) which may be augmented by increased release of catecholamines (Tenney, 1956b; Sechzer et al. 1960).

Secondly we have to assume that either the effects of the sympathetic discharge are not homogeneous or that in some beds the direct dilator response of carbon dioxide is predominant. For example, the skin blood flow which in other circumstances is markedly affected by sympathetic vasoconstrictor activity increases with carbon dioxide as sympathetic activity is presumed to increase. In muscle, on the other hand, the vasoconstrictor action of carbon dioxide is enhanced by sympathetic activity, while in the kidney and gut, the evidence might suggest that the sympathetic nerves effectively oppose any direct dilatation from carbon dioxide.

A more convincing generalization is that a rise in P_{a,CO_2} whether from the inhalation of carbon dioxide or from asphyxia causes an integrated vascular response such that cardiac output is redistributed and certain beds, e.g. the coronary and cerebral are preferentially perfused. This may be the homologue of the response seen vividly in diving birds and mammals in which during apnoea, blood flow to muscle virtually ceases, blood flow to gut and kidney is reduced while blood flow to coronary and cranial vascular beds is increased (Feigl & Folkow, 1963).

This notion is not new since it was shown by earlier workers, e.g. Bronk & Gesell (1927) that in response to asphyxia, blood flow to muscle diminished while that to the brain increased. But whereas these workers implied that the brain, so to speak, stood aloof from the redistribution of cardiac output between peripheral vascular beds, and participated in the overall response merely by dilating its vessels, more recent evidence suggests that the cerebral vascular response is precisely controlled by the autonomic nervous system. In this sense, therefore, the cerebral vessels can be considered to take part in the overall vascular response to a rise in P_{a,CO_2}. This evidence will now be considered.

7. Regulation of cerebral vessels by carbon dioxide

Neural control of the cerebral vascular response to carbon dioxide

As will be discussed more fully in Chapter 10, there is substantial evidence that cerebral blood vessels are under neural control although under physiological limits with respect to blood gas tensions, pH and systemic pressure, tonic vasoconstrictor activity mediated by the cervical sympathetic nerves is small. In a review in 1936, Wolff claimed that the dilatation of pial vessels or increase in blood flow in the frontal cortex in response to the inhalation of carbon dioxide was unaffected by section of the spinal cord, VIth, VIIth or VIIIth cranial nerves, cervical sympathetic nerves or by decerebration. It is particularly unfortunate that no documentary evidence is given for this claim, the results being listed as unpublished. In particular, it would be important to know how quantitative these measurements were, for although it is perfectly conceivable that cerebral vessels dilate in the absence of all nerve supply in response to carbon dioxide, it is clearly pertinent to know whether the response was reduced or modified. No further studies of this topic had been carried out until recently when it was shown that the vascular response to carbon dioxide in the sheep foetus and newborn lamb (Purves & James, 1969), and in the adult baboon (James, Millar & Purves, 1969), was considerably modified when the extrinsic nerves were stimulated. The results were similar at both ages and are summarized in Fig. 7.11. This shows that section of the cervical sympathetic nerves has a small effect upon blood flow at $P_{a,O_2} < 40$ mmHg but that as P_{a,CO_2} is increased, the effect becomes more pronounced. From this it is reasonable to suppose that, in the intact animal, sympathetic vasoconstrictor activity increases with P_{a,CO_2} and its effect is to limit the vasodilatation caused directly by carbon dioxide and other influences.

This is consistent with the observation that a rise in P_{a,CO_2} causes an increase in activity in the cervical sympathetic nerve (Biscoe & Millar, 1968) while further experiments confirm that the effect of stimulating the cervical sympathetic nerve is dependent upon the level of P_{a,CO_2}, Fig. 7.12.

Fig. 7.13 shows the effect of cutting the vagus nerves and VIIth cranial nerve upon the vascular response to carbon dioxide. It is apparent that after section of the VIIth cranial nerve, the

Carbon dioxide and cerebral vascular smooth muscle

response to carbon dioxide is reduced while after vagotomy the response is very considerably reduced and in some cases abolished (James, Millar & Purves, 1969). This suggests that a considerable proportion of the vasodilator response to carbon dioxide is accounted for by active neurally induced vasodilatation; and

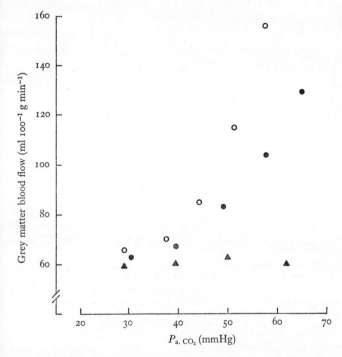

Fig. 7.11. The relation between cortical blood flow and P_{a,CO_2} in a baboon, anaesthetized with pentobarbitone, paralysed with gallamine and ventilated mechanically. Control (open circles); following ipsilateral cervical sympathectomy (open circles), and during sympathetic stimulation at $5-10\ s^{-1}$, $15\ V$ and $100\ \mu s$ duration. (From James, Millar & Purves (1969), *Circulation Res.* **25**, 77–93, Fig. 2A.)

when either the vagi or VIIth nerve was stimulated, in most experiments cerebral blood flow increased and the magnitude of this increase was independent of carbon dioxide levels.

These results taken together raise some points of interest. First, they demonstrate that the vasoconstrictor activity mediated by the sympathetic is not constant but varies with carbon dioxide

7. Regulation of cerebral vessels by carbon dioxide

level. This one fact may explain many of the discrepant results obtained by early workers who measured the effect of section or stimulation of the sympathetic nerves in animals whose P_{a,CO_2} was unknown, but may well have been low since it tends to fall during long experiments.

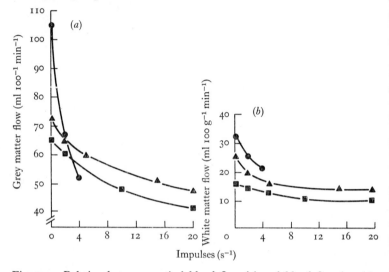

Fig. 7.12. Relation between cortical blood flow (a) and blood flow in white matter (b) and frequency of stimulation of the cervical sympathetic at three levels of P_{a,CO_2} (mmHg): solid circles, $P_{a,CO_2} = 54$; triangles, $P_{a,CO_2} = 40$; and closed squares, $P_{a,CO_2} = 34$. (From James, Millar & Purves (1969), *Circ. Res.* **25**, 77–93, Fig. 4.)

Secondly, the results indicate that vasodilator activity mediated through a reflex arc which includes vagi, possibly the carotid sinus nerves and VIIth cranial nerves, is of considerable importance in determining dilatation of cerebral vessels, and that the final level of flow at any level of P_{a,CO_2} is determined by the interaction of neural constrictor and dilator activity, and this interaction is most probably determined by afferent activity from aortic and carotid body chemoreceptors.

Thirdly, these results are consistent with those of Shalit *et al.* (1967 b) who showed that the cerebral vascular response to carbon dioxide could be greatly reduced or abolished if certain areas of the mid-brain were cooled. The validity of these experiments is discussed in more detail in the next chapter.

Carbon dioxide and cerebral vascular smooth muscle

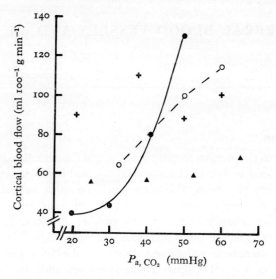

Fig. 7.13. Relation in one baboon between cortical blood flow and P_{a,CO_2} with nerves intact (solid circles), following division of the intracranial roots of seventh and eighth cranial nerves (open circles), following subsequent vagotomy (triangles) and during stimulation of the distal cut end of the seventh and eighth cranial nerves at a frequency of 10 Hz, 20 V and 100 μs duration (crosses). (From James, Millar & Purves (1969), *Circ. Res.* **25**, 77–93, Fig. 9.)

Finally, the results dispose of the idea that the brain takes no part in the integrated vascular response to a rise of P_{a,CO_2} whether by inhalation of carbon dioxide or asphyxia. How far the vessels dilate in response to carbon dioxide when bereft of all neural supply has not been determined, but such dilatation as occurs by local action of carbon dioxide is both augmented and limited by the action of extrinsic nerves.

8 CEREBRAL BLOOD VESSELS AND pH

Introduction

In the previous chapter, an account was given of the overall response of cerebral blood vessels to changes in P_{a,CO_2} and it is evident that as in other vascular beds, this response is complex and almost certainly the resultant of a number of factors which under different conditions may enhance or oppose one another. Of these factors, the more important are the direct action of carbon dioxide upon vascular smooth muscle, reflex vasomotor control which is probably initiated by peripheral arterial chemoreceptors and the action of catechol amines or other vasoactive substances. The attempt to assess the relative importance of these factors has only just begun.

In the present chapter, we shall consider in more detail one of these factors, namely the direct action of carbon dioxide upon cerebral vascular smooth muscle. Carbon dioxide could act in molecular form or it could act by altering plasma pH; but there is now some evidence that it acts by altering the pH of intra- or extracellular fluids of smooth muscle. If this is so then the question of control of cerebral vascular smooth muscle becomes at once more complex and biologically more interesting: complex because the pH of e.c.f. can be altered in other ways, e.g. by changes in the level of neuronal or glial metabolism or by changes in the level of cerebral blood flow. It is more interesting because there is now abundant evidence that the level of alveolar ventilation can also be affected by changes in the pH of e.c.f. or c.s.f. (Leusen, 1950a, b; 1954a, b; Mitchell, Loeschke, Severinghaus, Richardson & Massion, 1963). Thus the rates of alveolar ventilation and cerebral blood flow may be linked by their common dependence upon the pH of e.c.f. or c.s.f.

Cerebral vascular smooth muscle and pH

The effect of changes in plasma pH

In the least complicated experiments quoted in the last chapter, it was concluded that the frequency and force of the isolated perfused guinea-pig heart was more closely related to changes in pH than to changes in the carbon dioxide level of the perfusing fluid (Vaughan Williams, 1955). In the more complex situation *in vivo*, this distinction has not been so clearly made. In the cerebral vascular bed, early experiments suggested that the calibre of pial vessels or the level of blood flow was closely related to changes in plasma pH (Wolff & Lennox, 1930; Geiger & Magnes, 1947; Lubin & Price, 1942). However, in these experiments, little attempt was made to control P_{a,CO_2} while blood flow was calculated from the A–V oxygen difference or was measured as part of a preliminary attempt to perfuse the vascularly isolated cerebral vessels of the cat. Certainly, in later experiments, this relation between cerebral blood flow and plasma pH has not been confirmed. Kety, Polis, Nadler & Schmidt (1948) were unable to show any increase in cerebral blood flow in patients with diabetic acidosis; Shieve & Wilson (1953a) could not find any change in blood flow in man during non-respiratory acidosis or alkalosis. More recently, Harper & Bell (1963) have shown in dogs that blood flow is independent of plasma pH over the range shown in Fig. 8.1 while in a further series of experiments, also in dogs, McDowall & Harper (1968) have confirmed that if a non-respiratory acidosis is induced at constant P_{a,CO_2} there is no significant change in cortical blood flow, Fig. 8.2.

Other modes of action of carbon dioxide

Carbon dioxide could act upon vascular smooth muscle in molecular form. Since there has been no direct experimental work on this point, the idea cannot be dismissed except to point out that evidence for such a direct action is rare in biological systems; the formation of carbamino-complexes is one example and the anaesthetic action of carbon dioxide in high concentrations is possibly another.

An alternative action of carbon dioxide could be through its effect in altering the pH of intra- or extracellular fluid of vascular smooth muscle: such an action is reasonable since carbon dioxide

8. Cerebral blood vessels and pH

is known to diffuse freely between plasma and c.s.f. (Robin, Whaley, Crump, Bickelmann & Travis, 1958; Bradley & Semple, 1962; Mitchell, Carman, Severinghaus, Richardson, Singer & Schnider, 1965).

It must be said at the outset that the evidence which indicates that carbon dioxide may act upon cerebral vascular smooth muscle by altering the pH of i.c.f. or e.c.f. is mostly indirect. Some evidence

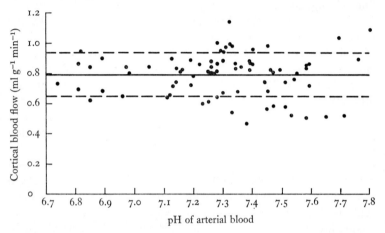

Fig. 8.1. The relation between cortical blood flow (ml g^{-1} min^{-1}) and pH of arterial blood in dogs anaesthetized with thiopentone, paralysed with suxamethonium and ventilated mechanically with a gas mixture of nitrous oxide and oxygen. P_{a,CO_2} was held constant at 30–40 mmHg while pH$_a$ was altered by infusing lactic acid or sodium bicarbonate intravenously until steady flow was obtained. Values from ten dogs. (From Harper & Bell (1963), *J. Neurol. Neurosurg. Psychiat.* **26**, 341–4, Fig. 1.)

derived from experiments with other excitable tissue may be relevant. Thus Hutter & DeMello (1966) and Strickholm, Wallin & Shrager (1969) have shown that a rise in carbon dioxide causes an increase in chloride conductance of the membrane of frog sartorius muscle. These observations have been extended by Walker & Brown (1970) who have shown in the abdominal cells of *Aplysia californica* that the rise in chloride conductance of the membrane with carbon dioxide is due solely to the fall in extracellular pH; that is, an increase in the gas level was ineffective if the pH of the perfusing fluid was held constant while other evidence indicated that the response was not due to changes in intracellular pH or [HCO_3^-]. If these results can be confirmed

Cerebral vascular smooth muscle and pH

with respect to mammalian vascular smooth muscle, they would be consistent with the more circumstantial evidence which suggests that the pH of e.c.f. is an important controller of vessel calibre and thus of flow. An additional and attractive consequence of such a relation, that is between $pH_{e.c.f.}$ and the calibre of cerebral

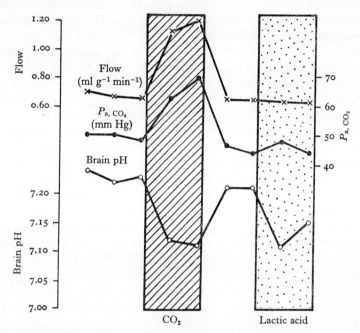

Fig. 8.2. The effect of altering P_{a,CO_2} and intravenous infusion of lactic acid on cortical blood flow and pH measured with a glass electrode on the surface of the cortex. When pH was altered independently of P_{a,CO_2} cortical blood flow correlated more closely with changes in P_{a,CO_2}. (From McDowall & Harper (1968), in *Blood flow through organs and tissues*, ed. Bain and Harper. Livingstone: Edinburgh, Fig. 1.)

vessels, would be that the level of cerebral blood flow could be affected by and adjusted to changes in local tissue metabolism as originally suggested by Roy & Sherrington (1890).

Evidence for a barrier to H^+ between plasma and brain

An important feature of the proposition that carbon dioxide exerts its effects upon vascular smooth muscle by altering $pH_{e.c.f.}$ is that whereas carbon dioxide diffuses freely from plasma to e.c.f., H^+

8. Cerebral blood vessels and pH

does not. This could be most simply tested by measuring $pH_{e.c.f.}$ as the pH of plasma is varied at constant P_{CO_2}. This is technically difficult because of the small volume of brain e.c.f. and because the necessary H^+ sensitive microelectrodes have not been totally satisfactory. This problem may be overcome with the introduction of suitable electrodes (Thomas, 1971). In the absence of measurements of $pH_{e.c.f.}$, evidence for a barrier to H^+ between plasma and brain rests almost entirely upon simultaneous measurements of the pH of plasma and c.s.f. and the first question that such evidence raises is how far the composition of c.s.f. sampled (as it is usually) from the large cavities is representative of that of e.c.f.

This distinction between the composition of c.s.f. and e.c.f. must be drawn since it is possible (a) that part of the known concentration gradients between plasma and c.s.f. may be accounted for by trans-ependymal gradients, and (b) that the composition of e.c.f. is more directly and rapidly affected by blood flow and the composition of plasma than of c.s.f. Further, it is known that the composition of c.s.f. is considerably modified between its site of origin where it is similar to that of a plasma ultrafiltrate to the site of sampling, usually in a large cavity. The most notable changes are in $[K^+]$ (4.5 to 2.8 m-equiv $kg^{-1} H_2O$), $[H^+]$ (40×10^{-6} to 50×10^{-6} m-equiv $kg^{-1} H_2O$) and $[HCO_3^-]$ (28 to 22 m-equiv $kg^{-1} H_2O$) (Pappenheimer, 1965). There is also evidence of close control of these ionic concentrations in c.s.f. which cannot be accounted for simply by passive diffusion between c.s.f. and e.c.f. Thus c.s.f. $[K^+]$ varies by only a fraction of a m-equiv l^{-1} as plasma $[K^+]$ is varied over the range 2.7 to 8.4 m-equiv l^{-1} (Bekaert & Demeester, 1954; Bradbury & Kleeman, 1967) although it is known that $^{42}K^+$ exchanges rapidly between plasma and c.s.f. (Cserr, 1965). The mechanism of this remarkable stability of c.s.f. has been partly explained by the demonstration that the efflux of K^+ from c.s.f. to brain and plasma (under conditions where c.s.f. $[K^+]$ is increased) is controlled by sodium–potassium pump which can be inhibited by ouabain (Bradbury & Stulcova, 1970).

With respect to the gradient for $[HCO_3^-]$ between plasma and c.s.f., the evidence is much less direct. As a result of experiments in which the ventricles were perfused with mock-c.s.f. of differing $[HCO_3^-]$, Pappenheimer (1965) has concluded that this gradient is maintained by active transport and that the pump is located at or near the capillary–glial boundary. If this is so, then [HCO_3^-] of interstitial fluid will be similar to that of c.s.f. in the steady state, i.e. when the net flux of HCO_3^- between plasma and c.s.f. is zero. From this and from the fact that carbon dioxide diffuses freely between plasma, e.c.f. and c.s.f., it follows that under these conditions [H^+] of interstitial fluid and c.s.f. will be similar. However, if the steady state is altered, e.g. by a change in P_{a, CO_2} then c.s.f. [H^+] cannot be assumed to represent e.c.f. [H^+] until equilibrium is restored. Precisely how long it takes for equilibrium to be restored between [H^+] and [HCO_3^-] of plasma and c.s.f. has never been satisfactorily determined; nor is this a realistic possibility since the factors which tend to stabilize $pH_{c.s.f.}$ in the face of a disturbance of plasma pH all proceed at very different rates, e.g. changes in alveolar ventilation, cerebral blood flow, renal excretion of HCO_3^-. This may be illustrated by the observation of Robin et al. (1958) that the injection of acid into plasma is followed by paradoxical changes in $pH_{c.s.f.}$ which may be explained as follows. Addition of acid to blood causes a rapid fall in plasma [HCO_3^-] with an equally rapid passive flux of HCO_3^- from e.c.f. to blood. The fall in e.c.f. [HCO_3^-] at constant P_{CO_2} will cause an increase in e.c.f. [H^+], hyperventilation and a fall in P_{a, CO_2}. As the [HCO_3^-] of c.s.f. in large cavities equilibrates with the low P_{a, CO_2}, c.s.f. [H^+] will fall. The extent to which c.s.f. [H^+] will fall will depend upon the buffering capacity of c.s.f.: the rate at which c.s.f. [H^+] will alter will depend on the degree to which c.s.f. [HCO_3^-] is affected on the one hand by changes in P_{a, CO_2} and on the other by the relatively slow changes in carbon dioxide stores of brain tissue with which c.s.f. is also in equilibrium. According to Robin's data, $pH_{c.s.f.}$ could be in fresh equilibrium with the pH of plasma minutes or hours after the disturbance, while the equilibrium between HCO_3^- in the two compartments could be even longer delayed.

There is therefore no simple answer at present to the two related questions: (a) how representative of e.c.f. is a sample of c.s.f. taken from a large cavity some minutes or even hours after disturbance of acid-base variables in plasma or c.s.f. and (b) how accurately can one deduce the composition of e.c.f. which might affect the calibre of cerebral blood vessels from a sample of c.s.f.

Failure to answer these questions also makes it difficult to

8. Cerebral blood vessels and pH

estimate whether a significant barrier to H^+ exists between plasma and e.c.f. or c.s.f. Part of the evidence has been derived from studies in which the composition of c.s.f. has been measured as plasma pH has been altered by carbon dioxide or by the injection of acids or alkalis. It has been known for fifty years that $pH_{c.s.f.}$ remains remarkably constant (Collip & Backus, 1920) and this has been confirmed many times. Thus $pH_{c.s.f.}$ which in many normally varies between 7.309 units (Bradley & Semple, 1962) and 7.349 (Schwab, 1962a; Mitchell et al. 1965) varies only slightly in non-respiratory alkalosis (Mitchell et al. 1965; Bradley & Semple, 1962), on exposure of subjects to altitude (Severinghaus & Carcelan, 1964), during pregnancy (Mitchell et al. 1965) or in certain clinical conditions with chronic acid–base disturbances, heart failure (Schwab, 1962a), emphysema (Schwab, 1962b).

However, there are a number of conditions in which the pH of c.s.f. has been shown to be significantly altered – in subjects with hypercapnia (Merewarth & Siecker, 1961; Buhlmann, Scheitlin & Rossier, 1963), in patients with chronic bronchitis and hypercapnia (Alroy & Flenley, 1967) in whom $[H^+]_{c.s.f.}$ was observed to increase 0.21 nmol l^{-1} per mmHg rise in P_{a,CO_2} and in unanaesthetized goats following intravenous injection of acid (Pappenheimer, 1965). Thus if a barrier to H^+ diffusion exists between plasma and brain or c.s.f., it appears to be a relative one and in the absence of any direct evidence on this point, it is probably more proper to consider a diffusion barrier to H^+ as one of a number of mechanisms which regulate $pH_{c.s.f.}$. Other mechanisms will be considered in later sections.

Evidence for the regulation of cerebral vascular smooth muscle by the pH of brain e.c.f.

The evidence which might support the idea that pH of brain e.c.f. regulates cerebral vascular smooth muscle is derived principally from studies in which cerebral blood flow has been correlated with simultaneously measured $pH_{c.s.f.}$. In view of the points discussed in the last section, such evidence depends mainly upon the assumption that equilibrium with respect to H^+ exists between c.s.f. and e.c.f. at the time of measurement. A small part of the evidence is derived from studies in which cerebral blood flow or the calibre of pial vessels has been measured following direct alterations of $pH_{c.s.f.}$ or of perivascular fluid.

Cerebral vascular smooth muscle and pH

Correlation between $pH_{c.s.f.}$ and cerebral blood flow. Most of the experiments in this group have been carried out by changing P_{a,CO_2} in experimental subjects or by measuring $pH_{c.s.f.}$ and cerebral blood flow simultaneously in patients with chronic respiratory disorders. In experiments involving hypocapnia for up to six hours, McDowall & Harper (1968) in monkeys and Marshall, Gleaton, Hedden, Dripps & Alexander (1966) in goats, have shown that as plasma pH and $pH_{c.s.f.}$ rise from normal levels, the latter to > 7.46 units, these changes are accompanied as expected by an initial fall in cerebral blood flow. In the subsequent 3–4 hours, $pH_{c.s.f.}$ fell towards normal levels reaching approximately 7.40 units: however, no corresponding increase in cerebral blood flow was observed. This lack of response of cerebral blood vessels to a steady fall in $pH_{c.s.f.}$ might be explained in terms of damage due to tissue hypoxia associated with the severe degree of hypocapnia induced (P_{a,CO_2} < 20 mmHg) were it not for the demonstration by McDowall and Harper that after the period of hypocapnia and when P_{a,CO_2} was restored to normal levels, cerebral blood flow increased to and overshot control levels in the same way as has been observed in other types of cerebral ischaemia (Kjallquist, Siesjö & Zwetnow, 1969).

Contrary results were obtained from experiments in cats in which P_{a,CO_2} was raised and maintained at high levels for rather longer periods, 8–11 hours (Agnoli, 1968). Cerebral blood flow increased initially as plasma and $pH_{c.s.f.}$ fell but on this occasion, as $pH_{c.s.f.}$ subsequently returned towards control levels, cerebral blood flow also fell. Similar evidence that a relation exists between blood flow and $pH_{c.s.f.}$ has been given by Skinhøj (1968) who studied patients with long standing respiratory or non-respiratory acidosis. In patients with a high plasma P_{CO_2} (47–86 mmHg), cerebral blood flow was found to be within normal limits provided that renal or other types of compensation had occurred and $pH_{c.s.f.}$ had returned to within or near normal limits (7.304–8.310 units). If this compensation had not occurred and $pH_{c.s.f.}$ remained low, i.e. < 7.28 units, cerebral blood flow was invariably elevated. Similarly, in patients with non-respiratory acidosis, cerebral blood flow was found to vary with $[H^+]_{c.s.f.}$ although P_{a,CO_2} was uniformly low (< 30 mmHg) in all patients. These observations have been confirmed by Betz (1968) in cats and by Fencl, Vale & Broch (1968) in human subjects.

8. Cerebral blood vessels and pH

In another type of study, cerebral blood flow (calculated from the ratio, mean cerebral oxygen uptake–A–V oxygen difference) and $pH_{c.s.f.}$ were measured in human subjects during acclimatization to altitude (Severinghaus, Chiodi, Eger, Brandstatter & Hornbein, 1966). The results are summarized in Table 8:1 and show that such changes in cerebral blood flow as were observed were principally due to arterial hypoxia. Changes in P_{a,CO_2} and pH_a were modest compared to those quoted above but if they had been induced by hyperventilation at sea level (Wasserman & Patterson, 1961; Reivich, 1964), then, on average, cerebral blood flow would have been some 6 ml 100 g^{-1} min^{-1} lower than that observed after 6–12 hours at altitude and 9–10 ml 100 g^{-1} min^{-1} lower than the values observed after 3–5 days at altitude. The only close correlation was between the levels of cerebral blood flow and $pH_{c.s.f.}$.

Although this last group of experiments provide some circumstantial evidence that $pH_{c.s.f.}$ or $pH_{e.c.f.}$ is an important factor in the regulation of cerebral blood flow, an explanation is still

Table 8.1. *Cerebral blood flow in man at sea level and at altitude* (from Severinghaus *et al.* 1966. *Circulation Res.* **19**, 274–82)

	pH_a	P_{a,CO_2} (mmHg)	P_{a,O_2}	$pH_{c.s.f.}$	c.b.f. (ml 100 g^{-1} min^{-1})	c.b.f. (% of control)
1. Sea level						
Mean	7.414	40.8	85.2	7.317	41.6	—
S.E. ±	0.005	0.8	1.5	0.007	1.8	—
2. At 3810 m, 6–12 hours						
Mean	7.453	35.0	43.5	7.317	50.7	124
S.E. ±	0.01	0.9	1.4	0.01	4.0	—
3. Breathing added oxygen for 10 min						
Mean	7.458	35.1	79.8	—	41.1	100
S.E. ±	0.025	1.5	1.5	—	3.5	—
4. 3810 m, 3–5 days						
Mean	7.467	29.7	51.2	7.308	47.2	113
S.E. ±	0.007	0.7	1.3	0.013	2.8	—
5. Breathing added oxygen for 10 min						
Mean	7.450	30.9	95.5	—	42.0	107
S.E. ±	0.010	0.8	1.0	—	3.0	—
6. Breathing added oxygen and carbon dioxide for 10 min						
Mean	7.428	35.2	170.0	—	53.0	133
S.E. ±	0.006	0.9	10.0	—	4.5	—

required for the apparently discrepant results obtained by McDowall & Harper (1968) and by Marshall et al. (1966). It should be noted first that in these experiments, the hypocapnia was induced for comparatively short periods and was severe, i.e. well below the level of carbon dioxide which would abolish phrenic motor neurone activity in most species. At this level of P_{a,CO_2}, it has been shown that cerebral blood flow changes little with carbon dioxide (Reivich, 1964). A quantitative study relating cerebral blood flow and either $pH_{c.s.f.}$ or $pH_{e.c.f.}$ has not yet been carried out in this range, but it would be reasonable to suppose that a similar degree of insensitivity of blood vessels would be observed with c.s.f. in the highly alkaline range, i.e. > 7.40 units. If future experiments show this to be the case, then the findings of McDowall & Harper (1968) and Marshall et al. (1966) are readily explained. If, however, the experiments of McDowall & Harper (1968) and Marshall et al. (1966) are repeated with a more modest degree of hypocapnia induced, e.g. with a P_{a,CO_2} of 25–35 mmHg in either goats or monkeys where the sensitivity of cerebral blood vessels to carbon dioxide is greatly increased and cerebral blood flow is still shown to be unaffected as $pH_{c.s.f.}$ drifts back towards normal levels, then this would constitute an important objection to the proposition that the effects of carbon dioxide are mediated by changes in $pH_{c.s.f.}$ and that $pH_{c.s.f.}$ is an important regulator of cerebral blood flow.

An attempt has been made by Alexander, Marshall & Agnoli (1968) to resolve these differences. In a control group of goats, these workers first confirmed the results reported by Marshall et al. (1966): that is, if P_{a,CO_2} was held at 19 mmHg for six hours, cerebral blood flow fell as expected but remained unchanged although $pH_{c.s.f.}$ after rising to 7.466 units subsequently fell to 7.401 units. In a second group of goats, P_{a,O_2} was additionally reduced to 41 mmHg and cerebral blood flow measured during short periods when the hypoxia was abolished by giving the goats high oxygen to breathe. These workers found that after 45 min of combined hypoxia and hypocapnia, cerebral blood flow was substantially lower than in the control group but that after a further $3\frac{1}{2}$ hours, it had risen by a mean of 6 ml $100g^{-1}$ min^{-1}. The authors interpreted the differences between the two groups as being due entirely to the effects of cerebral hypoxia and have concluded that this would explain the adaptation of cerebral

8. Cerebral blood vessels and pH

blood flow during acclimatization to altitude. This may be so: but it is particularly unfortunate that these authors did not report the values for $pH_{c.s.f.}$ in the second group of goats since it is probable that under these severe conditions of hypoxia and hypocapnia, the c.s.f. lactate concentration would have been higher than control and the $pH_{c.s.f.}$ lower. It would not then be surprising that after $\frac{3}{4}$ hour, cerebral blood flow was higher than control nor that after a further 3–4 hours cerebral blood flow steadily increased. It is therefore possible that the results obtained from these groups of experiments in which a severe degree of hypocapnia was induced and maintained represent an extreme position. Certainly, with more modest deviations of carbon dioxide tension from the normal range, cerebral blood flow closely follows changes in c.s.f. $[H^+]$. But it must be emphasized that this relation is circumstantial and a more causal relationship is considered in the next section.

Experiments involving alterations of $pH_{c.s.f.}$. The first attempt to assess the effect of altering $pH_{e.c.f.}$ or $pH_{c.s.f.}$ upon cerebral vessels arose out of an investigation into the most suitable kind of fluid with which to irrigate the brain surface during prolonged exposure to air during neurosurgical operations (Elliott & Jasper, 1949) for it had been shown that under such conditions, oedema of the brain was common (Prados, Strowger & Feindel, 1945a, b). The pH of fluid on the surface of the cortex was found to be between 7.07 and 7.34 units depending on how close to a pial vessel the measurement was made. Unbuffered (bicarbonate-free) isotonic solutions applied to the surface of the brain rapidly assumed pH levels of 6.0 units or less and under these conditions, the pial vessels were observed to dilate and the cortical surface to flush. If similar solutions with bicarbonate added were applied, the pH of the fluid on the surface of the brain was approximately 7.4 units or more and the pial vessels became constricted and the pial surface blanched. Similar results were obtained by Gotoh, Tazaki & Meyer (1961) who altered $pH_{c.s.f.}$ by blowing carbon dioxide across the surface of the cortex.

Some caution is required in interpreting the results of experiments in which the cortex is exposed for any length of time and in which the pH or P_{CO_2} of overlying fluid is measured by relatively large electrodes apposed to the surface of the brain.

Cerebral vascular smooth muscle and pH

In such situations, there is, as described above, a real danger of oedema of the brain and disruption of the microcirculation. This may for example explain the observation made by McDowall & Harper (1968) and illustrated in Fig. 8.2 that the pH of fluid overlying the cortex fell not only during the inhalation of carbon dioxide (which would be expected) but also during the intravenous infusion of lactic acid. This finding could be interpreted as demonstrating that long exposure of the cortex leads to oedema of the brain and the normal mechanisms which regulate the transport of H^+ are disrupted. Similarly Betz & Kozak (1967) measured cortical blood flow (using heat clearance methods) and the pH of fluid overlying the cortex (which was used as an index of the pH of brain e.c.f.). Changes in blood flow were brought about by altering carbon dioxide or by brief periods of severe hypoxia and, according to Betz & Kozak, correlated most closely with changes in cortical fluid pH. From these results, a substantial hypothesis which implicated the pH of brain e.c.f. as an important regulator of cerebral vascular smooth muscle was constructed. Neither the results nor, for reasons outlined above, the methods used really substantiate this. The most important objection is that, with their experimental design, a direct action of carbon dioxide or hypoxia upon cerebral vessels cannot be excluded.

Preliminary attempts to alter $pH_{c.s.f}$ alone and directly have been made by Siesjö, Kjällquist, Ponten & Zwetnow (1968) who used the ventricular perfusion technique introduced by Pappenheimer, Heisey, Jordan & Downer (1962) in dogs ventilated mechanically and in whom P_{a,CO_2} was held constant. The results shown in Fig. 8.3 indicate that cerebral blood flow follows changes in $[H^+]_{c.s.f.}$ closely. Further evidence that alterations of the pH of perivascular fluid can affect the calibre of pial vessels has been provided by Wahl, Deetjen, Thurau, Ingvar & Lassen (1970) who have repeated the experiments of Elliott & Jasper (1949) under more carefully controlled conditions. They injected solutions of various composition (Table 8.2(a)) locally by means of a micropipette into the fluid immediately surrounding pial vessels on the cortical surface and whose calibre could be measured and photographed. They showed that if an unbuffered solution but one otherwise similar in composition to c.s.f. was locally injected, the pial vessels dilated, see Plate 5. The changes in pial calibre in response to the injection of various solutions is sum-

8. Cerebral blood vessels and pH

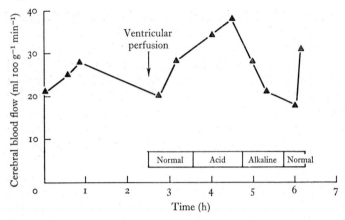

Fig. 8.3. The relation between cerebral blood flow and the ventriculo-cisternal perfusion of artificial cerebrospinal fluid. 'Normal' fluid contained 25, 'Acid' contained 10 and 'Alkaline' contained 40 m-equiv HCO_3^- l^{-1}. The chloride concentrations were varied reciprocally to give a constant osmolarity. Lactate–pyruvate ratio and lactate concentration in c.s.f. was continually monitored to check that the redox state of brain tissue did not alter. (From Siesjö, Kjällquist, Ponten & Zwetnow (1968), in *Progress in brain research*, **30**, 93–8, Fig. 4.)

Table 8.2. *Effects of perivascular microinjection of fluids of various compositions upon the calibre of pial vessels in the rat* (from Wahl et al. (1970). *Pflüg. Arch.* **316**, 152–63)

(a) Composition of solutions used for perivascular injection

Solution	NaCl	KCl	CaCl$_2$	NaHCO$_3$	HCl
			(m-equiv l^{-1})		
A	144	2.5	3.0	—	—
B	119	5.0	3.0	—	—
C	119	5.0	3.0	25	—
D	94	5.0	3.0	50	—

(b) Frequency of vascular reactions (% of those observed)

Solution (bicarbonate)	Vasoconstriction		Vasodilatation		No response	
	rat	cat	rat	cat	rat	cat
A (0)	0	0	84.6	40	15.4	60*
B (minus 25)	0	0	100	100	0	0
C (plus 25)	93.5	100	0	0	6.5	0
D (plus 50)	100	100	0	0	0	0

* During respiratory alkalosis (pH 7.6–7.8).

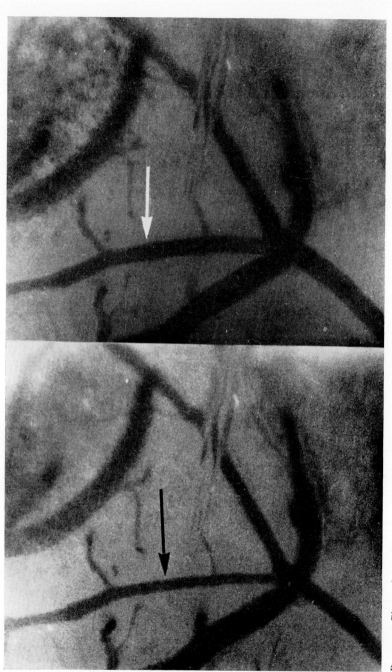

PLATE 5. Photomicrographs showing pial vessels, indicated by arrows, on the surface of the rat brain. Left: control. Right: arteriolar dilatation induced by the close perivascular application of 'bicarbonate-free' solution A, NaCl 144, KCl 2·5, CaCl$_2$ 3·0 m-equiv l^{-1}. (From Wahl et al. (1970). Pflüg. Arch. **316**, 152–63, Fig. 1.)

Cerebral vascular smooth muscle and pH

marized in Table 8.2(b) and although the results are qualitative and although no local measurements of pH were made, it is reasonable to suppose from the observations of Elliott & Jasper (1949) that the unbuffered solutions assumed a low pH and that the solutions with bicarbonate had a pH of 7.4 or greater. This group of experiments deserves to be repeated and extended using accurate methods of measuring local perivascular pH.

Summary

In the matter of whether the pH of brain e.c.f. is an important regulator of cerebral vascular smooth muscle, there is an abundance of hypotheses and singularly little concrete evidence. Clearly, the day of experiments in which cerebral blood flow and $pH_{c.s.f.}$ are measured as P_{a,CO_2} is altered is over since this method cannot establish the means whereby carbon dioxide affects vascular smooth muscle. Similarly, because there is uncertainty as to whether a plasma–brain barrier for H^+ exists or indeed how H^+ is transported, experiments in which plasma pH is altered are also likely to give ambiguous results. The best approach is therefore the most direct, namely to alter the pH of c.s.f. or e.c.f. by local or by ventricular perfusion and it would be important to check by implantation of H^+ sensitive microelectrodes that the changes in $pH_{c.s.f.}$ were in fact reflected in $pH_{e.c.f}$

A difficulty which must be recognized at the moment is that we do not know for certain which lengths of the pial arteriole effectively regulate blood flow; whether they are the pial vessels which course over the surface of the cortex and which are bathed in c.s.f. or whether they are the penetrating arterioles which might be more affected by the composition of e.c.f. This additional uncertainty makes it even more important to measure possible gradients for H^+ between c.s.f. and e.c.f. for it may ultimately be found that cerebral blood flow (as with the activity of respiratory neurones) is not a function of either plasma or $pH_{c.s.f}$ but of some point on the gradient for H^+ between c.s.f. and plasma.

The regulation of pH in brain e.c.f. and c.s.f.

The ionic composition of c.s.f. represents a balance between a number of factors which include the bulk secretion of fluid

8. Cerebral blood vessels and pH

by the choroid plexus and passive diffusion of ions from the cells of brain tissue and e.c.f. and across the blood–brain barrier. Other important factors which affect $pH_{c.s.f.}$ are summarized in somewhat speculative fashion in Fig. 8.4. The principal factor controlling $pH_{c.s.f.}$ may well be $P_{c.s.f., CO_2}$ and this, in turn, is regulated by

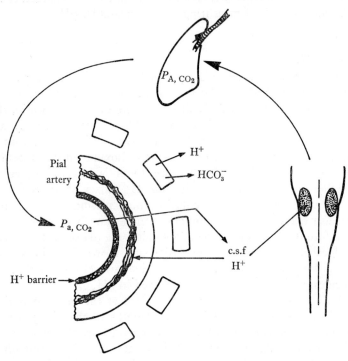

Fig. 8.4. A diagrammatic representation of some of the known and possible factors which link the levels of cerebral blood flow and alveolar ventilation. The cycle includes a section of pial vessel with a putative barrier for H^+ interposed between plasma and smooth muscle cells; rectangular blocks representing neurones and/or glial cells; the kidney-shaped H^+-sensitive areas on the ventro-lateral surface of the medulla and the lung representing the level of alveolar ventilation.

(a) the level of cerebral blood flow and (b) the level of alveolar ventilation and P_{a, CO_2}. As discussed in the previous section, the level of cerebral blood flow may be affected by the pH of e.c.f. or c.s.f. while the evidence that alveolar ventilation is affected by $pH_{c.s.f.}$ is briefly considered in a later section. The pH of

The regulation of pH in brain e.c.f. and c.s.f.

c.s.f. is also determined by the level of metabolism of neurones and glia, that is by the efflux of H^+ and HCO_3^-, the products of metabolism and by HCO_3^- of c.s.f. Except under a few conditions which are discussed in detail in Chapter 11, the level of metabolism of the brain remains remarkably constant so that the factors which are of immediate importance in the regulation of $pH_{c.s.f.}$ are the rate of cerebral blood flow, the level of alveolar ventilation and HCO_3^- of c.s.f. These will now be discussed.

Regulation of $pH_{c.s.f.}$ by cerebral blood flow

At a constant level of brain metabolism, the arterio-venous difference for carbon dioxide and hence the level of $pH_{c.s.f.}$ will be a function of cerebral blood flow. It has been suggested by Lambertsen, Semple, Smyth & Gelfand (1961) and by Bradley & Semple (1962) that this gradient – between P_{a, CO_2} and $P_{c.s.f., CO_2}$ – affords an index of cerebral blood flow. In practice, this does not appear to be so. Thus, under normal conditions, a gradient of approximately 10 mmHg for P_{CO_2} between c.s.f. and arterial blood has been observed (Mitchell *et al.* 1965; Bradley & Semple, 1962; Pauli, Vorburgher & Reubi, 1962). As cerebral blood flow increases, this gradient should diminish: in fact, it either does not change or else increases. Under conditions where cerebral blood flow is known to increase, e.g. in patients with emphysema and carbon dioxide retention, the c.s.f.–arterial P_{CO_2} gradient was found to be 17.3 mmHg (Mitchell *et al.* 1965), 9.1 mmHg (Schwab, 1962*b*), 14.0 mmHg (Buhlmann *et al.* 1963) or in subjects at altitude with hypoxaemia and hypocapnia, 10 mmHg (Mitchell *et al.* 1965). Similarly the finding that this gradient diminishes in non-respiratory acidosis and increases in non-respiratory alkalosis (Mitchell *et al.* 1965; Bradley & Semple, 1962) suggests that cerebral blood flow might be dependent upon arterial pH: yet, as has been shown in a previous section, this does not appear to be the case. Alternatively, it could be argued that the gradient for P_{CO_2} and cerebral blood flow were affected by the changes in plasma P_{CO_2} which accompany non-respiratory acidosis and alkalosis rather than changes in plasma pH. If this were so, then the effect would be to make $pH_{c.s.f.}$ less rather than more stable.

These inconsistencies which seriously limit the usefulness of the c.s.f.–arterial P_{CO_2} gradient as an index of blood flow may most simply be explained on the basis that the acid–base variables

8. Cerebral blood vessels and pH

of large cavity fluid do not accurately reflect those of e.c.f. which will be principally affected by changes in blood flow. That is, it may take much longer for equilibrium to be restored between these compartments than is usually recognized. Thus it is only after some days of application of a stimulus that changes in the c.s.f.–arterial P_{CO_2} gradient are consistent with measured changes in cerebral blood flow. Severinghaus, Mitchell, Richardson & Singer (1964) showed that the per cent reduction in this gradient in subjects after some days at altitude was similar to the increase in blood flow, while McCall (1953) has shown that in the last trimester of pregnancy, although P_{a,CO_2} is reduced, the P_{CO_2} gradient is unaffected as is the level of cerebral blood flow.

In view of the uncertainties mentioned above, it is difficult to give quantitative estimates of the importance of cerebral blood flow in regulating c.s.f. P_{CO_2} and hence $pH_{c.s.f.}$. From the data of Lambertsen et al. (1961), it can be estimated that changes in cerebral blood flow can account for not less than 10 per cent of the total change in $pH_{c.s.f.}$ induced by the inhalation of carbon dioxide and a change in P_{a,CO_2}. It is however possible that the size of this contribution may vary at different levels of P_{a,CO_2}.

Regulation of $pH_{c.s.f.}$ by changes in alveolar ventilation

It is abundantly clear that alveolar ventilation can be affected by changes in P_{a,CO_2} independently of reflex activity from peripheral arterial chemoreceptors and the problem which has yet to be solved is whether changes in ventilation are brought about by a non-specific action of carbon dioxide or H^+ upon respiratory neurones or whether there is a specific group of sensitive neurones or receptors which mediates the response.

This question has been answered in part by the demonstration that a variety of areas on the surface of the brain are sensitive to the local application of drugs or of mock-c.s.f. of altered $[H^+]$ and these include the lateral recesses of the fourth ventricle (Loeschke, Koepchen & Gertz, 1958; Loeschke & Koepchen, 1958a, b, c), the ventrolateral surface of the medulla (Mitchell, Loeschke, Massion & Severinghaus, 1963), the cranial nerve roots at the medullary pontine angle (Loeschke, Mitchell, Katsaros, Perkins & Konig, 1963). Although stimulation of these areas has been shown to cause changes in alveolar ventilation, it is still a matter of speculation as to how or whether they are stimulated

The regulation of pH in brain e.c.f. and c.s.f.

under physiological conditions and since there has been no accompanying histological study which has shown the presence of specific receptors, it is probably best that the use of the term 'central' or 'intracranial' chemoreceptors is postponed until this point is settled. Further, the use of such a term may distract attention from the possibility that neurones located somewhere along the H^+ gradient between c.s.f. and plasma may respond to changes in the H^+ of c.s.f. or plasma. Pappenheimer, Fencl, Heisey & Held (1955) for example, have concluded from experiments in which the composition of c.s.f. was altered in goats, that alveolar ventilation is a function of $[H^+]$ of interstitial fluid at a point three-quarters of the c.s.f.–plasma gradient for HCO_3^-. If these conclusions are accepted, and modified to include the possibility that neurones may be affected at different points of this gradient, it would be reasonable to suppose that the superficially placed H^+-sensitive areas quoted above consisted of neurones which responded to changes at the c.s.f. rather than at the plasma end of the gradient.

Whichever view is accepted, it is acknowledged that alveolar ventilation can be altered by changes in $[H^+]$ or $[HCO_3^-]$ of c.s.f. and that, as is shown in Fig. 8.4, the resulting changes in P_{a,CO_2} will tend to oppose the alterations in composition of c.s.f. and so lead to stability of $pH_{c.s.f.}$. This mechanism is enhanced by the fact that c.s.f. contains no buffers other than HCO_3^- so that any change in P_{a,CO_2} will be almost exactly reflected in c.s.f. Mitchell (1966) has shown that $\Delta P_{c.s.f.,CO_2} = 1.16\ \Delta P_{a,CO_2}$ and that *in vitro* the H^+ change in c.s.f. is approximately 95 per cent of the theoretical maximum (Van Heijst, Visser & Mass, 1961; Mitchell, Herbert & Carman, 1965). The question of tissue buffering of c.s.f. is considered in a later section (p. 227).

These points indicate the $pH_{c.s.f.}$ both affects and is affected by changes in the level of alveolar ventilation and P_{a,CO_2}. It is difficult to be certain in quantitative terms how important these effects are for it is clear that *in vivo*, $pH_{c.s.f.}$ is finely regulated to constancy by other mechanisms. This may be illustrated as follows. A rise in P_{a,CO_2} of 10 mmHg (as in respiratory acidosis) should lead to a rise in $P_{c.s.f.,CO_2}$ of 11.6 mmHg and a fall in $pH_{c.s.f.}$ of at least 0.07 units. However, in a group of patients with long term hypercapnia (P_{a,CO_2} 49.2 mmHg), the mean value for $pH_{c.s.f.}$ was found to differ from control by only 0.012 units (Mitchell *et al.* 1965).

8. Cerebral blood vessels and pH

The distribution of H^+ and HCO_3^- between plasma, e.c.f. and c.s.f.

The size and direction of ionic fluxes in general between plasma and c.s.f. have been studied in considerable detail by means of radioisotope techniques and the results of such studies have recently been reviewed by Davson (1967). Because these methods cannot be satisfactorily applied to H^+ and HCO_3^-, we remain uncertain about the nature and size of fluxes of these ions across the blood–brain barrier, the size of the H^+ efflux from glial cells and neurones, the precise gradients for these ions between plasma, e.c.f. and c.s.f. and the ways in which these gradients are affected by chronic disturbances of acid–base balance of plasma or by hypoxia. It follows that the precise environment with respect to these ions to which cerebral vascular smooth muscle and the neurones which are involved in the regulation of respiration is unknown.

It is known that under resting conditions, $[HCO_3^-]$ is lower and $[H^+]$ is higher in c.s.f. than in arterial plasma and also that c.s.f. is a few mV positive to blood (Held, Fencl & Pappenheimer, 1964). This indicates that H^+ and HCO_3^- are not in electrochemical equilibrium across the blood–brain barrier and this difference could be explained by (a) the active transport of H^+ or HCO_3^- between blood and c.s.f. or (b) by the metabolic production of H^+ and the flux of HCO_3^- to titrate it producing carbon dioxide and water. In the steady state it has proved difficult to estimate which of these processes predominates and most workers have therefore examined the changes which occur in response to disturbance of acid–base balance or to hypoxia, it being assumed that under these conditions, the energy requirements for active transport, the permeability of membranes to H^+ and HCO_3^- and the rate of H^+ production remain constant. The electrochemical potential differences for H^+ and HCO_3^- ($\Delta\mu H^+$ and $\Delta\mu HCO_3^-$) are derived (in volts)

$$\Delta\mu H^+ = \psi + 6.15\,(pH_{c.s.f} - pH_{pl}),$$

$$\Delta\mu HCO_3^- = \psi + 6.15 \log \frac{[HCO_3^-]\,c.s.f.}{[HCO_3^-]\,pl},$$

where ψ is the d.c. potential difference and which assumes that there are no appreciable hydrostatic gradients and no significant solvent drag effects (Ussing, 1960). The first step has therefore been to measure ψ under resting and other clearly defined conditions.

The regulation of pH in brain e.c.f. and c.s.f.

ψ, the electrical potential difference between c.s.f. and plasma. There has been almost complete unanimity among workers who have studied the problem that under resting conditions c.s.f. is a few (average 4) mV positive with respect to plasma or extrameningeal tissues (Lehmann & Meesmann, 1924; Loeschke, 1956; Held et al. 1964; Welch & Sadler, 1965; Goodrich, 1965; Finn, Kao, Mei & Harmel, 1968; Kjallquist & Siesjö, 1968; Kjallquist,

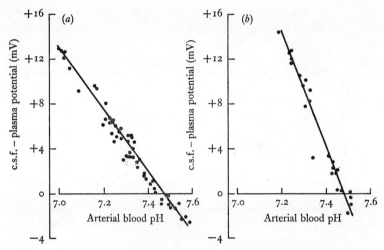

Fig. 8.5. The effects of acidosis and alkalosis on the c.s.f.–plasma potential. (a) pH altered by carbon dioxide inhalation or passive ventilation. (b) pH altered by intravenous infusion of 0.15 mM HCl or NaHCO$_3$. (From Held, Fencl & Pappenheimer (1964), *J. Neurophysiol.* **27**, 942–59, Fig. 1.)

1970; Sørensen & Severinghaus, 1970). Tschirgi & Taylor (1958) found the opposite but their results may most probably be explained by the use of NaCl bridges at liquid junctions. Similarly, most workers have shown that ψ increases as plasma pH falls and representative plots of this relation are shown in Figs. 8.5 and 8.6. In most studies, this relation has been shown to be linear with an average slope of 30.5 mV per pH unit change although Kjallquist (1970) has recorded a curvilinear relation in the rat.

The electrical potential difference is unaffected by changes in [H$^+$] of the fluid in artificial perfusion of the ventricles (Held et al. 1964; Welch & Sadler, 1965). These workers have shown that

8. Cerebral blood vessels and pH

the relation is affected by changes in [K^+] of the perfusing fluid and Cameron & Kleeman (1970) have demonstrated a clear reduction in the slope of the ψ–pH_{pl} curve. However, since [K^+]$_{c.s.f.}$ remains remarkably constant under a wide variety of conditions (Cserr, 1965; Bradbury & Kleeman, 1967; Cohen, Gershenfeld & Kuffler, 1968), it is probable that under the conditions tested, changes in [K^+]$_{c.s.f.}$ were small and of little consequence.

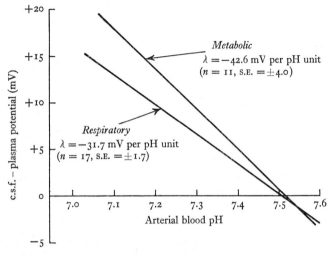

Fig. 8.6. The relation between c.s.f.–plasma potential and arterial pH. Average steady state values from twenty-eight dogs. (From Held, Fencl & Pappenheimer (1964), *J. Neurophysiol.* **27**, 942–59, Fig. 2.)

Site of the origin of the electrical potential difference. There has been much speculation as to the site of the potential difference between c.s.f. and plasma. It is logical to consider most strongly the capillary–glial interface because of the known existence of tight junctions between capillary endothelial cells (Brightman & Reese, 1969) shown by their resistance to the passage of tracers such as horseradish peroxidase and the probable site of resistance of smaller ions (Cohen, Gershenfeld & Kuffler, 1968).

Evidence for the active transport of HCO_3^-. Apart from the theoretical calculations by Held *et al.* (1964), evidence for the existence of active transport of HCO_3^- from c.s.f. to plasma has

The regulation of pH in brain e.c.f. and c.s.f.

almost entirely been derived from experiments in human subjects acclimatized to high altitude (Severinghaus et al. 1964). These workers observed that as a result of the hyperventilation which developed at altitude, P_{a,CO_2} fell from 40 to about 30 mmHg, $P_{c.s.f.,CO_2}$ fell from 49 to 38 mmHg and arterial pH rose from 7.43 to 7.48 units. From the amount by which $P_{c.s.f.,CO_2}$ fell, it could be calculated that if $[HCO_3^-]_{c.s.f.}$ had remained constant, $pH_{c.s.f.}$ would have risen from 7.32 to 7.43 units. No change in $pH_{c.s.f.}$ was however noted in any subject exposed to altitude for up to one week. The calculated fall in $[HCO_3^-]_{c.s.f.}$ was 4–5 m-equiv of which 0.2 m-equiv could be explained by hyperventilation and the fall in $P_{c.s.f.,CO_2}$. In seeking an explanation for the fall in HCO_3^-, Severinghaus et al. (1964) rejected the possibility that HCO_3^- has been excreted by the kidney since the plasma standard bicarbonate had fallen by only 1–2 m-equiv and the relatively slow rate of renal compensation would be unlikely to explain the early changes observed. The possibility that the arterial hypoxia caused an increase in lactic acid production which caused a fall in $[HCO_3^-]$ was also rejected because although blood lactate increased by 1–2 m-equiv, c.s.f. lactate was unchanged from control in the period up to 10 days at altitude. Further, it was found that hypoxaemia induced a 12 per cent increase in cerebral blood flow which by narrowing the A–V difference for P_{CO_2} (and assuming an unchanged rate for tissue metabolism) might have caused a greater fall in capillary HCO_3^- than that observed in arterial blood. It was calculated that this factor could account for a fall of only about -0.3 m-equiv, l^{-1}. Further experiments in which the subjects were given $NaHCO_3$ or NH_4Cl by mouth at altitude showed that the fall in $[HCO_3^-]_{c.s.f.}$ was unaffected by changes in $[HCO_3^-]_{pl}$.

By this process of exclusion and after making certain assumptions, Severinghaus et al. (1964) inferred that the reduction in $[HCO_3^-]_{c.s.f.}$ was due to active transport of HCO_3^- from c.s.f. to plasma but in a further paper (Severinghaus, 1965), some more direct evidence was given from measurements in man and in the dog of the c.s.f.–plasma potential together with measurements of the concentrations of HCO_3^-, Cl^-, Na^+ and K^+ and activity of H^+ from which calculations of the net electrochemical potential difference for these ions could be calculated.

Severinghaus reported that ψ returned to control levels and

8. Cerebral blood vessels and pH

his calculations of $\Delta\mu H^+$ were based upon this assumption although no precise data were given. These results confirmed those of Held et al. (1964), namely that Na^+ and Cl^- were close to electrochemical equilibrium but that H^+, K^+ and HCO_3^- were out of equilibrium and that if the subjects were hyperventilated and given $NaHCO_3$ to maintain a constant $[HCO_3^-]_{pl}$, $\Delta\mu H^+$ and $\Delta\mu HCO_3^-$ were stabilized yet further from equilibrium. The results of these experiments taken with those of Mitchell, Carmen, Severinghaus, Richardson, Singer & Schnider (1965) showing that $pH_{c.s.f.}$ was regulated to constancy under a wide variety of conditions constitute a suggestive case for the hypothesis that the active transport of HCO_3^- from c.s.f. to plasma is an important determinant of $\Delta\mu H^+$.

Difficulties with the active transport of HCO_3^- hypothesis.
In 1965, Goodrich reported that in the rat, ψ was upheld in chronic acidosis and alkalosis and this has subsequently been confirmed by Kjällquist & Siesjö (1967, 1968). In a further study (Kjällquist, 1970) ψ was measured over the range 7.15 to 7.55 pH_a under control conditions and when the animals were made acidotic or alkalotic with or without hypercapnia. The relation obtained between ψ and pH_a is shown in Fig. 8.7: it is not linear and the change in slope between acid and alkaline ends may reflect the respiratory and non-respiratory means adopted to obtain the values. In this case, the relation is similar to those obtained by Held et al. (1964), Figs. 8.5 and 8.6. When $\Delta\mu H^+$ and $\Delta\mu HCO_3^-$ were calculated, they were found to be unaffected by $pH_{c.s.f.}$ but they were affected by P_{a, CO_2}, Figs. 8.8 and 8.9. It should however be borne in mind that the data are derived from comparatively short periods of exposure to carbon dioxide (6 hours) or acid–base disturbances (24 hours) while the changes in $pH_{c.s.f.}$ wrought by non-respiratory methods were comparatively small compared to the changes when carbon dioxide was added.

However, Kjällquist's results make it clear that the active transport of HCO_3^- hypothesis is unlikely for if HCO_3^- was transported from c.s.f. to plasma, then $\Delta\mu HCO_3^-$ should diminish in acidosis and increase in alkalosis. In non-respiratory acid–base disturbances, $\Delta\mu HCO_3^-$ was found to be virtually unaltered. However, the changes seen in $\Delta\mu HCO_3^-$ with hypocapnia do support the hypothesis that HCO_3^- is transported actively since

The regulation of pH in brain e.c.f. and c.s.f.

$\Delta\mu HCO_3^-$ does fall with $pH_{c.s.f.}$ but there is no evidence that this regulates $pH_{c.s.f.}$ to constancy. Indeed, since $\Delta\mu HCO_3^-$ increases with hypercapnia, transport of HCO_3^- from c.s.f. to plasma would aggravate the situation. Further evidence has been obtained by Sørensen & Mines (1970), Sørensen (1970) and Mines & Sørensen

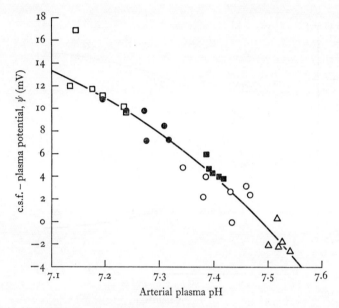

Fig. 8.7. The relation between arterial plasma pH and the c.s.f.–plasma potential (ψ) of rats made acidotic (filled circles) or alkalotic (triangles) by means of intraperitoneal injections of NH_4Cl or $NaHCO_3$. The control group (open circles) was given a simulated fluid intraperitoneally. In the last two groups, $NaHCO_3$ or NH_4Cl were given but carbon dioxide was simultaneously given to keep either the plasma pH (filled squares) or plasma (HCO_3^-) (open squares) constant. The curve was obtained by best visual fit. (From Kjällquist (1970), *Acta Physiol., Scand.* **78**, 85–93, Fig. 1.)

(1971) who related $\Delta\mu H^+$ with $pH_{c.s.f.}$ from data obtained in human and dog experiments in which simultaneous measurements were made of $pH_{c.s.f.}$ and pH_{pl} under varying conditions. They found no correlation between $\Delta\mu H^+$ and $pH_{c.s.f.}$ whereas according to the hypothesis, $\Delta\mu H^+$ should increase when $pH_{c.s.f.}$ increases and should fall when $pH_{c.s.f.}$ falls. These workers found that the closest correlation was between $\Delta\mu H^+$ and $[HCO_3^-]_{pl}$. This con-

8. Cerebral blood vessels and pH

clusion differs from that of Siesjö & Kjällquist (1969) and Kjällquist (1970) and further studies will be required to resolve this difference.

The present position then appears to be that the electrochemical potential difference cannot be explained solely in terms of the active transport of H^+ or HCO_3^- between plasma and c.s.f. The

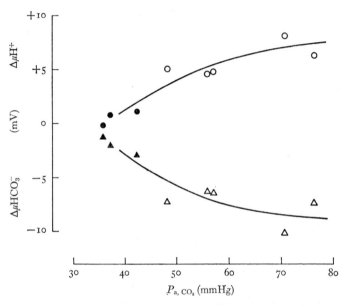

Fig. 8.8. The relation between the calculated electrochemical potential differences for H^+ ($\Delta\mu H^+$, circles) and for HCO_3^- ($\Delta\mu HCO_3^-$, triangles) between c.s.f. and plasma, and the arterial carbon dioxide tension. The normocapnic groups are denoted by filled symbols. Note the increasing values for $\Delta\mu$ with P_{a,CO_2}. (From Kjällquist (1970), *Acta Physiol., Scand.* **78**, 85–93, Fig. 4.)

position with respect to the efflux of H^+ from brain tissue is also not settled although Mines & Sørensen (1971) have shown that hypoxia causes a marked increase in $\Delta\mu HCO_3^-$. Finally, some points require emphasis which may be relevant to the interpretation of the studies quoted and may explain some of the discrepancies which have been observed. The more obvious differences obtained for ψ can most probably be explained by the fact that the recording electrodes were placed in fluids of imperfectly known composition. Further, the calculation of electrochemical

The regulation of pH in brain e.c.f. and c.s.f.

potential differences requires values for [HCO_3^-] and [H^+] in *capillary* plasma and these are usually estimated from arterial plasma values on the assumption that differences between arterial and capillary values are linear. Kjällquist (1970) for example, used the estimates of capillary plasma [HCO_3^-] suggested by Gibbs, Gibbs, Lennox & Nims (1942) and Kety & Schmidt (1948b). Other assumptions are discussed in the following section in the context of calculating intracellular pH.

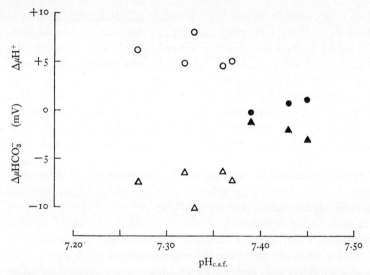

Fig. 8.9. The relation between the calculated electrochemical potential difference for H^+ ($\Delta\mu H^+$) and HCO_3^- ($\Delta\mu HCO_3^-$) between c.s.f. and plasma and the pH of c.s.f. Symbols as in Fig. 8.8. There is no simple relation between the calculated $\Delta\mu$ values and those of $pH_{c.s.f.}$. (From Kjällquist (1970), *Acta Physiol., Scand.* **78**, 85–93, Fig. 2.)

These assumptions together with those involved in the use of pK' values for the calculation of [HCO_3^-] and the more purely technical question of differences of junction potential between c.s.f. and blood may explain why, under the same conditions, Severinghaus (1965) reported a value for $\Delta\mu H^+$ of $+7$ mV and Kjällquist (1970) one of $+0.2$ mV. They may also explain why Kjällquist's values for $\Delta\mu H^+$ and $\Delta\mu HCO_3^-$ do not add up to zero. These observations would thus seem to suggest that at the moment deductions based on absolute values for $\Delta\mu H^+$ should

8. Cerebral blood vessels and pH

be treated with caution: on the other hand, changes in these values which occur in acid–base disturbances or with hypoxia are likely to be of greater value.

Intracellular pH of the brain

Measurement of $pH_{i.c.f.}$. Intracellular pH of fibres in the mammalian C.N.S. has not yet been measured but it is possible to make some estimates of probable values. Although Donnan (1911) implied that hydrogen ions inside a cell should be in equilibrium with those outside, it has been found in practice that the simple relationship which he proposed applies only under special conditions in biological systems (Caldwell, 1958). This is due to the relatively small number of hydrogen ions which pass across membranes and this is due in turn either to the relatively small concentrations of H^+ involved or to permeability barriers. If a range of likely resting membrane potential of -50 to -90 mV is taken for mammalian C.N.S. fibres and if equilibrium is assumed, from the Nernst equation, intracellular pH will vary from 5.9 to 6.5 units. Such a value, average 5.99 units range 5.77 to 6.29 units has been obtained experimentally in rat muscle cells (Carter, Rector, Campion & Seldin, 1967). Other workers have obtained values for intracellular pH of between 6.5 and 7.2 units in cut crab muscle (Caldwell, 1958) and mollusc ganglia (Sorokina, 1965); clearly the latter values are not consistent with a passive equilibrium across the cell membrane. Although Carter et al. (1967) have suggested that the high values obtained by other workers for intracellular pH were due to technical factors such as the length of sensitive tip of the electrode, the same arguments can be applied to their results since it is possible that the very high resistance which they observed in their electrodes was due to faulty insulation. Certainly recent workers using similar electrodes to those of Carter have obtained values for intracellular pH in the crab and rat muscle which are similar to those obtained by Caldwell (1958) (Paillard, Sraer, Leviel & Claret, 1971).

It is clear that development in this field is likely to be rapid over the next year or so especially with the introduction of new types of ion-sensitive microelectrodes (Thomas, 1971) so that values for intracellular pH should be established before long which will be tolerably free from technical qualifications.

The regulation of pH in brain e.c.f. and c.s.f.

Calculation of intracellular pH. Intracellular pH can be calculated thus:

$$\mathrm{pH}_i = 6.12 + \log \frac{[\mathrm{HCO}_3^-]_i}{P_{t,\mathrm{CO}_2} \cdot 0.0314},$$

where P_{t,CO_2} is tissue tension of carbon dioxide and, in brain, is assumed to be the same as $P_{\mathrm{c.s.f.,CO}_2}$: 6.12 is the apparent pK' of carbonic acid; 0.0314 is the solubility coefficient for carbon dioxide in intracellular water (m-mol kg^{-1} mmHg^{-1}); $(\mathrm{HCO}_3^-)_i$ is the intracellular bicarbonate concentration and pH_i is intracellular pH (Siesjö, 1962 a, b; Siesjö, Brzezinski, Kjällquist & Ponten, 1967). On the assumption that blood and extracellular fluid compartments occupy respectively 3 and 12 per cent of the tissue weight of brain (Rall, Oppelt & Patlak, 1962; Woodward, Reed & Woodbury, 1967), intracellular bicarbonate concentration has been calculated:

$$[\mathrm{HCO}_3^-]_i = \frac{[\mathrm{HCO}_3^-]_t - 0.03[\mathrm{HCO}_3^-]_{bl} - 0.12[\mathrm{HCO}_3^-]_{c.s.f.}}{0.64},$$

where the subscripts i, t, bl. and c.s.f. refer to the concentration of bicarbonate in intracellular water, in wet tissue (m-equiv kg^{-1}), in whole blood and c.s.f. respectively. The 0.64 is the intracellular water content (g intracellular water per g of wet tissue).

Using these calculations, Kjällquist, Nardini & Siesjö (1969) have shown that at a plasma P_{CO_2} of 40 mmHg in the rat, intracellular pH is 7.10 units assuming a 12 per cent extracellular volume, Fig. 8.10. This figure also shows that the buffer capacity of the intracellular space ($\beta = \Delta \log P_{\mathrm{CO}_2}/\Delta \mathrm{pH} = 2.3$) exceeds that in both c.s.f. ($\beta = 1.5$) and in blood *in vivo* ($\beta = 1.3$). As plasma P_{CO_2} is raised above 40 mmHg, the buffer capacity is constant in these three phases but as P_{CO_2} is lowered below 30 mmHg, the intracellular pH approaches constancy.

These results are of interest for a number of reasons. First, they confirm the greater buffering capacity of c.s.f. compared to blood *in vivo* (Schwartz & Relman, 1963). This may at first sight be surprising because of the lack of haemoglobin and protein in c.s.f. but it may be explained by the exchange of bicarbonate between plasma and extracellular fluids in hyper- and hypocapnia and by the pH-dependent changes in c.s.f. plasma potential which will cause an increase in c.s.f. bicarbonate in hypercapnia and a decrease in hypocapnia (Kjällquist & Siesjö, 1967, 1968, 1969).

8. Cerebral blood vessels and pH

Secondly, the results indicate a tissue buffering capacity which is not only higher than that of both c.s.f. and whole blood *in vivo* but higher than the previously obtained values by Ponten (1966). They are however consistent with those determined by Roos (1965) who estimated intracellular pH from the distribution of

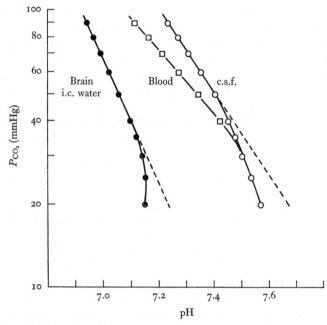

Fig. 8.10. The relation between the tissue (and c.s.f.) carbon dioxide tension (logarithmic scale) and the pH values derived for the c.s.f. blood and intracellular spaces assuming a 12 per cent extracellular volume. In hypercapnia, the apparent buffer capacity of the c.s.f. ($\beta = \Delta \log P_{CO_2}/\Delta$ pH $= 1.5$) and the intracellular space ($\beta = 2.3$) exceeded that for blood *in vivo* ($\beta = 1.3$). At low carbon dioxide tensions, the $\Delta \log P_{CO_2}/\Delta$ pH changes were similar in all three phases although the intracellular phase showed a greater tendency to pH constancy. (From Kjällquist, Nardini & Siesjö (1969), *Acta Physiol., Scand.* **76**, 485–94, Fig. 4.)

5,5-dimethyloxazolidine-2,4-dione-2-^{14}C (DMO) between brain, plasma and c.s.f. in the cat. It is possible that the discrepant results obtained by Ponten (1966) were due to the fact that his animals were anaesthetized with pentobarbitone; the animals in the other series quoted were all anaesthetized with nitrous oxide.

A third point which emerges from the results obtained by

The regulation of pH in brain e.c.f. and c.s.f.

Kjällquist, Nardini & Siesjö (1969) is the change in buffer capacity with hypocapnia. These results also differ from those of Ponten (1966) who found only small changes in the concentration of lactate. Kjällquist, Nardini & Siesjö (1969) also observed substantial changes in both lactate and pyruvate and in the lactate–pyruvate ratio as P_{a,CO_2} was lowered below 30 mmHg.

If the calculated values for intracellular pH can be confirmed by direct measurement, then it is clear that there cannot be a passive equilibrium for hydrogen ions across the cell membrane. It is most probable that hydrogen ions are actively transported from cells to brain e.c.f. and blood. Alternatively, hydrogen ions could combine with pyruvate to form lactate and since the pK of lactic acid is so low, 3.8, it can be regarded as almost completely ionized in tissue. There is no direct evidence that lactate anions can leak out of cells but there is circumstantial evidence that this may be so since a close correlation between c.s.f. and intracellular lactate has been observed over the range 20–90 mmHg P_{a,CO_2} in the rat (Kjällquist, Nardini & Siesjö, 1969). Hydrogen ions may also combine with HCO_3^- and be dehydrated by carbonic anhydrase to give carbon dioxide which can diffuse freely from the cell. If this is an important route for hydrogen ion transport, then because of the larger carbonic anhydrase concentration in glia than in nerve cells, intracellular pH would be likely to be lower in nerve cells than the average for brain and that in glia rather higher. Finally, the lactate ion can combine with intracellular protein and so alter the buffer base concentration. Kjällquist, Nardini & Siesjö (1969) have shown that a fall in [HCO_3^-] occurs both in intracellular water and in c.s.f. in rats but the change in c.s.f. is greater since c.s.f. lacks buffers other than bicarbonate. In the intracellular water, the reduction in the concentration of bicarbonate is less marked.

The results of this series of experiments then indicate that the intracellular fluid of brain cells is more completely buffered than either blood or c.s.f. and that the buffering capacity is virtually ideal when P_{a,CO_2} is low. It is also clear that there cannot be a passive equilibrium of H^+ across the membrane between cell and e.c.f. and various possible modes of H^+ transport have been proposed. The rate at which H^+ moves from cell to e.c.f. is likely to be considerably increased in the presence of tissue hypoxia which accompanies hypocapnia.

8. Cerebral blood vessels and pH

Summary

The evidence presented and discussed in this second half of the chapter concerning the distribution of H^+ and HCO_3^- between plasma, e.c.f. and c.s.f. is, like that in the first half, indirect and far from decisive. It certainly is not clear enough for us to establish with any certainty what the H^+/HCO_3^- environment is for cerebral vascular smooth muscle, and therefore to conclude whether, or how far under physiological conditions or in chronic hypoxia or acid–base disturbances this environment contributes to the regulation of cerebral blood flow.

The one certain fact is that under physiological conditions, H^+ and HCO_3^- are not in electrochemical equilibrium across the blood–brain and blood–c.s.f. barriers. This disequilibrium has been variously described as depending on plasma pH (Fencl et al. 1964), $pH_{c.s.f.}$ (Severinghaus et al. 1963; Mitchell et al. 1965), plasma P_{CO_2} (Siesjö & Kjällquist, 1969; Kjällquist, 1970) and plasma $[HCO_3^-]$ (Mines, Morill & Sørensen, 1971) and the mechanisms which have been proposed include active transport of H^+ and HCO_3^- between plasma and c.s.f. and the effects which follow production of H^+ by actively metabolizing neuronal or glial tissue.

There are probably three reasons why the literature in this field is so confused. First, estimates of electrochemical potential difference have been derived from very different types of experiment, in only a few of which has it been possible to control factors such as plasma P_{CO_2} or pH which must certainly be taken seriously as affecting the potential difference. This is particularly true of experiments in man and there has been a tendency to extrapolate values for electrochemical potential differences *as if physiological conditions obtained*: or else to assume certain values in the dog, e.g. that ψ varies with plasma pH by 30.5 mV per pH unit and apply these to the analysis of changes seen in man.

Secondly, the estimates of electrochemical potential difference are based on numerous assumptions which are very far from being established. It will have been observed, for example, that when measurements necessary for calculating the electrochemical potential difference are made in chronic hypoxia or acid–base disturbances, it has been assumed that if H^+ or HCO_3^- is transported actively, the energy required is the same as in physiological

Summary

conditions and that the permeability for H^+ and HCO_3^- remain unchanged. Similarly the calculations strictly require measurement of acid–base variables in *capillary* blood and it will be obvious from the calculations of Siesjö & Kjällquist (1969) how many assumptions are implicit in their estimate of capillary variables and intracellular buffering capacity – all of which could materially affect the actual values obtained and therefore estimates of the electrochemical potential difference.

Thirdly, and following from this, it is most probable that confusion exists because analyses in thermodynamic terms are being applied to a system or set of systems whose properties are imperfectly understood. If there is difficulty in understanding the mechanisms of transport across the red cell membrane, how much more complex is likely to be the system which includes an uncertain number of membranes interposed between which is tissue whose level of metabolism is difficult to measure and is likely to be far from homogeneous. In addition to this, there is difficulty in obtaining samples of fluid which either in time or in content represent the true distribution of ions and a complex and highly efficient regulating process which tends to ensure the stability of concentration of the ions in question.

In the face of these difficulties and assumptions, it seems to the author that the approach adopted by authors quoted in this chapter, that is the calculation of the electrochemical potential difference under normal and disturbed conditions in order to determine the mechanisms of transport at individual sites, is an exercise of limited usefulness. In order to avoid endless arguments as to whether H^+ or HCO_3^- is actively transported and if it is, at what site and under what conditions, what seems to be required is an attempt to analyse in greater detail and depth, the properties and mechanisms of transport at each of the sites interposed between blood and c.s.f. The techniques for such an analysis are now becoming available with the introduction of ion-sensitive microelectrodes though it should be recognized as it was by Kuffler & Nicholls (1966) in their study of glial mechanisms, that systems which are simpler than the mammalian brain may first have to be studied.

9 REGULATION OF CEREBRAL VESSELS BY OXYGEN

Just as there appear to be compensating mechanisms which ensure an adequate perfusion pressure when arterial pressure falls, so we might expect a system which ensures that venous P_{O_2} is kept as high as possible if P_{a,O_2} falls. In Chapter 2, it was shown that such an assumption was implicit in the models for oxygen exchange in the tissues and evidence was quoted which showed that, in the long term, the effects of a reduced venous P_{O_2} were offset by a reduction in the mean intercapillary distance. In the present chapter, the question is asked, 'How do changes in P_{a,O_2} affect cerebral vessels and blood flow?': and in answering this, the question arises whether local changes in P_{O_2} as a result of altered metabolic rate can affect cerebral vessels.

The effects of arterial hypoxia upon pial vessels was studied in the nineteenth century (e.g. by Donders, 1859) but in these and in later studies, the effects of hypoxia were not separated from those due to changes of P_{a,CO_2} which either rose, as in asphyxia (Roy & Sherrington, 1890): or fell in response to the inhalation of low oxygen mixtures. Similar difficulties have attended the measurement of the effect of high oxygen levels upon cerebral vessels and blood flow. The evidence which indicates that changes in P_{a,O_2} have a specific action upon blood vessels, the mechanism of that action and its homeostatic significance will now be considered.

Effect of arterial hypoxia upon cerebral blood vessels and flow

The inhalation of low oxygen mixtures causes an increase in the calibre of pial vessels (Binet, Cachera, Fauvert & Strumza, 1937; Wolff & Lennox, 1930) and in retinal vessels (Cusick, Benson & Boothby, 1940; Hickam, Shieve & Wilson, 1953; Sieker & Hickam, 1953) whether the arterial P_{CO_2} is controlled or not. Increases in blood flow with arterial hypoxia have been observed

Arterial hypoxia and cerebral blood vessels

in numerous animal experiments when cerebral blood flow has been measured during natural perfusion of the brain (Noell & Schneider, 1942a; Schmidt, 1928a), during artificial perfusion of the brain (Geiger & Magnes, 1947; Schmidt, 1928a, b), by the insertion of thermocouples into brain tissue (Norcross, 1938; Schmidt, 1934; Schmidt & Pierson, 1934) or cerebral vessels (Noell & Schneider, 1941) and from changes in arterio-venous oxygen difference (Courtice, 1941; Irving & Welch, 1935; Nöell & Schneider, 1942b). In man, similar evidence has been obtained by insertion of a Gibbs thermoelectric flow meter in the internal jugular vein (Lennox, Gibbs & Gibbs, 1935) and from measurement of the arterio-venous oxygen difference (Gibbs, Gibbs, Lennox & Nims, 1942, 1943; Lennox & Gibbs, 1932; Lennox, Gibbs & Gibbs, 1938). The majority of these methods are indirect or are at best semi-quantitative, and assume that no changes in cerebral metabolism have taken place. More direct and quantitative measurements of the effects of hypoxia upon cerebral blood flow have been carried out in the monkey (Dumke & Schmidt, 1943; Schmidt, Kety & Pennes, 1945), in the cat (Hansen, Sultzer, Freygang & Sokoloff, 1957) and in man (Fazekas, Alman & Bessman, 1952; Heyman, Patterson & Duke, 1952; Heyman, Patterson, Duke & Battey, 1953; Kety & Schmidt, 1948b; Lambertsen, Kough, Cooper, Emmel, Loeschke & Schmidt, 1953; Patterson, Heyman & Duke, 1952; Turner, Lambertsen, Owen, Wendel & Chiodi, 1957); and these have all confirmed that arterial hypoxia causes an increase in cerebral blood flow. In few of these studies was the complication of hypocapnia avoided. For example, in the study of Kety & Schmidt (1948b) when healthy young men were studied, the average increase in cerebral blood flow on inhaling 10 per cent oxygen in nitrogen was 19 ml 100 g^{-1} min^{-1} (+32 per cent of control); P_{a,CO_2} fell by an average of 4 mmHg and no measurements of P_{a,O_2} were recorded. In the experiments reported by Dumke & Schmidt (1943) in which blood flow was measured by the insertion of a bubble flow meter into the internal carotid artery of monkeys, it is probable that nerves involved in reflex activity from the carotid body chemoreceptors had been damaged during surgery, for in these monkeys, there was no increase in ventilation when low oxygen mixtures were administered. Similar criticisms can be raised concerning most of the experimental results cited above.

9. Regulation of cerebral vessels by oxygen

More precise but qualitatively similar results have been obtained when cerebral blood flow has been measured using the clearance of radioisotopes. McDowall (1966) measured cortical blood flow in dogs using ^{85}Kr clearance and confirmed that flow increased at constant P_{a,CO_2} when P_{a,O_2} was reduced to approximately 50 mmHg. The data are derived from the pooled results of an unspecified number of dogs and it is not certain how many measurements were made on each dog. But examination of the responses seen in individual dogs (McDowall, personal communication) indicates that, at constant P_{a,CO_2}, blood flow was virtually unchanged over the range 60–140 mmHg and, as P_{a,O_2} was lowered below this range, blood flow increased sharply by amounts varying between +60 and +120 per cent of control (Fig. 9.1). Similar results have been obtained in the adult baboon (James, Millar & Purves, 1969). Above the physiological range of P_{a,O_2}, the reduction in cortical or white matter flow was too small to be detected and was within the error of measurement. As P_{a,O_2} was reduced to c. 50–60 mmHg, blood flow increased and the overall relation between cortical blood flow and P_{a,O_2} was best described by a rectangular hyperbola. The average asymptote at infinitely high blood flow was 11.0 mmHg P_{a,O_2} and at infinitely high P_{a,O_2} was 55 ml 100 g^{-1} min^{-1}.

Variations in the flow response to P_{a,O_2} with age

Changes in blood flow with P_{a,O_2} have been measured in the sheep foetus and newborn lamb (Lucas, Kirschbaum & Assali, 1966; Purves & James, 1969). In the former study, 'cephalic' blood flow was measured by placing electromagnetic flowmeters on the carotid artery. The results obviously have the limitation that an unknown proportion of the measured flow was destined for the brain. Some evidence was obtained from the study of Purves & James (1969) that extracranial flow in the sheep foetus is small in comparison with intracranial flow: so that the observation by Lucas et al. (1966) that arterial hypoxia in the foetus, induced by giving the ewe low oxygen mixtures to breathe, was accompanied by a rise in blood flow is probably valid. Because P_{a,O_2} in the foetus is low by adult standards (15–22 mmHg) and because foetal P_{a,O_2} varies only slightly with changes in maternal P_{a,O_2} (Parker & Purves, 1967), it is difficult to construct a flow–P_{a,O_2} curve as in

Arterial hypoxia and cerebral blood vessels

the adult. But with a reduction of some 5–9 mmHg P_{a,O_2} in the foetus, an increase of cortical blood flow of 25–37 ml 100 g^{-1} min^{-1} above control was observed when blood gas tensions and pressure were within the physiological range. The pattern of response and magnitude of change in blood flow in the newborn lamb with alterations in P_{a,O_2} was found to be similar to that in the adult baboon.

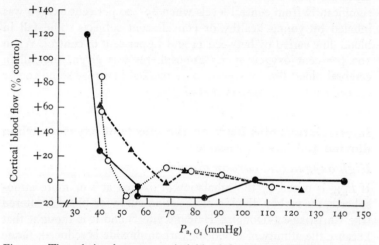

Fig. 9.1. The relation between cortical blood flow (expressed as per cent of control) and P_{a,O_2} (mmHg), when the animals breathed air. Results from three dogs anaesthetized with pentobarbitone sodium, paralysed with suxamethonium and ventilated mechanically. P_{a,CO_2} maintained within physiological limits throughout. (Abstracted from McDowall, 1966, *Oxygen measurements in blood and tissues*. Churchill: London, Fig. 1.)

Data on the response to hypoxia in the aged are equally meagre. One study by Fazekas, Alman & Bessman (1952) in elderly patients with clinical evidence of arteriosclerosis shows that first, the control level of blood flow, measured by the nitrous oxide method, was just over half that found in healthy young men: the increase in blood flow when 10 per cent oxygen in nitrogen was inhaled was on average however similar to the younger age group, +33 per cent of control: but in this study, a direct comparison may not be valid since no measurements of the accompanying changes in blood gas tensions were made.

9. Regulation of cerebral vessels by oxygen

The effects of high oxygen tension upon cerebral blood flow

In the majority of the studies quoted in the previous section, the effect of a rise in P_{a,O_2} induced by the inhalation of 50 to 100 per cent oxygen upon cerebral blood flow was also measured. Thus high P_{a,O_2} causes constriction in pial and retinal arteries. In a group of studies in man, summarized by Sokoloff (1959) cerebral blood flow, measured by the nitrous oxide method was reduced significantly from control levels when 85–100 per cent oxygen was inhaled by young healthy or convalescent subjects. The fall in blood flow varied by between 11 and 14 per cent of control. When 100 per cent oxygen at 3.5 atmospheres was given, the fall in cerebral blood flow was even more marked being by some 20 per cent of control (Lambertsen *et al.* 1953).

Inter-relationships between the effects of oxygen, carbon dioxide and blood pressure

High oxygen and carbon dioxide

If P_{a,O_2} is raised by the inhalation of oxygen at 1 or more atmospheres, the carbon dioxide dissociation curve for blood is altered (the Christiansen–Douglas–Haldane effect) and it is thought that because the affinity of blood for carbon dioxide is reduced, tissue carbon dioxide and [H$^+$] increase. Such an increase has been held to be responsible for the relative hyperventilation, fall in P_{a,CO_2} and cerebral vasoconstriction which accompany prolonged oxygen inhalation compared to the hypoventilation and rise in P_{a,CO_2} which is associated with transient inhalation of oxygen. Certainly a rise in cerebral venous P_{CO_2} and a fall in P_{a,CO_2} have been demonstrated in subjects inhaling oxygen at 1 or more atmospheres (Lambertsen *et al.* 1953) and the hypothesis outlined above has received further support by the observation that if alveolar P_{CO_2} is held constant while subjects successively inhale air and high oxygen, there is no significant change in cerebral blood flow (Turner, Lambertsen, Owen, Wendel & Chiodi, 1957).

These results would suggest that a rise in P_{a,O_2} has no direct effect upon cerebral vessels and similar difficulties of interpretation arise from other studies in which subjects inhaled high oxygen (Kety & Schmidt, 1948*b*, Huckabee, 1961; Jacobsen, Harper & McDowall, 1963*b*). There is some evidence that the effects of

Inter-relationship of oxygen and blood pressure

inhaling high oxygen depend on the level of P_{a,CO_2}. Thus at low P_{a,CO_2} (19 mmHg) caused by voluntary hyperventilation, subjects inhaled successively 6 per cent oxygen in nitrogen, and 100 per cent oxygen at 2 atmospheres (Rievich, Dickson, Clark, Hedden & Lambertsen, 1968). Cerebral blood flow was measured from the arterio-venous oxygen difference on the assumption that cerebral metabolic rate for oxygen did not change (Alexander, Cohen, Wollman, Smith, Rievich & Vander Molen, 1965; Kety & Schmidt, 1948b; Lambertsen et al. 1953). At this low and maintained level of P_{a,CO_2}, cerebral blood flow fell by 22 per cent when the subjects inhaled oxygen. If, as the authors suggest, this fall in flow was a consequence of the rise in P_{a,O_2}, it follows that high oxygen has a direct vasoconstrictor effect or by removing the hypoxic vasodilator effect when the subjects inhaled 6 per cent oxygen, the full vasoconstrictor effect of low P_{a,CO_2} was revealed. Somewhat different results were obtained by McDowall (1966) in dogs and these are summarized in Table 9.1. This shows that at constant P_{a,CO_2}, blood flow progressively fell as P_{a,O_2} rose and this constitutes the best evidence so far that high oxygen has a constrictor effect on cerebral blood vessels which is independent of changes in P_{CO_2}.

Table 9.1. *Blood flow through the cerebral cortex of the dog during ventilation with air, oxygen at 1 atmosphere and oxygen at 2 atmospheres at constant arterial carbon dioxide tension. The last column gives the mean value for flow during ventilation with air after exposure to 2 atmospheres of oxygen* (from McDowall, 1966. *Oxygen measurements in blood and tissues*, ed. Payne, J. P. & Hill, D. W. London: Churchill)

	Air	Oxygen (1 atmosphere)	Oxygen (2 atmospheres)	Air
Cortical blood flow (ml 100 g^{-1} min^{-1})	0.96 ± 0.20	0.86 ± 0.16*	0.76 ± 0.16†	0.83 ± 0.20
Arterial P_{CO_2} (mmHg)	36.1	36.7	36.9	36.8

* $P < 0.05$.   † $P < 0.005$.

9. Regulation of cerebral vessels by oxygen

Low oxygen and carbon dioxide

A similar difficulty arises in distinguishing the effects of low P_{a,O_2} from the accompanying changes either in P_{a,CO_2} due to the hyperventilation initiated by the peripheral arterial chemoreceptors or the presumed changes in tissue carbon dioxide and [H^+] as a result of a change in the carbon dioxide dissociation curve for blood. It has been clearly shown that hypoxia which is severe enough to cause a fall in P_{a,O_2} to < 60 mmHg causes an increase in blood flow, and this occurs whether P_{a,CO_2} is controlled (as in the experiment illustrated in Fig. 9.1) or allowed to fall. There are to date no experiments in which the rise in blood flow under these two conditions has been quantitatively compared: it is not therefore known whether the vasoconstrictor effects of low P_{a,CO_2} should merely be subtracted from the vasodilator effects of low P_{a,O_2} or whether, as is the case with the level of alveolar ventilation (Cormack, Cunningham & Gee, 1958), or of carotid body chemoreceptor discharge (Hornbein, Griffo & Roos, 1961; Biscoe, Purves & Sampson, 1970) under these conditions, there is evidence of inhibition. Some evidence on this point may be derived from the experiments of Severinghaus, Chiodi, Eger, Brandstater & Hornbein (1966) who studied the changes in cerebral blood flow during the early phases of acclimatization to altitude, Fig. 9.2. This shows that eight hours after ascent to 3810 m, the increase in cerebral blood flow was abolished by the inhalation of oxygen which raised P_{a,O_2} from an average of 43.5 to 79.8 mmHg. On the fourth day at altitude, cerebral blood flow had fallen by some 3 ml 100 g^{-1} min^{-1} which may have been due in part to the rise in P_{a,O_2} over this period from 43.5 to 51.2 mmHg; but, when the subjects inhaled oxygen sufficient to raise P_{a,O_2} to 95.5 mmHg, cerebral blood flow fell by a mere 5 ml 100 g^{-1} min^{-1}. There are other more complex reasons for the apparent reduction in the effectiveness of hypoxia in maintaining a high level of cerebral blood flow with time and as P_{a,CO_2} falls at altitude and these are discussed in Chapter 10. But it would seem from this evidence that the antagonistic effects of low P_{a,O_2} and low P_{a,CO_2} cannot be described as a simple arithmetical relationship. Indeed, from Fig. 8.2 it might be possible to predict that after a longer period at altitude, hypoxia played little or no part in maintaining an elevated blood flow, i.e. the inhalation of added oxygen would

Inter-relationship of oxygen and blood pressure

not significantly affect blood flow. This has certainly been found to be true with respect to alveolar ventilation in subjects who have lived for long periods at high altitude (Bainton, Carcelan & Severinghaus, 1965). The mechanism of this insensitivity to

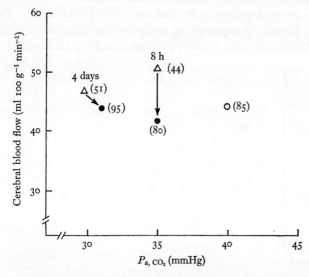

Fig. 9.2. The relation between cerebral blood flow (ml 100 g^{-1} min^{-1}) and P_{a,CO_2} (mmHg) during acclimatization to altitude (3810 m). The sea level value is shown by an open circle and the open triangles represent values for cerebral blood flow at 8 h and 4 days at altitude. At each of these and the other points, the figures in parenthesis give the P_{a,O_2} (mm Hg). The arrows represent the change in cerebral blood flow when the subjects inhaled high oxygen for 10 min toward the final values (closed circles). (Redrawn from Severinghaus et al. (1966), Circulation Res. **19**, 274–82, Fig. 1.)

oxygen was attributed to failure of the carotid body chemoreceptors. If this is true, it may be a response peculiar to man for the chemoreceptors in the cats native to high altitude have been found to respond normally.

On the other hand, the effect of hypoxia upon cerebral blood flow has been found to be similar whether P_{a,CO_2} is controlled at physiological levels or at levels 7–11 mmHg higher as is shown in Fig. 9.3 (James, Millar & Purves, 1969). These results and those of Shapiro, Wasserman & Patterson (1966), provide no support for the concept that carbon dioxide potentiates the vascular response to hypoxia.

9. Regulation of cerebral vessels by oxygen

Oxygen and blood pressure

The finding by McDowall (1966) that hyperbaric oxygen had a constrictor effect upon cerebral vessels formed part of a study of the clinical usefulness of hyperbaric oxygen in maintaining adequate tissue P_{O_2} levels in patients with hypotension. The relation between changes in P_{a,O_2} and arterial pressure upon cerebral blood flow was studied in dogs (Harper, Ledingham &

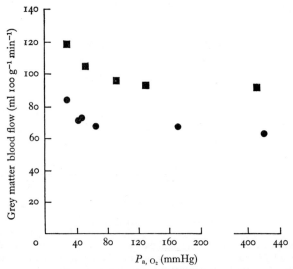

Fig. 9.3. The relation between cortical blood flow (ml 100 g^{-1} min^{-1}) and P_{a,O_2} in a baboon lightly anaesthetized with pentobarbitone, paralysed with gallamine and ventilated mechanically. P_{a,CO_2} was held at 41–44 mmHg (closed circles) and at 53–57 mmHg (closed squares) (From James, Millar & Purves (1969). *Circulation Res.* **25**, 77–93, Fig. 5.)

McDowall, 1965) and the results are summarized in Tables 9.2 and 9.3. These show that when the dogs breathed air and when arterial pressure was reduced to 50 mmHg by bleeding, there was the expected fall in cortical blood flow. There was also an accompanying fall in cortical oxygen uptake, a 17 per cent fall in cerebral venous oxygen saturation and a fall of approximately 0.2 pH units in arterial blood. P_{a,CO_2} and P_{a,O_2} were unchanged. While still hypotensive, the dogs were ventilated with oxygen at 2 atmospheres and while cortical flow was unaffected, cortical oxygen uptake was restored to the level seen under control

Table 9.2. *Cerebral cortical blood flow in the normotensive and hypotensive situations* (from: Harper, Ledingham & McDowall, 1965. *Hyperbaric oxygenation*, ed. Ledingham, I. McA. Edinburgh: Livingstone)

	Blood flow (ml g^{-1} min^{-1})		
Experiment	Control (Air)	Hypotension (Air)	Hypotension (2 atmospheres oxygen)
1	0.71	0.43	0.43
2	0.85	0.59	0.62
3	0.64	0.45	0.40
4	0.76	0.52	0.51
5	0.54	0.42	0.39
6	0.66	0.36	0.29
7	1.25	0.64	0.64
8	1.13	0.64	0.62
9	0.92	0.41	0.43
Mean and s.d.	0.83 ± 0.24	*0.50 ± 0.11	*0.48 ± 0.12

* $P < 0.005$.

Table 9.3. *Oxygen uptake values in the normotensive and hypotensive situations* (from Harper, Ledingham & McDowall, 1965. *Hyperbaric oxygenation*, ed. Ledingham, I. McA. Edinburgh: Livingstone)

	Oxygen uptake (ml g^{-1} min^{-1})		
Experiment	Control	Hypotension (Air)	Hypotension (2 atmospheres oxygen)
1	0.061	0.033	0.062
2	0.040	0.035	0.076
3	0.048	0.041	0.053
6	0.050	0.037	0.034
7	0.059	0.038	0.063
8	0.046	0.043	0.059
9	0.059	0.041	0.065
Mean and s.d.	0.052 ± 0.007	*0.038 ± 0.003	0.059 ± 0.009

* $P < 0.01$.

9. Regulation of cerebral vessels by oxygen

conditions, while P_{a,CO_2} rose on average by 6 mmHg. The possibility that a further fall in blood flow was masked by the vasodilator properties of a rise in carbon dioxide may probably be discounted since it has been shown that with severe hypotension, carbon dioxide is without effect (Harper & Glass, 1965). It was suggested by these authors that the apparent failure of hyperbaric oxygen to constrict cerebral vessels under these conditions was because the presumed tissue hypoxia caused local vasodilatation which offset it. However, there is no evidence that tissue hypoxia, as defined by a rise in venous lactate concentrations and/or a rise in tissue NADH causes vasodilatation (Granholm, Lukjanova & Siesjö, 1968): nor in the experiments of Harper et al. (1965) was there any significant fall in cerebral venous oxygen saturation, when the dogs were either made hypotensive or were ventilated with hyperbaric oxygen, to indicate significant cerebral hypoxia. An alternative possibility in these animals was that the very low levels of pH, average 6.97 units associated with a low cardiac output and hypotension rendered the cerebral vessels unreactive. Such a sequence of events has been observed in the baboon (James, Millar & Purves, unpublished results) and in the foetal lamb (Purves & James, 1969) under comparable conditions and it has been reported that under conditions of hypoxia, cerebral blood flow is no longer independent of arterial pressure (Freeman & Ingvar, 1968). The possibility arises that where the oxygen supply to the brain is impaired for some time either by a reduction in P_{O_2} or in blood flow, the reactivity of vessels is affected. This loss of reactivity may be complex. Thus Harper & Glass (1965) have shown that the vascular response to high carbon dioxide is lost with progressive reduction in arterial pressure. On the other hand, a dissociation between the vascular response to changes in pressure, which is usually lost, and the vascular response to hypocapnia which is preserved following periods of hypoxia or ischaemia has been reported (Häggendal, 1968; Baldy-Moulinier & Frerebeau, 1968; Easton & Palvölgyi, 1968).

Finally, it should be noted that the values obtained by Harper et al. (1965) for cortical oxygen uptake are low by comparison with other series, e.g. Lübbers, Ingvar, Betz, Fabel, Kessler & Schmahl (1964) in which, with comparable methods of measurement in the dog, values for cortical oxygen consumption of between 8 and 9 ml O_2 100 g^{-1} min^{-1} were obtained. The possi-

Inter-relationship of oxygen and blood pressure

bility arises that the low values obtained by Harper et al. (1965) were due to the anaesthetic used, trichloroethylene which has been shown to reduce oxygen uptake (McDowall, Harper & Jacobson, 1964). Thus although the results are of some interest in demonstrating that hyperbaric oxygen can reverse the fall in oxygen consumption brought about by severe hypotension, the possible artifacts introduced by anaesthesia and the grossly abnormal acid–base balance limit the significance of the results and it would certainly be unwise to extrapolate from these results to the clinical setting without further studies.

The homeostatic role of the cerebrovascular response to changes in P_{a,O_2}

The rise in cerebral blood flow as a consequence of progressive hypoxia undoubtedly helps to maintain not only a relatively high cerebral venous P_{O_2} but a relatively high level of oxygen supply. These changes are shown in Fig. 9.4. Cerebral venous P_{O_2} is plotted against P_{a,O_2} as the latter was altered randomly in steps over the range 35 to 120 mmHg in a lightly anaesthetized baboon. The departure of the $P_{v,O_2}/P_{a,O_2}$ curve from linearity reflects (a) the effect of increasing blood flow and (b) the fact that P_{a,O_2} has to fall by some 30–40 mmHg below normal limits before this change takes place. Similarly, the figure illustrates the fact that due to the shape of the oxygen dissociation curve and to the rise in blood flow, the rate at which available oxygen (oxygen content × blood flow) falls as P_{a,O_2} is reduced from the physiological range slowly diminishes.

Thus it is clear that a number of factors effectively ensure an adequate venous P_{O_2} as P_{a,O_2} is reduced to 35 mmHg. By 'adequate' is meant a P_{v,O_2} which is still above the 'critical threshold' of 17–19 mmHg at which humans lose consciousness (Thews, 1963). Further, the dilatation of vessels will ensure a greater surface area for gas exchange and the probability that adequate oxygen diffusion to the periphery of the tissue cylinder will be maintained.

It may be recalled that all these effects will be offset to a certain extent if the subject is allowed to hyperventilate in response to the hypoxia: on the other hand, if the fall in P_{a,O_2} is accompanied by a rise in P_{a,CO_2} as in any form of alveolar hypoventilation, the maintenance of an adequate P_{v,O_2} will be enhanced.

9. Regulation of cerebral vessels by oxygen

There are however clear examples where the factors which bring about homeostasis either fail or else are compromised. Thus it has been shown in the dog (Harper *et al.* 1965) and in the sheep foetus (Purves & James, 1969) that if the circulation fails, cerebral vessels do not respond to hypoxia or hypercapnia. It would then be expected that venous P_{O_2} would fall progressively and that oxygen consumption would be reduced. This would be of particular

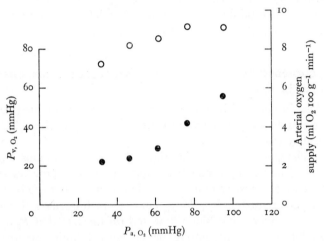

Fig. 9.4. The relation between venous P_{O_2} (mmHg, closed circles) and the oxygen supply (cerebral blood flow × oxygen content of arterial blood (ml O_2/100 g^{-1} min^{-1}, open circles) and P_{a,O_2}. P_{a,CO_2} held constant at 42 mmHg. (From James, Millar & Purves, unpublished observations.)

importance at the moment of birth for if there is evidence of circulatory distress *in utero* as shown by a fall in pH_a and even allowing for the fact that there are other protective mechanisms in the foetus and newborn (Chapter 11), the cerebral cortex will be particularly vulnerable to the effects of birth asphyxia. The incidence of cerebral injuries attributable to birth asphyxia suggests that further studies in this field would be profitable.

These examples illustrate the failure of homeostasis. In slighter degree, it is clear that the failure to maintain a constant P_{v,O_2} during hypoxia also represents a failure of true compensation, and a progressive reduction in P_{a,O_2} leads to signs of cerebral hypoxia and loss of consciousness. The level of P_{v,O_2} below which consciousness is invariably lost has been given as 15–20 mmHg

The homeostatic role of cerebrovascular response to P_{a,O_2}

by Lennox, Gibbs & Gibbs (1935) and it is at this level that the effects of hypoxia are signalled by a fall in the frequency of cortical potentials (Lennox, Gibbs & Gibbs, 1938) (Fig. 9.5). This figure emphasizes that the fall in the frequency of cortical potentials occurs within a narrow range of cerebral oxygen saturation or tension. A similar critical level of P_{v,O_2} was observed in dogs at which consciousness was lost (Noell & Schneider, 1944) and in

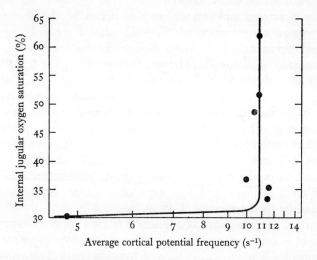

Fig. 9.5. The relation between frequency of cortical potentials to the oxygen saturation of internal jugular venous blood in one human subject. The average frequency per second was calculated by counting the duration of all the waves in a given length of electrogram from the occipital lead (indifferent electrode on the ear) and is represented on the abscissa. The effect of low oxygen levels on cortical frequency is critical but is manifest only when the venous oxygen saturation has fallen to about 30 per cent which approaches the level at which consciousness is lost. (From Lennox, Gibbs & Gibbs (1938), *Res. Publs Ass. Res. Nerv. Ment. Dis.* **18**, 277–97, Fig. 119.)

further studies in man (Gibbs, Williams & Gibbs, 1940; Kety & Schmidt, 1948b). In this latter paper, the authors also found evidence of mental confusion before consciousness was lost and that at this stage when P_{v,O_2} was 25–30 mmHg, there was no measurable change in cerebral uptake. Similarly, with slightly lower inspired oxygen concentrations, 8 per cent and at constant alveolar carbon dioxide levels, Turner et al. (1957) were unable to detect any change in cerebral oxygen uptake.

9. Regulation of cerebral vessels by oxygen

The dissociation between changes in cerebral oxygen consumption and levels of consciousness and mental symptoms associated with progressive arterial hypoxia may be explained in a number of ways. First, it is probable that the sensitivity of the method of measuring overall oxygen consumption of the brain is insufficient to detect changes of less than 10–15 per cent, for the error of blood flow measurement is of the order of 10 per cent, to which must be added the errors involved in measuring oxygen content of arterial and venous samples. Secondly, the changes in total oxygen consumption of the brain almost certainly conceal regional variations in both blood flow and in oxygen consumption and therefore in tissue P_{O_2}. Hansen, Sultzer, Freygang & Sokoloff (1957) have shown that the changes in flow in the brain of the cat in response to the inhalation of 10 per cent oxygen are far from uniform, and this has been confirmed recently by the demonstration that the absolute or percentage change in cortical and white matter blood flow response to hypoxia varies considerably (James, Millar & Purves, 1969). The probability therefore is that different parts of the brain become hypoxic at different rates and to different degrees and this type of response would not be detected by measurements of mean venous P_{O_2}.

Thirdly, there is the vexed question of how far the changes in the level of consciousness are a function of the hypocapnia which accompanies hypoxia rather than of hypoxia itself. This question was raised by Gibbs, Gibbs, Lennox & Nims (1942) who showed that cortical electrical activity was more sensitive to changes in P_{a,CO_2} than to changes in P_{a,O_2}: compare the changes in activity with carbon dioxide illustrated in Fig. 9.6 with those in 9.5. In this connection it is possible that the tissue hypoxia caused by hypocapnia is more severe than that caused by hypoxia because of the lowering of blood flow which accompanies a reduction in P_{a,CO_2}. This topic is considered in more detail in Chapters 10 and 11.

Mechanism of action of changes in P_{a,O_2}

The evidence which has been considered above has all been derived from experiments in which the subject has inhaled or been ventilated with different concentrations of oxygen and the effective and measured stimulus has been the arterial saturation,

Mechanism of action of changes in P_{a,O_2}

content or tension of oxygen. Without being explicit on the point, it has been assumed that the oxygen affects the smooth muscle of the vessel walls; high concentrations of oxygen causing the vessels to contract and low concentrations causing them to dilate. Alternatively, in those experiments, which are considered

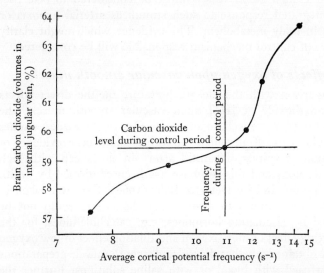

Fig. 9.6. The relation between the frequency of cortical potentials and carbon dioxide content of internal jugular venous blood. The frequency of cortical potentials calculated as in Fig. 9.5 and expressed as the reciprocal of wave duration. (From Lennox, Gibbs & Gibbs (1938), *Res. Publs Ass. Res. Nerv. Ment. Dis.* **18**, 277–97, Fig. 120.)

in Chapter 10, where the rate of oxidative metabolism of cerebral tissue has been altered, it has been assumed that the resulting changes in local P_{O_2} affect the smooth muscle of blood vessels and hence the rate of blood flow. These arguments are plausible because they not only describe the observed changes in blood flow and calculated changes in vascular resistance, but they also define a homeostatic mechanism whereby on the one hand, cerebral tissue is protected from the effects of arterial hypoxia and on the other blood flow is matched with metabolic requirements of cerebral tissue. The difficulties about these arguments are that there is singularly little evidence to back them and secondly, they entirely leave out the possibility that part at least of the control

9. *Regulation of cerebral vessels by oxygen*

of cerebral vessels is neurally determined. If the nerves which supply cerebral vessels are unimportant, then we might regard the control of blood flow in the brain as being self sufficient, and dependent on other vascular beds only in so far as blood is redistributed in favour of the brain. If the nerves are important, then intracranial blood vessels must be considered to take part in an integrated response to such stimuli as arterial hypoxia or changes in body metabolism. The evidence which might clarify the types of control mechanism responsible will be considered.

Local effects of oxygen upon vascular smooth muscle

By comparison with the meagre literature on the direct effects of carbon dioxide or [H^+] upon vascular smooth muscle, the effects of oxygen are well documented. One of the earliest studies was by Garry (1928) in which the records show clear evidence of contraction of spirals of carotid artery in the sheep with high oxygen tensions and this response has been confirmed by Smith & Vane (1966) and by Detar & Bohr (1968). Care was taken in the latter two studies to ensure that the effects could not be explained by circulating hormones, e.g. catecholamines, for the constricting effect of high oxygen and dilating effect of low oxygen could be obtained whether the isolated smooth muscle preparations were perfused with blood or with saline solutions. Further the effect of oxygen appeared to be a direct one since it was not antagonized by specific pharmacological antagonists such as hyosine, phenoxybenzamine, hexamethonium, bromolysergic acid, diethylamide or mepyramine. It is probable that the effect is complex *in vitro* and even more so *in vivo* since these workers showed the high oxygen potentiated the constrictor effect of catechol amines and 5-hydroxytryptamine while Skinner & Costin (1969) have shown in the isolated perfused gracilis muscle preparation in the dog, that the vasoconstriction caused by sympathetic stimulation can be modified by the level of hypoxia, and that this effect is further dependent on the potassium ion concentration.

Nothing is known about the mechanism of action of oxygen upon smooth muscle nor unfortunately were any of these studies quantitative and consequently we have no idea as to the sensitivity of these muscle preparations to changes in P_{O_2}. Further, these studies were carried out on smooth muscle from large arteries, carotid or aorta and in the absence of any specific studies we

Mechanism of action of changes in P_{a,O_2}

have to ask whether it is possible to make any conclusions about the reactivity of cerebral vessels.

There is some evidence that vascular smooth muscle at different sites respond differently to changes in P_{O_2}. The ductus arteriosus in the foetus (Kovalcik, 1963) and the umbilical arteries (Eltherington, Stoff, Hughes & Melmon, 1968) respond to high oxygen with vigorous constriction and this is apparently a direct response and not mediated by nerves. On the other hand, isolated pulmonary vessels are relatively little affected by changes in P_{O_2} (Smith & Coxe, 1951). But if these extreme examples are excepted and cerebral vessels are assumed to behave as other systemic vessels, it is reasonable to suppose that a fall in P_{a,O_2} will cause cerebral vessels to dilate and high oxygen will cause them to constrict. Further, insofar as it has been shown that the P_{O_2} of tissue immediately neighbouring pial vessels is not measurably different from that of blood within (Davis, McCullogh & Roseman, 1944) it is reasonable to suppose that cerebral vascular smooth muscle could be affected by tissue P_{O_2}. All these assumptions are consistent with the observed effects of altering P_{a,O_2} upon cerebral blood flow and calculated vascular resistance.

The idea that changes in tissue oxygen tension are a major factor in determining the level of blood flow has been most vigorously adumbrated by Guyton and his colleagues (Ross, Fairchild, Weldy & Guyton, 1962; Carrier, Walker & Guyton, 1964; Guyton, Ross, Carrier & Walker, 1964; Fairchild, Ross & Guyton, 1966). In a series of experiments, these workers showed that the relation between blood flow and the oxyhaemoglobin saturation of perfusing blood was approximately linear, the vessels dilating with hypoxia. The effect was not significantly altered in the spinal animal and further, they showed that the recovery from hyperaemic state following occlusion of the arterial supply only occurred when the muscles were perfused with oxygenated blood. These results confirm that the effects of changes in P_{a,O_2} can be obtained *in vivo* and that they can be distinguished from the neural and other chemical influences: and they provide suggestive evidence that the increase in flow after occlusion of arterial blood supply is related to depletion of oxygen stores and fall in tissue P_{O_2}. The precedent for the type of flow response seen in the brain is therefore established in muscle. Whether the mechanisms in cerebral vessels are the same can only be established after further studies.

9. Regulation of cerebral vessels by oxygen

Reflex control of vessels in hypoxia

The evidence discussed in the previous section provides a basis for believing that molecular oxygen can affect vascular smooth muscle directly although it is possible that its effect is mediated by alterations in the ionic composition of intra- or extracellular fluids. But in the absence of any quantitative data on the effect of oxygen *in vivo* upon vascular smooth muscle, it has to be asked whether a direct action of oxygen can account for the total vascular response. In particular, is there any evidence that the vasodilatation which occurs with hypoxia is due to reflex activity?

This question was never studied by earlier workers, presumably because of the widespread assumption that the neural control of cerebral blood vessels was negligible or absent. Ask-Upmark (1935) studied the effects of stimulating the sinus nerve and of altering the pressure within the carotid sinus. His results are ambiguous because of the variable changes in pressure which these manoeuvres caused and although he claimed that vasomotor changes in pial vessels were initiated by activity from the carotid sinus and the carotid body, they could equally well have been due to the effects of alterations in pressure described by Fog (1937, 1939b) and by Forbes, Nason & Wortman (1937).

The subject has recently been studied by James, Millar & Purves (1969) who measured the cerebral vascular response to hypoxia with all nerves intact and following division or stimulation of the vagus and cervical sympathetic nerves. The effect of sympathectomy is shown for one animal in Fig. 9.7 and is seen to consist of an elevation of both grey and white matter blood flow 40–50 per cent above control over the range of P_{a,O_2} tested, 35 to 440 mmHg. When the cervical sympathetic nerve was stimulated at constant frequency and intensity, cortical flow was reduced to control levels.

These results indicate that, in contrast to the response seen with changes in P_{a,CO_2}, the sympathetic appears to exert a uniform vasoconstrictor effect over the range of oxygen tension tested: although, with respect to white matter flow, its effects appear to be potentiated at low P_{a,O_2}. By contrast, as is shown in Fig. 9.8, the effect of bilateral vagotomy is most marked in the hypoxic range and the results suggest that part at least of the vasodilatation seen with hypoxia is reflexly determined.

Mechanism of action of changes in $P_{a,O}$

These results make it clear that as with the vascular response to other physiological variables, e.g. carbon dioxide and perfusion pressure, the reflex participation in the response to hypoxia is by no means negligible and in our present state of knowledge it is reasonable to state that the vascular response to changes in P_{a,O_2}

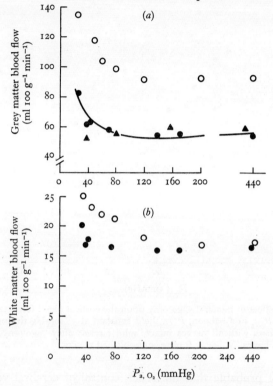

Fig. 9.7. The effect of ipsilateral cervical sympathectomy upon regional (parietal) cortical blood flow (a) and white matter blood flow (b) response to changes in P_{a,O_2} in a lightly anaesthetized baboon. P_{a,CO_2} held constant at 38 mmHg throughout. Control response, with all nerves intact, solid circles; following sympathectomy, open circles; during stimulation of the ipsilateral cervical sympathetic nerve at a frequency of 5–10 s^{-1}, 15 V and 100 μs duration, closed triangles. (From James, Millar & Purves (1969), *Circulation Res.* **25**, 77–93, Fig. 6.)

is due partly to direct effects of molecular oxygen and partly due to reflex vasoconstriction and vasodilatation. It would be naive to say that the reflex effects merely oppose one another. It is more probable that the proportion of vasoconstrictor and vaso-

9. Regulation of cerebral vessels by oxygen

dilator activity vary over the range of P_{a,O_2} tested but the net effect at low P_{a,O_2} is dilatation. To all these mechanisms must be added the probable local action of catechol amines and other vasoactive substances: but their actions upon cerebral vessels in hypoxia have never been studied *in vivo*.

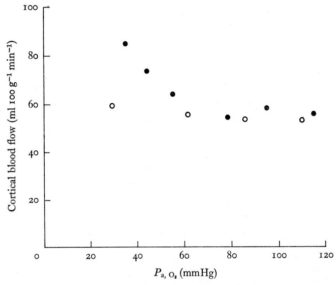

Fig. 9.8. The effect of bilateral vagotomy upon the cortical blood flow response to changes in P_{a,O_2} in a baboon, P_{a,CO_2} held constant at 38 mmHg throughout. Control responses with all nerves intact, solid circles; after vagotomy, open circles. (From James, Millar & Purves, unpublished results.)

It is most probable that the reflex control of cerebral vessels with changes in P_{a,O_2} is initiated by the carotid body and aortic group of chemoreceptors since, as far as is known, these are the only receptors known to be sensitive to oxygen. Further, it is tempting to speculate that the similarity between the flow–P_{a,O_2} and the carotid body chemoreceptor activity–P_{a,O_2} curves (Hornbein, Griffo & Roos, 1961; Biscoe, Purves & Sampson, 1970) is not accidental. These speculations however merely emphasize the need for studies which will determine the chemoreceptor participation in the reflex control of cerebral blood vessels.

10 THE NEURAL CONTROL OF CEREBRAL BLOOD VESSELS

Introduction

There can now be no doubt that cerebral blood vessels are plentifully supplied with nerves. The evidence, reviewed in Chapter 3 makes it clear that there are vasoconstrictor and vasodilator pathways and more recent studies have confirmed that adrenergic fibres can be seen on cerebral arterioles down to 10–15 μm both on the pia and within brain substance. Electron microscope studies have shown the presence of nerve terminals upon the adventitia of small cerebral vessels whose appearance is similar to adrenergic and cholinergic terminals elsewhere in the body.

The question may then be asked: what do these nerves do? This is never an easy question to answer in the cerebral or other peripheral vascular bed. For example, in skeletal muscle it has been shown that blood vessels dilate normally in response to muscular contraction after cholinergic vasodilator nerve block (Anrep & von Saalfeld, 1935; Ganter, 1928) and after extirpation of both sympathetic chains (Cannon, Newton, Bright, Menkin & Moore, 1929) and similarly that reactive hyperaemia occurs after blockade of all vasomotor nerves (Barcroft, 1963). Does this enable us to say that vasomotor nerves do not play some part in these responses in the intact animal; or do we have to suppose that vasomotor nerves are only or principally involved in the response to alterations in baroreceptor activity in order to maintain a constant circulating blood volume; or in the redistribution of cardiac output which occurs notably in diving birds or mammals during submergence?

When the cerebral circulation is considered in these terms, it is at once obvious that the necessary experimental evidence is almost completely lacking. Such evidence as there is, is mainly directed towards establishing whether or not the vasomotor nerves are involved in the maintenance of resting cerebral vascular tone –

10. *The neural control of cerebral blood vessels*

that is, from experiments in which cerebral blood flow or some index of it has been measured before and after section or stimulation of the cervical sympathetic or vagus nerves. Since many of these tests both in experimental animals and in man were carried out without adequate control of blood gas tensions or arterial pressure, it is not surprising that the estimates of the neural contribution to resting vascular tone have varied considerably. This confusion in the literature is unfortunate but it is even more unfortunate that a number of reviewers e.g. Sokoloff, 1960; Lassen, 1968a have dismissed the action of cerebral vasomotor nerves as negligible. Clearly, before this kind of statement can be made responsibly, we should have to know among other things how or whether vasomotor nerves are involved in the cerebral vascular response to alterations of blood gas tensions and arterial pressure and whether these nerves are involved in integrated vascular responses e.g. to altitude or active muscular exercise. So far from being dismissed, the problem has scarcely been tackled, as the evidence reviewed in this chapter will show.

The vasomotor contribution to resting cerebral vascular tone

Evidence from the study of pial vessels

It is clear that the question of the neural control of cerebral blood vessels occurred to the earliest investigators in this field. Brachet (1837), for example, observed congestion of the exposed cortical surface after removing the superior cervical ganglia. Callenfels (1855) noted that in a proportion of unanaesthetized rabbits, stimulation of the cervical sympathetic nerve caused constriction of the pial vessels on the same side while Ackerman (1858) and Donders (1859) showed that section of the sympathetic nerve caused constriction followed by dilatation.

Numerous similar observations were made over the following fifty years and the accumulated evidence indicated that the calibre of pial vessels could be affected not only by section or stimulation of sympathetic nerves (e.g. Nothnagel, 1867; Krauspe, 1874; Schuller, 1874; Vulpian, 1875) but also by stimulation of a variety of peripheral receptors. Thus Schuller (1874) observed that pial vessels constricted when he applied mustard to the animal's skin; Meyer & Pribam (1875) caused cerebral vasoconstriction by

The vasomotor contribution to resting blood flow

stimulating afferent fibres from the stomach and Spina (1898) observed cerebral vasodilatation after applying tetanic stimuli to the medulla following hind-brain section. On the other hand, a few workers using apparently the same methods were unable to confirm these results (e.g. Schultz, 1866; Reigal & Jolly, 1871) and it became apparent to these and to later workers that some at least of these discrepancies could be explained by the fact that the vascular response was in part determined by the changes in systemic arterial pressure which accompanied the various types of nerve stimulation.

Over the next few years, therefore, particular attention was paid to the changes in arterial pressure which occurred with each manoeuvre and it will be recalled that at that time, it was generally believed that cerebral blood flow followed changes in pressure passively. For proof of neurally mediated vasomotor activity, there had to be evidence of a change of blood flow in the opposite direction to the change in pressure or in the absence of any pressure change. A variety of methods of measuring cerebral blood flow or some index of it were used. Cramer (1873) used jugular venous pressure as an index of blood flow and, not surprisingly, obtained variable results when he stimulated the cervical sympathetic nerve. Gaertner & Wagner (1887) measured venous outflow directly and observed a large rise in blood flow during asphyxia; but, since there was also a rise in arterial pressure, they concluded that there was no need to postulate active changes in cerebral vessels. Roy & Sherrington (1890) used changes in brain volume as an index of blood flow. They also observed a rise in volume when the central end of the vagus was stimulated and this despite a fall in arterial pressure. However, they also noted a change in central venous pressure and, believing that this was the cause of the change in brain volume, were reluctant to ascribe vasomotor properties to cerebral vessels. Von Schulten (1884) recorded changes in cerebrospinal fluid pressure and internal carotid arterial flow using a thermoströhmur and observed that faradic stimulation of the cervical sympathetic nerve caused a fall in blood flow and constriction of the pial vessels without obvious change in internal carotid artery pressure. He therefore concluded that the vasoconstriction was active and neurally mediated.

Other workers used a similar method to that of Roy & Sherring-

10. The neural control of cerebral blood vessels

ton (1890) but with certain modifications and came to different conclusions. Thus, Müller & Siebeck (1907) in addition measured cerebral venous outflow. They also observed an increase in brain volume when the central end of the vagus was stimulated and they also observed a substantial increase in blood flow. Since this was associated with a fall in arterial pressure and no obvious change in jugular venous pressure, they concluded that a true vasodilatation had taken place and that in consequence cerebral vessels could be neurally regulated. These results were confirmed by Weber (1908). Jensen (1904) measured the velocity of carotid artery blood flow using a thermoströhmur and observed that this fell when the ipsilateral cervical sympathetic nerve was stimulated, and since systemic arterial pressure had risen by only a small amount, he concluded that active vasoconstriction had taken place.

A further attempt to demonstrate the neural control of cerebral vessels was undertaken by Anrep & Starling (1925) during the course of experiments in which the effect of changes in cerebral blood flow upon the medullary 'vasomotor centres' was measured. They used a cross-circulation technique whereby the head and neck of a dog was perfused with blood from a heart–lung preparation, the dog's body being perfused by its own heart. When pressure in the heart and aorta was raised by clamping the thoracic aorta, pressure in the cephalic circulation fell. But if blood pressure in the perfused head was held constant by means of a shunt, this manoeuvre caused a marked increase in cephalic blood flow – in the example quoted, from 335 to 500 ml min^{-1}. This response was abolished after vagotomy. When pressure in the cephalic circulation (which included the carotid sinuses) was raised or lowered, blood pressure in the femoral artery changed in the opposite direction and these responses were unaffected by vagotomy. At first sight, these results would seem to provide impressive evidence that a vasodilator neural pathway exists which includes fibres in the vagus and/or the aortic depressor nerve activity in which is initiated by the aortic group of baroreceptors. It is however unfortunate that these workers chose the dog as their experimental animal since an unknown proportion of blood perfusing the head was destined for the brain. It is therefore impossible to conclude as did the authors that an increase in aortic pressure and in depressor nerve activity caused vasodilatation of cerebral vessels only. Furthermore, it is not possible to make

The vasomotor contribution to resting blood flow

any statements on the basis of these results about the effects of changes in cerebral blood flow on the medullary vasomotor centres which would confirm or otherwise the results obtained by Francois Franck (1887). It was also unfortunate that Anrep & Starling gave no evidence of having read the published observations of Hering (1923, 1924) concerning the function of the carotid sinus: otherwise they would have interpreted their findings with respect to the changes in femoral artery pressure with changes in cephalic perfusion pressure very differently. Similar criticisms can be levelled at the experimental results and conclusions of Gesell & Bronk (1926) and Bronk & Gesell (1927) who observed an increase in carotid blood flow and reduced femoral arterial flow in response to hypoxia, hypercapnia or the injection of sodium cyanide in the dog.

The problem was taken up afresh by Forbes & Wolff (1928) using the skull window technique. They found that when the cervical sympathetic nerve was stimulated, the pial vessels on the same side constricted; when the central cut end of the vagus was stimulated, the pial vessels on both sides dilated and since this dilatation persisted after section of the sympathetic nerves, they concluded that the vagus effect was genuinely dilator and not merely inhibitory to sympathetic vasoconstrictor activity. Some of these responses are illustrated in Fig. 10:1 and in Figs. 5.4 & 5.5. By measuring the pressure changes in sagittal sinus and jugular veins, Forbes & Wolff were able to dispose of the suggestion made by Roy & Sherrington (1890) and repeated by Bayliss & Hill (1895) that the increase in brain volume on stimulating the vagus or with asphyxia was due to venous congestion. Such changes in pressure as did occur did not exceed $+4$ cm H_2O. Forbes & Wolff also showed that the changes in calibre of pial vessels with nerve stimulation occurred in vessels over a wide range of diameter, 110–340 μm. The average reduction in diameter with sympathetic stimulation was 7–8 per cent of control and the average increase in diameter with vagal stimulation was between 9 and 22 per cent. Two points may be made about these results. First, it should be recalled that since blood flow is related to the fourth power of the radius of a vessel, these changes in vessel calibre could conceal rather larger changes in blood flow. Thus, a very approximate estimate of changes in flow from these changes in calibre would imply that with sympathetic stimulation,

10. *The neural control of cerebral blood vessels*

blood flow was reduced by 30 per cent of control, and with vagal stimulation, blood flow doubled. It may be no more than a coincidence that these changes are similar to those found with similar manoeuvres when blood flow was measured more precisely (James, Millar & Purves, 1969). A second point to be made is that in the

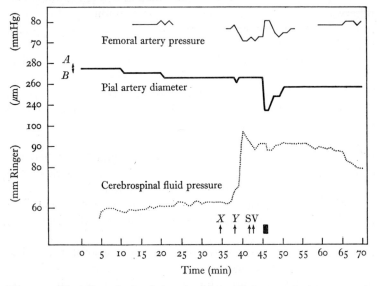

Fig. 10.1. The effect of stimulating the left cervical sympathetic nerve upon femoral artery pressure, pial artery diameter and cerebrospinal fluid pressure. At the mark X, artificial ventilation was started and shortly after this cerebrospinal fluid pressure rose abruptly and remained some 30 mm Ringer higher than control. The second mark, Y, indicates manipulation of the cervical sympathetic nerve for a few moments before section of the left sympathetic nerve (S) and vagus (V). The sympathetic nerve was stimulated as shown by the solid block and 15 s later the pial artery constricted and then slowly returned to its control size. (From Forbes & Wolff (1928), *Archs Neurol. Psychiat., Chicago*, **19**, 1057–86, Fig. 4. Copyright 1928, American Medical Association.)

experiments of Forbes & Wolff (1928) and in others using the same method, blood gas tensions or pH were rarely measured. Since P_{a,CO_2} and pH_a fall progressively during long experiments under general anaesthesia, it is probable they were low by control standards when the measurements were made. There is now some evidence that the magnitude of the vascular response to stimulation of the sympathetic nerve depends on the prevailing level of P_{a,CO_2} (James, Millar & Purves, 1969). Thus it should

The vasomotor contribution to resting blood flow

be borne in mind that the failure of many workers to observe significant changes in pial vessel diameter with sympathetic nerve stimulation may have been due to the presence of hypocapnia. The response of cerebral blood vessels to vagal stimulation on the other hand does not appear to depend upon the level of P_{a,CO_2}.

The findings of Forbes & Wolff have been confirmed and extended by Cobb & Finesinger (1932) and by Forbes & Cobb (1938). The study by Cobb & Finesinger (1932) was of particular interest since it was shown that the facial nerve carried dilator fibres to the cerebral vessels and thus confirmed in most details the morphological study of Chorobski & Penfield (1932). Cobb & Finesinger (1932) also demonstrated that the effect of facial nerve stimulation was limited to pial arteries on the same side and that this effect was independent of changes in arterial pressure. Further experiments at this time confirmed that changes in blood flow could occur with nerve stimulation at constant perfusion pressure. Finesinger & Putnam (1933) perfused the brains of monkeys whose own blood was pumped through one internal carotid artery, the other afferent arteries having been tied off. Blood flow was measured at constant pressure and was shown unequivocally to fall when the cervical sympathetic was stimulated and to rise when the vagus was stimulated.

It has been pointed out on a number of occasions, e.g. Forbes & Cobb (1938), Sokoloff (1959), that the changes in pial artery diameter with stimulation of the sympathetic nerve are small, i.e. approximately 10 per cent of those observed in skin or vessels of the rabbit's ear. Such comparisons require to be treated with caution. In the first place, electrical stimulation of the sympathetic nerve with a repetitive train of impulses differs considerably from the natural discharge which is either random or grouped. Further, the stimulating conditions may vary from experiment to experiment or even between observations so that the same vascular response can only be obtained by altering the intensity or frequency of stimulation. In the papers referred to by Forbes & Cobb (1938), none of the particulars of the recording conditions is recorded and in only a few, are details given of the threshold of stimulating intensity above which vascular responses were observed. In addition to these difficulties, there is the question as to whether the conditions with respect to blood gas tensions and arterial pressure were similar in the compared experiments. Finally, some of these

10. *The neural control of cerebral blood vessels*

comparisons were derived from animals of different species, i.e. pial vessels in the cat and blood vessels in the rabbit's ear, while few indications are given as to the different levels of anaesthesia. An example of this type of comparison is given in Fig. 10.2. This certainly shows that the change in calibre to sympathetic stimulation of the rabbit's ear vessels is very much greater than that

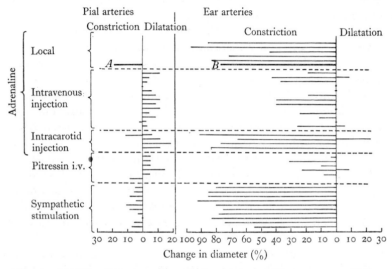

Fig. 10.2. Each fine line represents a single experiment. The percentage change in the diameter of the pial artery is directly opposite to that of the artery in the ear in the same experiment. In the case of the local application of adrenaline, *A* indicates the average constriction in twenty-eight cases. *B* indicates the average constriction of the five arteries the individual constrictions of which are shown. (From Forbes, Finley & Nason (1933), *Archs Neurol. Psychiat. Chicago*, **30**, 957–79, Chart 3. Copyright 1933, American Medical Association.)

in pial vessels. But it is also clear that the vessels respond to other stimuli in opposite directions. These differences should not occasion surprise since the function of the vessels is quite different; that is, we might suppose that the all-important function of cerebral vessels is to ensure a constant environment for the central nervous system while that of ear vessels is to lose heat from the body. For all these reasons, it is suggested that these types of comparisons are spurious and best avoided.

In general, the studies of Forbes & Wolff (1928) and others have been repeated on a number of occasions since (Koopmans,

1938; Wierzuchowski, 1938) and the results confirmed. Gurdjian and his colleagues (Gurdjian, Webster, Martin & Thomas, 1958; Gurdjian & Thomas, 1961) have however failed to confirm that stimulation of the sympathetic nerves causes pial constriction. In the monkey, these workers recorded changes in pial artery calibre on moving film and it was unfortunate that no indication was given of the sensitivity of this method, nor of the range of spontaneous changes in pial vessel calibre. In human curarized and anaesthetized subjects, these workers did observe spontaneous changes in pial vessel calibre and suggested that the results obtained by earlier workers were confused with these. This seems improbable since particular note was taken by Forbes & Wolff (1928) of such spontaneous changes during long experiments and the expected error of measurement was included in their published figures. It is also worth noting that in the experiments of Gurdjian & Thomas (1961), they reported 'beading' of the pial vessels – a feature which is associated with trauma to vessels in other species (Wentsler, 1936). Further when the pial vessels were mechanically irritated, they dilated: and this also contrasts with the constriction which is usual in normally responsive vessels (Florey, 1925; Echlin, 1942). These points suggest that the vessels whose changes were recorded by Gurdjian may have been damaged either during the surgery necessary to expose the vessels or by the pathological process which led to the surgical intervention. There is now clear evidence that extensive surgery to or exposure of pial vessels will lead to a marked reduction in vascular response (Sagawa & Guyton, 1961).

Evidence from measurements of cerebral blood flow

Using thermocouples to give an index of blood flow. The evidence so far reviewed indicates that under a variety of experimental conditions, the calibre of pial vessels can be markedly altered by changes in vasomotor neural activity. The technique of observing changes in the calibre of pial vessels is limited partly because it gives little and occasionally misleading information about changes in cerebral blood flow and partly because the areas of cortex accessible to direct study are limited. The introduction by Gibbs (1933b) of a heated thermocouple which could be used as an index of flow in organs made it possible to extend the study to different parts of the brain. This type of thermocouple

10. *The neural control of cerebral blood vessels*

was first used in the kidney and applied to the problem of cerebral blood flow in an attempt to measure the changes which occurred with the administration of convulsants. Schmidt & Pierson (1934) were the first to undertake a systematic study of the responses of blood flow in different parts of the brain to section or stimulation of the vasomotor nerves. When they measured changes in blood flow in the medulla with a thermocouple cooled with respect to the surrounding tissues, they were unable to demonstrate any change when the vagus, sympathetic or sinus nerves were stimulated. On the other hand, they observed large increases in blood flow with hypoxia or hypercapnia.

It is worth interpolating at this point that the reason why this group of experiments was carried out, was to test the hypothesis that an inverse relation existed between the level of cerebral blood flow and alveolar ventilation (Schmidt, 1928 *a*, *b*, *c*). Schmidt assumed that if other factors, notably the metabolic rate of neurones comprising the 'respiratory centre', remained constant the chemical environment of these neurones and hence the level of respiration would be entirely regulated by the rate of cerebral blood flow to the medulla. This view had to be modified following the observation by Heymans & Heymans (1927) that the level of alveolar respiration could be and was normally modified by reflex activity initiated by the carotid body group of chemo-receptors, and that the medulla appeared to have little control over its own blood supply. It was further shown that the changes in blood flow were not directly affected by chemoreceptor reflex activity but indirectly by the redistribution of cardiac output in favour of the head (Bouckhaert & Heymans, 1935).

Schmidt (1934) then studied other areas of the brain, and in the hypothalamus found no evidence of vasodilator activity when the vagi were stimulated, but a constant reduction in blood flow when the cervical sympathetic nerve was stimulated. This reduction in blood flow was peculiar. It lasted for many minutes after the stimulus was discontinued and was not completely reversible. Schmidt also noted that there was no vasodilatation on cutting the sympathetic nerves and this was in contrast to the response seen in the vessels of the tongue or temporal muscle. This unusual response to stimulation of the sympathetic raised the question as to whether an alternative mechanism, e.g. the release of catechol amines, could be involved. In all other respects,

The vasomotor contribution to resting blood flow

i.e. in response to hypoxia or hypercapnia, the sensitivity of the blood vessels of the hypothalamus was similar to that in other parts of the brain. In a further series of experiments, Schmidt (1936) studied the changes of blood flow within and on the surface of the parietal cortex. The changes observed here with stimulation of the cervical sympathetic nerves were most marked; those in sub-cortical tissue, less so. Schmidt was unable to confirm the findings of previous workers (e.g. Forbes & Wolff, 1928) that stimulation of the vagus caused vasodilatation; neither did he observe any change in flow when the sinus nerves were stimulated. In all other respects, the results obtained with the thermocouple were qualitatively similar to those obtained by direct inspection of the pial vessels.

A later study by Forbes, Schmidt & Nason (1939) reconciled the discrepant results obtained when the vagi or facial nerves were stimulated. These workers obtained unequivocal evidence of vasodilatation which was most marked in the cortex and to a lesser extent in sub-cortical areas. The fact that these types of nerve stimulation were not equally effective in various parts of the brain was confirmed in experiments by Ludwigs & Schneider (1954) and these workers extended their observations to show that the reduction in local cerebral blood flow which occurred with stimulation of the sympathetic nerves could be abolished by the administration of hexamethonium or other ganglion blockading agents.

Using direct methods. The evidence quoted so far indicates that, under experimental conditions in animals, there is clear evidence of a small but consistent neural contribution to resting vascular tone of cerebral vessels, and of changes in blood flow which vary in degree throughout the brain when the vasomotor nerves are stimulated. Inevitably, the literature contains some discrepant results and these together with the relatively small response seen when vasomotor nerves are cut have been seized upon by a number of reviewers as evidence that the neural contribution to the regulation of cerebral vessels is so slight as to be negligible. In view of this controversy, it is odd that the matter has never been systematically studied following the introduction of more direct methods of measuring cerebral blood flow. It is just possible that some workers have been dissuaded

10. *The neural control of cerebral blood vessels*

from such a study by the idea that the decisive experiments had already been carried out. The study of Dumke & Schmidt (1943) is often quoted by reviewers (e.g. Schmidt, 1960; Kety, 1960) as showing that section of the sympathetic nerves had no significant effect upon blood flow. Closer examination of the evidence shows that such a conclusion is not justified. In the experiments of Dumke & Schmidt, cerebral blood flow in monkeys was measured by a bubble method which required that a coil of 22 gauge tubing of unstated length be inserted into the internal carotid arteries. Certainly, when the sympathetic nerves were subsequently cut, there was a negligible change in blood flow. However, it was also noted that when these animals were given low oxygen to breathe, there was virtually no change in the level of respiration which suggests that during the preliminary surgery, the sinus nerves had been damaged or that the animals were very deeply anaesthetized. Secondly, Dumke & Schmidt observed that when the animals were given high carbon dioxide to inhale, mean cerebral blood flow increased from a resting value of 45 ml min^{-1} by only 5 ml min^{-1}. Under comparable conditions, other workers have found that cerebral blood flow almost doubles. These features raise serious doubts as to the sensitivity of the cerebral vessels in these experiments, for many workers, e.g. Echlin (1942), have shown that such a failure of cerebral vessels to respond to changes in P_{a,CO_2} usually points to damage by trauma or other factors. For these reasons, therefore, it is doubtful whether the results obtained by Dumke & Schmidt are helpful in determining whether sympathetic vasoconstrictor activity is important in regulating cerebral blood vessels.

This question has also been studied in man as part of an investigation into the therapeutic value of cervical sympathectomy for conditions where there was evidence of vascular 'spasm' or reduced cerebral blood flow. Some previous studies had shown that this might be valuable (Leriche & Fontaine, 1936; Mackey & Scott, 1938; Risteen & Volpitto, 1946). The stellate ganglion was blocked with local anaesthetic on both sides in two series (Harmel, Hafkenshiel, Austin, Crumpton & Kety, 1949; Shenkin, Cabieses & Van den Noort, 1951) and on one side only in a third series (Scheinberg, 1950). The subjects of these experiments formed a heterogeneous series, some with hypertension with or without evidence of cerebral atherosclerosis; some with normal blood pressure but with other diseases, e.g. diabetes mellitus and Parkinson's disease, which could well affect cerebral blood vessels. Only six normal subjects were included and complete data are available in four. When the stellate ganglion was blocked in these patients, no significant change in cerebral blood flow (measured by the nitrous oxide technique) was observed in the first series (Harmel *et al.* 1949) and the change in blood flow observed in the second series (Shenkin *et al.* 1951) could in part have been explained by the fall in oxygen content and arterial haematocrit associated with the operation. Similarly, no significant change in blood flow was observed following unilateral stellate ganglion blockade (Scheinberg, 1950).

The results of these experiments have often been used (e.g. Lassen, 1968a) to demonstrate that vasomotor nerves contribute negligibly to the control of cerebral blood flow; but it is clear, as was emphasized by Harmel *et al.* (1949), that it is difficult if not impossible to draw any conclusions about the physiological functions of sympathetic nerves from these studies. In the first place, the studies were carried out to answer a different question – namely, will stellate ganglion blockade materially affect cerebral blood flow in patients with various forms of vascular disease and with evidence of vascular spasm. The answer was no, and not unexpectedly so, since in the group of patients as a whole, the rate of blood flow was uniformly lower and cerebrovascular resistance higher than normal and, as was discussed in Chapter 4, it is known that the sensitivity of blood vessels is markedly reduced under these conditions. Any change might have been small and concealed in the errors of the method. A second objection to this type of experiment has become clear from the results of fluorescent histochemical studies which have shown that whereas the fluorescence of adrenergic nerves is completely abolished after removal of the superior cervical ganglia, the fluorescence after stellate ganglionectomy or blockade is reduced but by no means abolished (for references, see Chapter 3). The explanation for this is not known with certainty but the observation suggests that the post-ganglionic sympathetic fibres supplying cerebral blood vessels derive from the superior cervical rather than the stellate ganglia and that, in consequence, blockade of the stellate ganglia might be expected to be ineffective in denervating cerebral vessels.

10. *The neural control of cerebral blood vessels*

Other data in man have been obtained by less direct means. Bridges, Clark & Yahr (1958) measured changes in intracranial volume with a sensitive pressure transducer placed in a trephine hole in the skull of lightly anaesthetized patients. In addition to the expected changes in cerebral volume associated with obstruction of the jugular veins, and the inhalation of high carbon dioxide or low oxygen, they observed that procaine block of the superior cervical ganglion caused a rise in brain volume, while stimulation of the ganglion caused a fall. By contrast, blockade or stimulation of the stellate ganglion caused very much smaller changes. These observations, besides indicating that in man, as in other animals, sympathetic vasoconstrictor nerves to cerebral vessels can affect vascular tone, raise the question as to whether post-ganglionic fibres from the superior cervical ganglion can be affected by preganglionic fibres other than those in the cervical sympathetic nerve. One possible pathway between vagus and superior cervical nerve has been described by Ranson & Billingsley (1918) (see also Chapter 3) and other such pathways may exist. If so, this would constitute an additional reason why blockade of the stellate ganglion is an ineffective way of interrupting sympathetic pathways to cerebral vessels.

Cerebral vasomotor nerves and the responses to changes in blood gas tensions and arterial pressure

Thus far, the experiments which have been described have been directed at determining how far autonomic nerves affect cerebral vascular tone under control conditions. More recently, attempts have been made to assess their participation in the cerebral vascular responses to changes in blood gas tensions and arterial pressure. In the sheep foetus and newborn lamb, it has been shown that section of the vagi abolishes the large rise in cerebral blood flow and oxygen consumption which occur when the umbilical cord is occluded and respiration starts and that the vascular response to carbon dioxide is enhanced following sympathectomy and reduced following vagotomy (Purves & James, 1969). These studies have been extended in the adult baboon in which the separation of intra- and extracerebral circulations is easier to achieve (James, Millar & Purves, 1969). In these experiments, it was observed first that the effect of section or stimulation of vasomotor nerves was most obvious in the vessels of the cortex. The results may be summarized as follows: Within physiological limits of blood gas tensions and arterial blood pressure, section of the cervical sympathetic nerve gave rise to a consistent rise in cerebral blood flow of up to +14 per cent of control and this difference increased as P_{a,CO_2} was raised, as P_{a,O_2} was lowered and as mean arterial pressure was raised. Following sympathectomy, the relation between blood flow and mean arterial pressure altered so that flow was no longer independent of pressure. When the cervical sympathetic nerve was stimulated electrically, blood flow always diminished and the amount by which it fell was proportional to the resting level and to the level of P_{a,CO_2}. This type of study has been repeated in two further series of experiments in which the responses of flow to changes in perfusion pressure have been measured in monkeys (Eklöf, Ingvar, Kågström & Olin, 1971), and in cats (Waltz, Yamaguchi & Regli, 1971) not less than two weeks following unilateral or bilateral cervical sympathectomy. The numbers involved in each series were small but as far as could be determined the responses observed under these conditions were essentially unchanged from control. This type of result is similar to that seen in other vascular beds, e.g. skeletal muscle with chronic denervation. The effects of chronic denervation upon the responses of various beds has been extensively reviewed by Shepherd (1963) and as he points out, the resumption of an approximately normal level of blood flow and vascular responses some days or weeks after section of the autonomic nerves makes it difficult, if not impossible, to assess the physiological importance of these nerves in the intact animal.

Qualitatively similar results were obtained whether the vagus and depressor nerve or depressor nerve alone was cut or stimulated. Section of the nerves gave rise to small and inconstant changes in blood flow; but the vascular response to carbon dioxide, low oxygen and hypotension was markedly reduced. In particular, following vagotomy, a modest reduction of mean arterial pressure to 70–80 mmHg caused a drastic fall in blood flow, cardiac irregularities and death if blood was not rapidly reinfused. Stimulation of the central cut end of the vagi caused substantial increases in blood flow which were independent of the level of P_{a,CO_2}, Fig. 10.3. Some confirmatory evidence was obtained in this series of experiments that vasodilator fibres were carried in the

10. *The neural control of cerebral blood vessels*

facial nerve but the results suggested that alternative efferent dilator pathways could exist.

These results are of interest since they indicate that the function of the autonomic nerves to cerebral blood vessels is almost exactly opposite to that which has been determined for the blood vessels

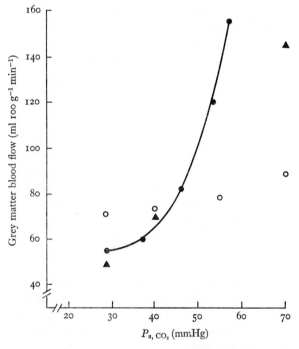

Fig. 10.3. The relation between cortical blood flow and P_{a,CO_2} in an anaesthetized baboon. With nerves intact, closed circles; following vagotomy, open circles and during stimulation of the central cut end of the vagus, triangles. (From James, Millar and Purves (1969), *Circ Res.* **25**, 77–93, Fig. 8.)

of skeletal muscle – that is, they contribute only slightly towards resting vascular tone but are of particular importance when blood gas tensions or systemic pressure deviates from the physiological range.

Similar results have been obtained by Kobayashi, Waltz & Rhoton (1971, personal communication) with respect to the effects of stimulation of the cervical sympathetic nerves. They noted, in addition, that although such stimulation gave rise to a fall in cerebral blood flow, measured with ^{85}Kr, there was not always a corresponding change in the calibre of pial arterioles with a diameter of $> 50\,\mu$m. These authors argue that the most prominent effect of sympathetic vasoconstriction is likely to be on the smaller precapillary arterioles of diameter $10-15\,\mu$m. If this can be confirmed, then it would provide further explanation for some of the early discrepant results in which the responses of the larger pial vessels only were observed.

Some further important evidence in this field has been provided by Mchedlishvili & Nikolaishvili (1970). They first confirmed that when mean arterial pressure is lowered, the pial vessels dilate and that they constrict when mean arterial pressure is restored (Fog, 1937). This response is shown against time in Fig. 10.4. When either atropine or other short acting cholinergic inhibitors were injected intravenously, the initial vasodilatation in response to a fall in arterial pressure was abolished although the subsequent passive dilatation and constriction when blood was replaced was unaffected. This response – that is a failure of pial vessels to dilate as pressure is lowered – is consistent with the observations of James, Millar & Purves (1969) and provides further evidence that the maintenance of a high rate of cerebral blood flow as mean arterial pressure is reduced depends upon intact cholinergic vasodilator nerves. As a consequence, the whole concept that the independence of blood flow and arterial pressure in the intact animal is a manifestation of 'autoregulation' requires reappraisal.

Other aspects of the neural reflex arcs

The evidence which has so far been reviewed has dealt principally with the efferent or motor side of the proposed reflex pathways and, in order to substantiate that these pathways exist, it is equally important to consider evidence from experiments in which (a) mid-brain structures and (b) peripheral receptors can be shown to be involved.

Effect of stimulating mid-brain structures

The first experiments in which mid-brain structures were stimulated occurred at a time when there was considerable controversy as to whether active vasomotor responses in cerebral blood vessels

10. *The neural control of cerebral blood vessels*

could be observed at all. Further, difficulties arose in interpreting the results because different methods of stimulation were employed, and different indices of blood flow were used. Thus, Gaertner & Wagner (1887) injected strychnine intravenously and concluded that any changes observed in pial vessels could be accounted for by the concomitant changes in systemic arterial pressure

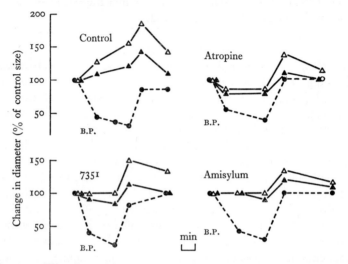

Fig. 10.4. The effect of haemorrhage and subsequent replacement of blood upon the diameter of pial arteries in a rabbit. Vessels larger than 90 μm, closed triangles; smaller than 90 μm, open triangles. The control curve shows that as blood pressure is lowered by haemorrhage, the pial vessels dilate. When blood is replaced and blood pressure rises, there is a further dilatation which is probably passive; the vessels then return to normal size. The remaining curves show the vascular response to the same procedure following intravenous injection of atropine, (0.2 mg kg^{-1}), '7351', (diethylaminopropyl diphenyl-α-isopropoxyacetate; 2–3 mg kg^{-1}) or 'Amizylum' (Benzactizine; 2–3 mg kg^{-1}) in isotonic saline. (From Mchedlishvili & Nikolaishvili (1970), *Plügers Arch. ges. Physiol.* **315**, 27–37, Fig. 1.)

Florey (1925) stimulated the 'vasomotor' centre faradically and observed no change in the calibre of pial vessels. Spina (1898, 1900) stimulated the medulla and observed dilatation of pial vessels, an effect which was unaffected by section of the hind-brain. These and other studies at this time gave rise to the concept that a reciprocal relation existed between cerebral blood flow and activity in the 'vasomotor' centres (in much the same way as Schmidt (1928a) had proposed a relation between blood flow and the

Other aspects of the neural reflex arcs

'respiratory' centres). If the vasomotor centre became 'anaemic' or 'ischaemic', then it initiated a sequence of cardiovascular changes which included a redistribution of cardiac output in favour of the head. In general, it was thought that the cerebral vessels and the nerves supplying them took little part in this response (Heymans & Bouckhaert, 1932; Gesell & Bronk, 1926).

The matter was first studied systematically by Stavraky (1936) and two features, at least, make this study an important one. First, he took advantage of the immense amount of careful work on the anatomy of the diencephalon in the preceding fifty years to define as precisely as possible the areas which he stimulated. Secondly, Stavraky took considerable care to dissociate the changes in pial artery calibre as a result of stimulation from those which occurred in response to changes in systemic pressure etc. He found that stimulation of the posterior region of the hypothalamus caused in addition to other features of sympathetic stimulation, e.g. pupillary dilatation, piloerection, hypertension, a bilateral constriction of pial vessels. This constriction occurred rapidly and was complete some seconds before there were any changes in blood pressure. He further showed that the vasoconstriction was not dependent upon changes in blood gas tensions associated with respiration; that it could be reduced or abolished by cervical sympathectomy and that any residual constriction of pial vessels was abolished by transection through the pons.

Stimulation of the ventral part of the hypothalamus caused bilateral dilatation of the pial vessels together with a fall in blood pressure and slowing of the heart. Again, this effect could be temporally distinguished from the systemic changes in pressure. Stimulation of the thalamus caused dilatation of the pial vessels but variable changes in arterial pressure. Stavraky also showed that stimulation of the central end of the sciatic nerve could cause either constriction or dilatation of pial vessels but he did not analyse this response further. Nevertheless, this observation together with that in which he showed that cervical sympathectomy did not always abolish the vascular response to hypothalamic stimulation suggests that the sympathetic fibres to cerebral vessels may be included in pathways other than the cervical sympathetic nerve and the question of precisely which vasomotor nerves are involved when peripheral nerves are stimulated requires further study.

10. The neural control of cerebral blood vessels

The question of mid-brain control of cerebral vessels has been taken a stage further by Ingvar & Soderberg (1956, 1958). They showed that regional cerebral blood flow was increased when the reticular formation was stimulated. They also showed that this type of stimulation caused desynchronization of the EEG pattern. Assuming that this pattern could be used as an index of local cerebral metabolic rate, Ingvar & Soderberg (1958) postulated that the primary response to brain stimulation was a rise in metabolic rate and that the blood flow changes were secondary to this. There are obvious difficulties in this type of interpretation. In the first place, as is discussed in more detail in Chapter 11, the relation between changes in EEG pattern and local metabolic rate in the brain is very far from being established and hypotheses based upon such a relation are unsupported by adequate evidence. Secondly, the speed of response of pial vessels to hypothalamic or reticular formation stimulation makes it unlikely that an intermediate chemical reaction is involved. This is especially true if the mechanism of the vascular response to changes in metabolism proposed by Wahl *et al.* (1970) is evoked, for it is noticeable that in these experiments, the change in calibre of pial vessels in response to alterations in pH of the perivascular fluid occurred in not less than 20–30 seconds.

An alternative but less direct method of approaching this problem was adopted by Cross & Silver (1962) who measured cerebral tissue P_{O_2} with a microelectrode and found that changes in tissue P_{O_2} could be used as an index of brain capillary flow. When these workers stimulated the hypothalamus, they obtained clear evidence of a widespread sympathetico-adrenal response, a rise in brain tissue P_{O_2} and, by contrast, a fall in tissue P_{O_2} in the testis. The changes in brain tissue P_{O_2} consisted of an early fall followed by a later rise and the time course of this response is similar to that seen when adrenaline was administered, Fig. 10.5. This type of response is difficult to interpret partly because there may have been changes in tissue oxygen consumption which would affect tissue P_{O_2} and partly because no details of the accompanying blood pressure changes were given. On the other hand, direct stimulation of the cervical sympathetic nerve gave rise to a fall in tissue P_{O_2} which, if other factors remained constant could have indicated a fall in capillary blood flow.

The relation between stimulation of various brain centres and

Other aspects of the neural reflex arcs

cerebral blood flow has also been studied by Molnar & Szanto (1964). They measured local changes in blood flow using implanted thermistors and also recorded EEG changes. Unfortunately no stereotaxic co-ordinates for their stimulating positions are given. Stimulation of the area in the hypothalamus which gave rise to an increase in systemic pressure and desynchronization of the electrical pattern also caused a fall in cortical blood flow. In a number of experiments, these workers were able to distinguish the changes in blood flow from those in blood pressure. However

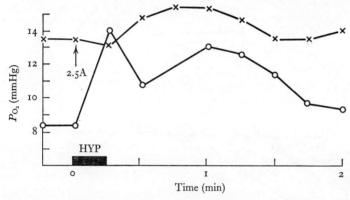

Fig. 10.5. A comparison of the effects of electrical stimulation of the perifornical area of the hypothalamus (HYP) and the intravenous injection of 2.5 μg adrenaline (2.5A) on the indicated level of tissue oxygen tension in the contralateral side of the hypothalamus. Note the biphasic response to stimulation. (From Cross and Silver (1962), *Proc. Roy. Soc.* (B) **156**, 483–99, Fig. 5.)

they noted, in contrast to Stavraky (1936), that the blood flow responses were unaffected by cervical sympathectomy. Stimulation of the hypothalamic area which gave rise to a fall in systemic pressure caused variable changes in cortical and sub-cortical blood flow as did stimulation of other bulbar areas in which virtually complete dissociation of changes in cortical blood flow and blood pressure were observed.

The possibility that selected areas in the brain stem are involved in the reflex regulation of cerebral blood flow has also been tested by Shalit, Reinmuth, Shimojyo & Scheinberg (1967b). They showed that lesions in certain defined areas of the brain stem (Fig. 10.6) caused a large reduction in the vascular response to

10. *The neural control of cerebral blood vessels*

carbon dioxide, Fig. 10.7, compared with the control response or with that seen when lesions were made in other parts of the brain stem or brain. Although these experiments can be criticized on the grounds that cerebral blood flow was not actually measured during the course of the experiments but derived from the reciprocal of the A–V oxygen difference it being assumed that

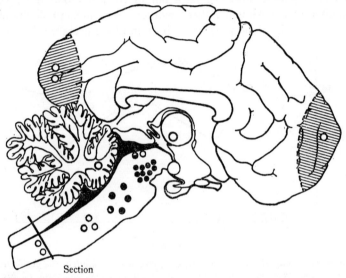

Fig. 10.6. Distribution and effectiveness of lesions on the cerebral vascular response to changes in P_{a,CO_2}. The dots indicate the approximate position of the lesions: solid circles, lesions which effectively reduced the vascular response to carbon dioxide; open circles, lesions which were ineffective; semi-solid circles, lesions which were only partially effective; hatched areas, sites of lobectomies. (From Shalit, Rienmuth, Shimojyo & Scheinberg (1967a), *Archs Neurol. Psychiat., Chicago*, **17**, 337–41, Fig. 1. Copyright 1967, American Medical Association.)

cerebral (or more accurately, cortical) oxygen consumption had not changed, such criticism cannot by itself invalidate the results since it applies to the responses to all the brain stem lesions, effective or otherwise. Similarly, it could be objected that the reduced vascular response to carbon dioxide was an artifact depending on initial surgical manipulation rather than to the brain stem lesion: this objection, again, could not explain why some lesions were effective and others only a few millimetres away were not so. Further, these workers showed that if the brain stem lesion was caused by local cooling, the normal vascular response to carbon dioxide was restored on rewarming the tissue. For these reasons, the experimental results may be interpreted as showing that part of the cerebral vascular response to carbon

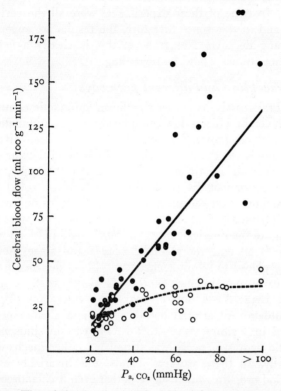

Fig. 10.7. The relation between cerebral blood flow and P_{a,CO_2}. Solid circles, control measurements from thirty-five dogs; open circles, measurements from fourteen dogs with effective lesions (as shown in Fig. 10.6). (From Shalit, Rienmuth, Shimojyo & Scheinberg (1967a), *Archs Neurol. Psychiat.*, Chicago, **17**, 337–41, Fig. 3. Copyright 1967, American Medical Association.)

dioxide is probably reflex in origin and that the pathway involved can be interrupted by localized brain stem lesions.

The evidence from the experiments discussed in this section provides some support for the proposition that neural pathways possibly involving a variety of neurones in the brain stem are involved in the regulation of cerebral blood flow. Technically,

10. *The neural control of cerebral blood vessels*

these are difficult experiments to carry out since apart from the accurate and reproducible production of lesions, great care is required in controlling such variables as respiration, blood gas tensions and systemic pressure which may alter with brain stem stimulation or lesions and themselves cause changes in cerebral blood flow. In none of these experiments were all the variables controlled and in this sense therefore, the results are suggestive without being decisive. Nevertheless, this is clearly a field in which further studies would be rewarding.

Possible receptors and afferent pathways

If the cerebral vessels are under significant autonomic control, it is likely that cerebral blood flow could be affected by stimulation of any peripheral receptors. Examples have already been given to show that this is so; the changes in cerebral blood flow or some index of it when the sciatic nerve was stimulated (Stavraky, 1936) or when the ear or foot was pinched (Cross & Silver, 1962).

The most important receptors which are likely to be involved are the peripheral arterial baroreceptors and chemoreceptors since they respond to the physical and chemical changes in blood which are known to affect cerebral blood vessels and since they are strategically placed to monitor changes in blood destined for the brain. In early experiments, (Forbes & Wolff, 1928; Cobb & Finesinger, 1932; Ask-Upmark, 1935), the vasomotor response to the stimulation of arterial baroreceptors and chemoreceptors was studied in a simple way – that is, the carotid sinus nerves or the aortic depressor nerve was stimulated either mechanically or faradically. In these workers' hands, this invariably caused a fall in blood pressure which suggests that even if chemoreceptor afferent nerves were being stimulated, the dominant effect was from baroreceptor stimulation. In the majority of these experiments, this manoeuvre caused pial vasodilatation although, in a number of Ask-Upmark's experiments, pial vasoconstriction was observed: no further analysis of this difference in response was carried out. In 1937, two papers appeared by Forbes, Nason & Wortman and by Fog in which it was shown that if systemic pressure was reduced, for example by stimulating the central cut end of the vagus nerves, the pial arteries dilated: and that this dilatation could be abolished or reduced if the nerve was stimulated but systemic pressure was held constant by infusing Ringer's solution into the mesenteric artery. This suggested that the pial artery dilatation was a consequence of the fall in systemic pressure and not of the vagal stimulation. Similar observations were made with respect to the carotid sinus and depressor nerve stimulations. It is also unlikely that the changes in pial vessel diameter observed by these workers was reflex since, in response to changes in systemic pressure, the calibre of the vessels changes after a comparatively long latent period following the onset of stimulation – 20–30 seconds. This type of response contrasts with the rapid response observed by the same group of workers (Forbes, Nason, Cobb & Wortman, 1937) when they stimulated the geniculate ganglion of the VIIth cranial nerve. The results of this group of experiments yield little information about the possible role of peripheral arterial chemoreceptors and it would be easy to conclude that the baroreceptors had little direct effect upon cerebral blood vessels. However, before this can be substantiated, it would be necessary to repeat these experiments preferably by stimulating the baroreceptors physiologically i.e. by means of an isolated carotid sinus preparation and that the effects should also be repeated at different levels of mean arterial pressure. Further, there were some indications from the experiments of Forbes, Nason & Wortman (1937) that the relation between changes in pial vessels and blood flow was not a simple one. They noted that at the same time as the pial vessels dilated in response to a fall in systemic pressure, the velocity of blood flow was reduced. This might not be unexpected in view of the reduced perfusion pressure and a greater cross-sectional area of the vascular bed. This dissociation between blood flow and pial vessel diameter may explain, for example why it was that Schmidt (1936) observed that cortical blood flow (measured with a ströhmur) fell when the vagus was stimulated and, in general, interpretation of changes in blood flow when extrapolated from changes in pial vessel diameter may sometimes be misleading.

Further studies to determine the role of baroreceptors in the control of cerebral blood flow were carried out by Rein (1931), Heymans & Bouckhaert, (1932) and, more recently, by Yoshida, Meyer, Sakamoto & Handa (1966). These three studies had in common the use of carotid arterial blood flow as an index of cerebral blood flow. The difficulty about this is that blood destined for the brain constitutes only a proportion of carotid arterial

10. *The neural control of cerebral blood vessels*

flow – 45 per cent according to Ishikawa, Handa, Meyer & Huber (1965) and further, there is no guarantee that this proportion is constant under all conditions. It is not therefore surprising that the results of these experiments varied considerably. Rein found that as carotid sinus pressure was reduced, cerebral vessels dilated. Heymans & Bouckhaert found that section of the carotid sinus nerve made no difference to the volume of blood flow in the artery on the other side while Yoshida *et al.* found that the pressure–flow relation could be altered by section of the carotid sinus nerves but that the effect of this was negligible within the physiological range of pressure.

An alternative approach was adopted by Rapela, Green & Denison (1967) who measured the cerebral venous outflow in the dog. The method which they used has been discussed in detail in Chapter 5; here, it may be remarked that this method involves considerable preliminary surgery to close the numerous anastomoses between intra- and extracerebral circulations while the values for blood flow obtained are so low compared to other methods that they raise questions as to the sensitivity of the cerebral vessels and the actual volume of brain being drained by the jugular veins. These workers altered vascular conductance by raising and lowering carotid artery pressure and found that although denervation of the carotid sinus baroreceptors caused a rise in cerebral vascular conductance and stimulation of the carotid sinus nerves reduced it, denervation of the baroreceptors did not affect the normal pressure–flow relation in which flow is independent of pressure. These results are difficult to interpret because the aortic group of baroreceptors was intact throughout and presumably responded to the changes in systemic pressure which occurred with changes in carotid sinus pressure. Furthermore, the results are considered by the authors only in terms of the baroreceptors while the possible consequences of altering carotid artery pressure upon the carotid body chemoreceptors are ignored (Floyd & Neil, 1952). As with many other of the problems discussed in the present chapter, decisive answers are unlikely to be obtained until experiments in which precise physiological stimuli are applied, direct methods of measuring cerebral blood flow are used and variations, notably of systemic arterial pressure, are controlled, are carried out.

The possible physiological role of vasomotor nerves to cerebral blood vessels

It is worth emphasizing that the bulk of evidence considered so far in this chapter has been derived from experiments aimed at establishing whether or not the nerves to cerebral blood vessels contribute to the resting tone of vascular smooth muscle within the physiological ranges of gas tensions and arterial pressure. Although there is not yet unanimity on this point, this evidence suggests that under the usual experimental conditions, that is – general anaesthesia and mechanical ventilation – there is evidence of tonic vasoconstrictor activity and rather less evidence of tonic vasodilator activity. More consistent is the evidence where the vasomotor nerves have been sectioned or their action blocked and there is also some evidence which suggests that these latter responses are dependent upon the prevailing levels of blood gas tensions. What remains unclear is how important this neural control in quantitative terms is, compared to the other factors which include the intrinsic properties of the vascular smooth muscle and chemical factors in plasma or e.c.f. Some recent evidence (James *et al.* 1969) which requires to be confirmed indicates that the neural contribution becomes progressively more important as blood gas tensions or arterial pressure depart from the physiological range.

Although it is comparatively simple to describe the changes in blood flow or pial vessel diameter which occur when the vasomotor nerves are stimulated or sectioned, it is very much more difficult in the cerebral as in other peripheral vascular beds to assign a definite physiological role to the vasomotor nerves. On the evidence which is available, they could be important in at least three ways.

First, vasomotor activity could interact with and act as a fine control for other factors which are known to affect vascular smooth muscle e.g. carbon dioxide, oxygen and hydrogen ions in plasma or e.c.f. Thus from the experiments of James *et al.* (1969) it would appear that in response to carbon dioxide and following section of the sympathetic nerves, the action of vasodilator nerves is 'unopposed' and the response is enhanced. Conversely, if the dilator pathway is interrupted by section of the facial or aortic nerves, vasoconstrictor activity is unopposed and the vascular

10. *The neural control of cerebral blood vessels*

response is reduced. It is probable that in the intact animal, the direct response of vascular smooth muscle to a rise in P_{a,CO_2} is modified and regulated by activity mediated by *both* vasoconstrictor and vasodilator pathways. The participation of vasomotor nerves appears to be of particular importance with changes in systemic pressure since the normal relation between blood flow and pressure, and the maintenance of a relatively high cerebral blood flow as pressure is reduced appear to depend considerably upon intact vasomotor pathways. A further circumstance in which an intact vasodilator pathway is important is the moment of birth; certainly the remarkable changes in both cerebral blood flow and in oxygen uptake cannot be explained in terms of the changes in blood gas tensions which occur at this time.

A second way in which vasomotor nerves could be involved in cerebral vascular responses is in determining the rate at which these responses occur. There is now a substantial body of evidence which has been discussed previously in this chapter and which shows that the intrinsic response of vascular smooth muscle to such stimuli as a change in systemic arterial pressure (Fog, 1937) or in the pH of perivascular fluid (Wahl *et al.* 1970) is relatively slow being measured in tens of seconds. On the other hand, there is evidence which is admittedly scanty but which indicates that in the intact animal, cerebral blood flow starts to change in response to a step change of carbon dioxide with a latent period which is approximately that of the lung-to-brain circulation time (Severinghaus & Lassen, 1967). The possibility therefore arises, and one which could be tested relatively easily by direct experiment, that the response of cerebral vascular smooth muscle consists of two principal components: a fast component which is reflexly determined and which may be initiated by stimulation of peripheral arterial baroreceptors and chemoreceptors, and a slow component which represents either the intrinsic response of vascular smooth muscle or its response to an altered chemical environment.

The third and possibly most important function of vasomotor activity of cerebral blood vessels is to integrate the response of the cerebral vascular bed with those of other peripheral beds. Early writers such as Gesell & Bronk (1926) suggested that the cerebral vascular bed was not actively involved in the redistribution of cardiac output which occurred in response to asphyxia. The cerebral vessels, they suggested, merely dilated and the autonomic

The possible physiological role of vasomotor nerves

nervous system ensured that sufficient blood was diverted from other vascular beds to fill them. Such a passive role for cerebral vessels may in fact turn out to be true but it would not be consonant with more recent ideas concerning the mode of cardiovascular control. These would suggest that although the responses of individual vascular beds are exceedingly complex, representing the interaction of numerous intrinsic and extrinsic factors, there is an overall pattern of response dictated by the autonomic nervous system, for example to asphyxia, in which all vascular beds participate. Vivid examples of this pattern are to be seen as part of the cardiovascular responses of diving mammals and birds in which the normal responses of individual beds to asphyxia, e.g. in skeletal muscle, are over-ridden by intense sympathetic vasoconstrictor activity in order to ensure that oxygen uptake in these 'expendible' beds is cut to a minimum during a prolonged dive. This is the kind of situation in which it might be possible to test the general hypothesis of an integrated neural regulation of cardiovascular responses and see whether the increased blood flow and vasodilatation which presumably occur in the cerebral vascular bed of the seal or duck during a dive is materially altered following section or blockade of the appropriate normal pathways.

11 CEREBRAL BLOOD FLOW AND METABOLISM

Introduction
Historical

Roy & Sherrington (1890) are generally credited with being the first to propose that blood flow in the brain was regulated to meet the requirements of metabolism. It is of some interest to recall how they arrived at this idea. They used the technique described in Chapter 5 whereby changes in the volume of the brain were used as an index of blood flow and observed that brain volume increased markedly following the intravenous injection of various acids. This increase could not be accounted for by changes in systemic arterial or venous pressure. They recalled that Lagendorf & Seelig (1886) had shown that the reaction of brain rendered ischaemic became acid, and they therefore prepared a filtrate from emulsified brain tissue from a dog which had been bled to death, and on injecting the filtrate intravenously, observed a similar rise in brain volume. On the assumption that this represented an increase in blood flow to the brain, they concluded that cerebral vessels responded not only to changes in arterial pH but also to acidification of 'lymph' surrounding them as a result of altered cellular metabolic activity.

It is remarkable that Roy & Sherrington observed a change in brain volume since numerous workers since, including Bayliss & Hill (1895) who repeated their experiments exactly, have been unable to show that a change in arterial pH affects cerebral vessels or blood flow. Furthermore, it might be supposed that following the intravenous injection of an acid, there would be a brisk increase in ventilation and a fall in P_{a,CO_2} which would either reduce cerebral blood flow or offset an increase due to any other cause. It would therefore seem that Roy & Sherrington's conclusions, however perceptive, were based upon experimental results which have been unreproducible.

Introduction

As has been shown in Chapter 8, there appear to be good reasons why Roy & Sherrington's experiments should not work in the way they observed for there is an important barrier for H^+ between plasma and extracellular fluid and probably between plasma and the intracellular fluid of vascular smooth muscle. At this stage, therefore, we can only conclude that the filtrate from dog's brain contained some vasoactive substance other than H^+. On the other hand, it has now been clearly demonstrated that pial vessels respond to changes in pH of the perivascular fluid (Wahl *et al.* 1970) and this constitutes at least one mechanism whereby cerebral blood vessels and blood flow could be regulated by local changes in metabolism. It is worthwhile emphasizing, however, that other mechanisms, e.g. that proposed by Guyton *et al.* (1964) where blood flow is regulated by oxygen demand, should be entertained and actively explored.

Further studies in this field had to await the revival of the cranial window technique (Forbes, 1928) and the introduction of the heated thermocouple (Gibbs, 1933*b*) which with the qualifications discussed in Chapter 4, can give a qualitative measure of local changes in blood flow. The evidence which had accumulated by the end of the 1930s and which suggested that a relation existed between local changes in metabolism and blood flow was threefold. First there were isolated experiments which showed that blood flow altered locally in response to a presumed change in functional activity. An example of this is shown in Fig. 11.1 and shows the change in temperature in the lateral geniculate ganglion associated with presumed increase in activity associated with illumination of the eye (Gerard, 1938). Similar results were obtained in the olfactory areas when ammonia vapour was blown into the nostrils and in the sensory cortex when the animal's paw was pinched or massaged. A similar response was elicited by Schmidt & Hendrix (1938) from the occipital cortex when the eye was illuminated.

A second approach used was to measure the calibre of pial vessels when analeptics such as picrotoxin or metrazole, which were known to excite the cerebral cortex, were injected intravenously. The results reported by a number of workers will be considered later in this chapter: here it may be said that the responses were not clear cut. In some experiments the pial artery diameter increased, in others it decreased and in others it gave

11. *Cerebral blood flow and metabolism*

a biphasic response. Further, it was not possible to exclude a direct action of these drugs on pial vessels when applied locally. In addition, marked changes in systemic arterial pressure accompanied the induced convulsions and although they were not measured, it is likely that there were also considerable changes in blood gas tensions and pH. For all these reasons, it is difficult from these experiments, to show that any clear change in blood flow or blood vessel response accompanied the presumed local changes in cortical activity.

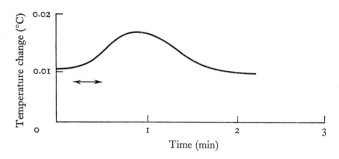

Fig. 11.1. The change in temperature in the lateral geniculate body of a cat following illumination of the eyes for the period indicated by the arrows. (From Gerard (1938), *Ass. Res. Nerv. Ment. Dis.* **18**, 316–45, Fig. 132.)

A third approach has been to correlate the vascularity of various areas of brain in terms of capillary density with known functional activity of that part. This topic has been considered in detail in Chapter 2 where it was shown that such a correlation existed. However, such a correlation is purely qualitative and yields no information as to the mechanisms involved.

General aspects of the relation between blood flow and tissue metabolism

It may be that the relation between blood flow to the brain and the level of metabolism is a special one. In other organs, e.g. skeletal muscle or exocrine glands, the level of metabolism varies very considerably and the problem of physiological interest is how blood flow is so precisely related to these variations. In the brain, however, all the measurements of metabolism so far made indicate a remarkably steady level under physiological conditions except that during sleep, there are striking changes. Moreover,

Introduction

the brain is, compared to other organs, remarkably dependent for its functioning on the constancy of blood flow since its oxygen consumption is high, its oxygen reserves are small and it has only a limited ability to oxidize substrates other than glucose. The problem of physiological interest here is therefore how blood flow is regulated to constancy and which ensures adequate flow for the metabolic requirements.

But the relation of blood flow to tissue metabolism is a problem of general physiological interest. For example, it has generally been assumed that blood flow is adjusted to metabolic requirements and this assumption follows from the experimental results of Pflüger (1872, 1875) who showed in isolated organs that the oxygen consumption of tissues was independent of its blood supply over a wide range of blood flow and oxygen saturation. Barcroft (1914) summed the matter up as follows: 'The issue before Pflüger may be stated in a few words. Is the quantity of oxygen taken up by the cell conditioned primarily by the cell, or by the supply of oxygen? The answer was clear: the cell takes what it needs and leaves the rest.'

More recent experimental results have questioned this generalization. For example, there are obvious physiological situations in which tissue metabolism is affected by blood flow, e.g. in limb muscles during a dive (Andersen, 1966) or in the kidney during muscular exercise (Radigan & Robinson, 1949). But there is now increasing evidence that in less extreme situations, the level of tissue metabolism may be considerably affected by blood flow. Thus although in the denervated limb or organ, oxygen consumption remains relatively constant despite changes in blood flow or pressure or oxygen saturation (as Pflüger showed), when the nerves are intact, the situation is quite different. It has been shown in the kidney (Dole, Emerson, Philips, Hamilton & Van Slyke, 1946) and in the carotid body (Purves, 1970 a, b) that as blood pressure is reduced, oxygen consumption is reduced and in the latter studies, oxygen consumption was also shown to be dependent upon P_{a,O_2} and P_{a,CO_2}. It is most probable that in all these situations, the uptake of oxygen by the tissues is a function of blood flow or its distribution through the organ. Thus, Pappenheimer (1941 a, b) showed that in the denervated and vascularly isolated hind limb, as blood flow was reduced (by reducing perfusion pressure) oxygen consumption was only slightly

11. Cerebral blood flow and metabolism

affected. But if blood flow was reduced by comparable amounts when the sympathetic nerves were stimulated, oxygen consumption was reduced by 70–80 per cent of control. Pappenheimer (1941 a) gives a number of reasons why the action of the sympathetic was less likely to be a direct one, i.e. upon the cells themselves, and more likely to be due to redistribution of blood flow such that blood was diverted from areas of high to low metabolism.

These somewhat fragmentary pieces of evidence are presented to suggest that in some organs and possibly in most, blood flow does not passively follow changes in metabolism but may be an important factor in regulating the level of tissue metabolism. This question has not been specifically studied in the brain except in a very small series (three animals) where cerebral metabolism was measured before and during stimulation of the cervical sympathetic nerve (James, Millar & Purves, 1969). Although cerebral blood flow was reduced by up to 50 per cent of control levels, the A–V oxygen difference varied inversely with blood flow. These studies could well be repeated and extended.

Aspects of brain metabolism

As for other organs, the metabolic rate of the brain may be derived from the product of blood flow and the arteriovenous difference in concentration of the substance to be studied. For oxygen, the arteriovenous difference is measured in ml O_2 100 ml of blood^{-1} and the metabolic rate expressed as ml O_2 100 g of brain^{-1} min^{-1}. (In order to compare the oxygen with other metabolites, it is more suitably expressed in molar units: thus

$$n \text{ ml } O_2 \text{ 100 g}^{-1} \text{ min}^{-1} = n.22.4 \text{ m-mol}^{-1} \text{ 100 g}^{-1} \text{ min}^{-1}.)$$

The A–V oxygen difference for oxygen is almost identical to that for carbon dioxide and the respiratory quotient (RQ) has been variously calculated in man as 0.95 (Lennox, 1931 a) or unity (for references, see McIlwain, 1966). This indicates that glucose is most likely to be the main substance oxidized. The arteriovenous difference for glucose in man is 9.8 mg 100 ml^{-1} and on the basis of a blood flow of 50 ml 100 g^{-1} min^{-1}, glucose utilization is at a rate of 30 μmol 100 g^{-1} min^{-1}. For complete oxidation this would require 180 μmol O_2 100 g^{-1} min^{-1}. From the most reliable data for young man (Kety & Schmidt, 1948a), cerebral blood flow was found to be 50 ml 100 g^{-1} min^{-1} and A–V oxygen difference

Introduction

6.6 ml 100 ml^{-1}. Oxygen utilization is therefore 147 μmol 100 g^{-1} min^{-1} and thus complete oxidation accounts for 147/180 or approximately 85 per cent of glucose removed by the brain. This implies that if all the carbon dioxide is derived from glucose, other substances requiring less oxygen for their formation must be produced: lactic and pyruvic acid are most likely and an example is given by Himwich & Himwich (1946) of the way in which carbon dioxide, lactic and pyruvic acids account for the glucose utilized. The A–V glucose difference was 10.2 mg 100 ml^{-1}, the carbon dioxide was found to be equivalent to 8.9 mg glucose 100 ml^{-1}; the lactic acid to 1.2 and pyruvic acid to 0.2, a total of 10.3 mg 100 ml^{-1}. This equivalence has been confirmed by numerous subsequent studies (Sokoloff, 1959) and the fact that lactic pyruvic acid rates can vary has been made the basis for a useful index of inadequate oxygen supply (Siesjö, Kaasik, Nilsson & Ponten (1968)).

It has long been held that glucose is the principal substrate of brain and that the ability of brain to use other substances is limited. This is shown in Table 11.1 in which the response of brain to hypoglycaemia induced by insulin is compared with that of muscle. In both tissues, there is a fall in glucose utilization but whereas oxygen uptake in muscle is virtually unimpaired, that in brain is severely reduced. This apparent vulnerability of the brain emphasizes the importance of maintaining an adequate level of blood flow and substrate supply in the short term.

In the long term however, as with glucose deprivation in pro-

Table 11.1. *Hypoglycaemia on oxidations in brain and muscle* (from Himwich & Fazekas, 1937. *Endocrinology* **21**, 800–7)

Measurement	Arterial blood glucose (mg 100 ml^{-1})	Arteriovenous difference	
		glucose (mg 100 ml^{-1})	oxygen (ml O$_2$ 100 ml^{-1})
Brain:			
Before experiment	90	13.1	9.3
Moderate hypoglycaemia	30	12.5	8.0
Intense hypoglycaemia	12	3.0	3.8
Muscle:			
Before experiment	90	7.6	6.9
Hypoglycaemia	20	1.7	6.0

11. Cerebral blood flow and metabolism

longed fasting, there is now abundant evidence that the brain utilizes other substances, notably keto-acids (for references, see Cahill, 1970). This adaptive property of the brain besides being of evolutionary interest suggests that the question of whether or how far glucose is the principal substrate in brain requires further scrutiny.

Measurement of cerebral metabolic rate *in vivo*
The whole brain

The measurements necessary for calculating cerebral metabolic rate are not technically difficult to make. Cerebral blood flow is measured using the nitrous oxide technique or one of its modifications. In some series, regional cerebral blood flow has been measured and mean cerebral blood flow derived (Obrist, Thompson, King & Wang, 1967). The concentrations of substrate, e.g. oxygen or glucose, are measured from representative samples of blood from artery and vein at the same time as measurements of flow are made and cerebral metabolic rate calculated according to the Fick equation.

Difficulties arise principally with venous sampling since there is the question of whether (a) the venous sample is contaminated with blood from extracerebral sources and (b) the blood is truly representative of the area of brain whose metabolism and whose blood flow is being measured. This particularly is the case when mean cerebral blood flow or metabolism is derived from measurements of regional blood flow. Some of the difficulties of interpreting venous samples have already been considered in Chapter 5 since they also apply to methods of measuring cerebral blood flow, including the nitrous oxide technique in which venous samples are taken. There it was shown that in the common laboratory animals, e.g. dog, cat and rabbit, it was virtually impossible to ensure that venous blood in jugular veins or the major sinuses was draining brain only and it is probable that this accounts for the very high estimates of cerebral oxygen consumption obtained in early experiments in this field, Table 11.2. On the other hand, if every conceivable anastomosis between intra- and extracranial circulations is tied or blocked, there is a real danger that the surgery and anaesthesia involved will have irreparably damaged the cerebral vessels and the rate of cerebral oxygen consumption affected (Sagawa & Guyton, 1961).

Measurement of cerebral metabolic rate in vivo

Table 11.2. *Values for cerebral oxygen consumption obtained in early experiments*

Experimental animal	Cerebral oxygen consumption ml O_2 100 g^{-1} min^{-1}	Reference
Dog	13.0	Alexander & Cserna (1913)
Dog (perfusion)	10.9	Handley et al. (1943)
Dog	10.0	Gayda (1914)
Dog	10.0	Rein (1941)
Dog (perfusion)	7.5	Schmidt (1928b)
Dog	5.8–7.8	Hou (1926)
Rabbit	9.4	Yamakita (1922)
Cat (perfusion)	3.3–5.0	Chute & Smyth (1939)

In spite of these difficulties, apparently satisfactory results were obtained by Geiger & Magnes (1947) in the cat in experiments in which the head was vascularly isolated and artificially perfused. They obtained venous blood from the transverse sinuses and calculated values for cerebral metabolism were > 4.0 ml 100 g^{-1} min^{-1} in those preparations in which there was evidence of cerebral activity, i.e. pupillary light reflexes and normal levels of respiration and blood pressure. Even in these experiments with the high degree of skill involved, it is not certain that the absolute values for oxygen consumption are overestimates because of contamination. However, it was possible for these authors to show several important features of cerebral metabolism *in vivo*; that cerebral metabolism diminished with time during the experiment and with the addition of sodium amytal to the perfusing blood, that cerebral metabolism increased – in fact, approximately doubled – when metrazol caused convulsions and that there appeared to be a close dependence of the level of metabolism on the rate of blood perfusion.

In primates, particularly, in man, the situation is more favourable because of the virtual separation of the intra- and extracranial circulations. There is good evidence that all but a very small proportion of jugular venous blood drains the cerebral hemispheres, this proportion being usually 3 per cent, maximally 7 per cent (Shenkin, Harmel & Kety, 1948): but in a proportion of cases in man, e.g. five of sixty-seven tests, gross contamination with

11. Cerebral blood flow and metabolism

extracerebral blood was found (Lassen & Lane, 1961). On the other hand, a number of workers have found definite differences in the oxygen content of internal jugular venous blood on both sides in man (Gibbs, Lennox & Gibbs, 1934; Ferris, Engels, Stevens & Logan, 1946) amounting to 1 ml 100 ml^{-1} and under certain circumstances, e.g. hyperventilation, this difference increases considerably. These observations suggest that apart from the question of contamination from extracerebral sources, the internal jugular vein drains different parts of the brain: samples taken arbitrarily from one jugular vein or the other may therefore be unrepresentative of mean cerebral venous blood. For example, Gibbs, Lennox & Gibbs (1934) found that the right internal jugular vein drained principally cortex and the left vein deeper structures. Because of the discrepancies in venous oxygen content, estimates of mean cerebral metabolic rate varied from 2.3 to 4.7 ml 100 g^{-1} min^{-1}.

These discrepancies may not be quantitatively very important and may do no more than account for the distribution of values obtained in any given series. However, it is clear that from time to time, subjects with gross degrees of contamination occur and, it must be presumed that there are intermediate degrees. It is often difficult during investigations to validate the origin of cerebral venous blood in every experimental animal or human subject: nevertheless, as the examples quoted above show, there can be marked variations in the origin of cerebral venous blood, and if cerebral metabolic rate is dependent on analysis of venous oxygen content *as well as* venous nitrous oxide for calculation of blood flow, the possibilities of error are considerable and should be borne in mind.

Measurement of regional metabolic rate

Early studies of brain metabolism *in vitro* have shown a large difference between the rates of oxygen uptake of cortex and white matter, e.g. Holmes (1930), who obtained values of 2.0 and 0.5 ml 100 g^{-1} min^{-1} respectively. These values, in keeping with others obtained *in vitro*, underestimate the rate of metabolism *in vivo*: nevertheless, the ratio between the rate of metabolism in the cortex and white matter has been confirmed by Homberger, Himwich, Etstein, York, Maresca & Himwich (1946). The derivation of cortical and white matter blood flow from the fast and slow components of the ^{133}Xe washout curve has meant that

Measurement of cerebral metabolic rate in vivo

values for cortical and white matter oxygen consumption can be calculated fairly easily if the A–V oxygen difference is also measured. But this form of calculation involves a number of assumptions concerning the relative weights of grey and white matter and how far the values have any meaning when they are extrapolated from volumes of brain which may be far from representative.

Attempts to measure the rate of oxygen uptake of even more localized parts of the brain using the Fick principle have been made by Gleichman, Ingvar, Lassen, Lubbers, Siesjö & Thews (1962) in the anaesthetized dog. They obtained values for circumscribed areas of the cerebral cortex which varied between 2.3 and 13.4 ml 100 g^{-1} min^{-1}, mean 7.0, S.D. ±2.3. This study illustrates most of the difficulties of the methods at present available and these are readily acknowledged by the authors. In the first place, there was wide variation in the values obtained for blood flow through the post-sigmoid gyrus, 19 to 106 ml 100 g^{-1} min^{-1}, mean 65, S.D. ±21.6 and these reflect not only the errors involved in measurement of blood flow but differences between animals with respect to blood gas tensions and levels of anaesthesia. In the unanaesthetized cat, for example, a value of 138 ml 100 g^{-1} min^{-1} was obtained for blood flow from the same area using the autoradiographic technique described by Kety, Landau, Freygang, Rowland & Sokoloff (1955). Secondly, the method of measuring regional blood flow employed – the desaturation method using ^{85}Kr (Lassen & Ingvar, 1961; Ingvar & Lassen, 1962) – measures average blood flow in the most superficial layers of cortical blood flow (depth about 1 mm) while the outermost layers influence the measurement more than the deeper ones.

Thirdly, there is the question already referred to above, as to how far blood sampled from the sagittal sinus is representative of that part of the cerebral cortex from which blood flow measurements were made. The authors (Gleichman, Ingvar, Lübbers, Siesjö & Thews, 1962) put their dilemma quite clearly. If they obtained samples of blood from a vein draining the circumscribed areas of cortex from which blood flow measurements were made, there would have been a real danger of damage to the vessels involved and alterations in the pattern of venous drainage. On the other hand, blood samples obtained from the sagittal sinus

11. Cerebral blood flow and metabolism

could include blood from (a) diplöic anastomoses and (b) sub-cortical tissue. Anastomotic blood flow was largely eliminated in these experiments by substituting polythene sheeting for the calvarium and although this leaves the possibility that other types of contamination, e.g. from ethmoidal veins occur, the evidence suggests that this is not important (Pfeiffer, 1928; Homberger *et al.* 1946; Himwich, 1951). At the moment, however, it is difficult to exclude the possibility that the sagittal sinus drains sub-cortical tissue and the proportion of this drainage may alter with anaesthesia or blood gas tension.

For all these reasons, absolute values for regional or local cerebral metabolism cannot be accepted as being valid at the moment and because of the inhomogeneity of rates of perfusion of the cortex (Wilkinson *et al.* 1969) it is impossible to generalize about the rate of 'cortical' metabolism from measurements made in one area of cortex. Probably the most that can be said is that cortical metabolic rate is substantially greater than that for white matter: but whether the difference is proportional to the difference in blood flow, i.e. three to fivefold, is a matter for conjecture.

Normal values for cerebral metabolism in man

For a variety of reasons, the most reliable and reproducible values for the cerebral metabolic rate for oxygen have been obtained in man. Thus it is possible to measure the metabolic rate in un-anaesthetized subjects under resting conditions or during sleep while the anatomical considerations which have been discussed in previous sections make it more certain than in other species that venous blood which is sampled drains the brain only. In addition, it has been possible to measure cerebral metabolism in certain important clinical circumstances, e.g. during various types of coma and mental deterioration which enlarge our understanding of the relation between cerebral activity and blood flow.

Most of the measurements in man were carried out in the years shortly after the introduction of the nitrous oxide technique by Kety & Schmidt (1948a) and the values which are listed in Table 11.3 were collected from this era. These values have been widely published and have been used as a reference with which other values are compared: but, clearly, they are only as reliable

Normal values for cerebral metabolism in man

Table 11.3. *Cerebral blood flow and respiratory rate in man*

Condition (adult subjects except as specified)	Cerebral blood flow (ml 100 g of tissue^{-1} min^{-1})	Cerebral respiratory rate (ml O_2 100 g of tissue^{-1} min^{-1})	References
Children, mean age 6.2 years	102	5.1	Kennedy (1956)
Normal resting subjects A	54	3.3	Kety & Schmidt (1948a)
Normal resting subjects B	58	3.2	Heyman et al. (1951)
Normal subjects C, supine	65	3.8	Scheinberg & Stead (1949)
Normal subjects C, erect	52	3.8	Scheinberg & Stead (1949)
Hyperventilation	34	3.7	Kety & Schmidt (1948b)
Breathing 5–7% CO	93	3.5	Kety & Schmidt (1948b)
Breathing 85–100% O_2	45	3.2	Kety & Schmidt (1948b)
Breathing 10% O_2	73	3.3	Kety & Schmidt (1948b)
Insulin hypoglycaemia* arterial glucose level 19 mg %	61	2.6	Kety, Polis et al. (1948)
Insulin coma* arterial glucose level 9 mg %	63	1.9	Kety, Woodford et al. (1948)
Irreversible insulin coma arterial glucose, 360 mg %	52	1.5	Fazekas et al. (1951)
Natural sleep	65	3.4	Sokoloff (1956)
Thiopentone anaesthesia, subjects D	52	1.9	Kety, Woodford et al. (1948) Wechsler et al. (1951)
Thiopentone; hyperventilation, subjects E	28	1.5	Pierce et al. (1962)
Schizophrenics	54	3.3	Kety, Woodford et al. (1948)
Uremic subjects	50	2.3	Heyman et al. (1951)
In diabetic acidosis	45	2.7	Kety, Polis et al. (1948)
In diabetic coma	65	1.7	Kety, Polis et al. (1948)
In myxedema	40	2.8	Scheinberg et al. (1950)
Extreme apprehensiveness	—	5.0	Sokoloff (1956)
Adrenaline perfusion	61	4.2	Sokoloff (1956)
In cerebral haemangioma	164	3.3	Schmidt (1950)

* Carried out with schizophrenic subjects.

as the values obtained for cerebral blood flow since cerebral metabolic rate was calculated from the Fick equation. Thus if ^{85}Kr is used as the indicator instead of nitrous oxide and cerebral blood flow calculated from the arterial and venous curves (as was

11. *Cerebral blood flow and metabolism*

discussed in Chapter 4) extrapolated to infinity rather than from the area at 10 min as was the case in many of the studies listed in Table 11.3, both cerebral blood flow and cerebral oxygen were approximately 15 per cent lower (Lassen, Fineberg & Lane, 1960). Representative examples obtained by this group of workers may be compared with those in Table 11.3. In normal young men, cerebral oxygen consumption was 2.975, S.D. $\pm$ 0.213; in normal old men, 2.716, S.D. $\pm$ 0.125 and in patients with various types of organic dementia, 2.281 ml 100 g^{-1} min^{-1}, S.D. $\pm$ 0.380.

The value of these figures for cerebral blood flow and oxygen consumption obtained under normal conditions lies first in the fact that they provide an estimate of the relative importance of the brain. Thus, the brain which is 2.5 per cent of the total weight of the body consumes between 20 and 25 per cent of the total oxygen taken up and receives approximately 15 per cent of the cardiac output. In conditions of asphyxia, particularly in diving mammals or birds, these proportions increase very considerably, while during muscular exercise, cerebral blood flow and oxygen consumption may fall to approximately 5 per cent of cardiac output and total body oxygen consumption respectively (Hedlund, Nylin & Regnström, 1962).

Changes with age

The general pattern which has emerged from species so far studied is that cerebral metabolism is perhaps half the adult level at birth, rises to a peak at or about the age of sexual maturity and then falls to adult levels. Thereafter cerebral metabolism falls to rather lower levels in old age.

Cerebral metabolism in the foetus and newborn. Cerebral metabolic rate *in vivo* has so far been measured only in the sheep foetus. In a preliminary study in which 'cephalic' oxygen consumption was measured from carotid artery blood flow and the A–V oxygen difference, Lucas, Kirschbaum & Assali (1966) obtained an average value of 0.26 ml O_2 min^{-1} kg of body weight^{-1}, or 1.5 to 2.0 ml 100 g^{-1} min^{-1} assuming 3.0 to 4.0 kg body weight for a mature sheep foetus. This can only give a very approximate estimate of cerebral blood flow since it is entirely a matter of speculation as to what proportion of carotid artery blood flow perfuses the brain.

Normal values for cerebral metabolism in man

A further attempt to measure cerebral oxygen consumption and blood flow in the foetus was made by Purves & James (1969). They used a modification of the ^{133}Xe desaturation method and obtained an average value of 2.00 ml O_2 100 g^{-1} min^{-1} in the mature foetus and 2.92 ml 100 g^{-1} min^{-1} in the newborn lamb under 48 hours of age. A similar increase was observed, approximately 50 per cent over 'birth' i.e. when the umbilical cord was clamped, and the lamb had been breathing regularly for half to one hour. Even greater increases in oxygen consumption were observed immediately after birth when the average rate was found to be 3.59 ml 100 g^{-1} min^{-1} but part of this could be accounted for by replenishment of oxygen reserves which had been depleted during the period between occlusion of the umbilical cord and the onset of respiration.

These values require to be treated with caution since (*a*) despite a number of control experiments, it was not possible to exclude some contamination with extracerebral venous blood, and (*b*) both a proportion of the ewes and all the newborn lambs had been anaesthetized with pentobarbitone sodium. The values obtained may therefore underestimate the true cerebral oxygen consumption. However, even if this is so, the changes in oxygen consumption and the accompanying changes in cerebral blood flow are of considerable interest.

In the first place, the low values in the foetus may reflect a low level of cerebral functional activity. In this connection it is of interest that there appears to be a difference between the foetus which has been exteriorized and the foetus *in utero*. In the former, numerous workers have found clear evidence of a generalized central nervous system inhibition, i.e. unresponsiveness to tactile or painful stimuli and reluctance to breathe until the umbilical cord is occluded: in the foetus, studied *in utero*, however, there is now evidence that respiratory-type movements are common and that these are associated with EEG tracings and ocular movements which resemble those seen in REM type of sleep (Merlet, Hoerter, Devilleneuve & Tchobroutsky, 1970; Dawes, Fox, Leduc, Liggins & Richards, 1970). *In utero* therefore, cerebral activity may be considerably higher and with it, the level of both cerebral oxygen consumption and blood flow. However, even if cerebral oxygen consumption is higher in the foetus *in utero*, the level is still substantially below that in the adult and this may be a factor of

11. Cerebral blood flow and metabolism

importance in enabling the foetus and newborn to withstand the effects of asphyxia very much more than can the adult (Boyle, 1670). The reason why cerebral oxygen consumption and blood flow increase at birth is not known. It would be tempting to relate this increase to the presumed increase in cerebral activity which occurs at the same time. It was observed that the increase in blood flow and oxygen consumption only occurred when the vagi were intact and it is therefore possible to argue either that vagal afferent activity is an important part of the process whereby the central nervous system is activated at birth; alternatively, the central nervous system could be activated by other means and activity in the vagus causes cerebral vasodilatation and an increase in blood flow commensurate with the increase in metabolic rate.

Some evidence of a different kind suggests that the oxygen consumption of the brain of the newborn in a number of species is low by adult standards. Thus in the rat, the formation of lactic acid in the brain approximately doubles between birth and three weeks (Himwich, Bernstein, Herrlich, Chesler & Fazekas, 1941; Himwich, Baker & Fazekas, 1939; Tyler & Van Harreveld, 1942), and numerous other studies show the same phenomenon in the dog (e.g. Himwich & Fazekas, 1941). This relation, between oxygen uptake and age is shown for the rat from data given by Tyler & Van Harreveld (1942) Fig. 11.2. From this figure and from Fig. 11.3, which show comparable values in the dog, it will be noted that the changes in oxygen consumption occur at different rates in different parts of the brain. Himwich (1951) pointed out, and it is worth emphasizing here, that these measurements can only be extrapolated to the situation *in vivo* if they are expressed in terms of wet weight of tissue for the percentage of dry weight increases with age (Donaldson & Hatai, 1931), being 12 per cent of moist weight at birth and 22 per cent in the mature rat. This may explain the finding of Reiner (1947) that the oxygen consumption in the rat, expressed as dry weight of tissue, developed the adult level within a month of birth and continued until the second year when it fell markedly.

These changes in the uptake of oxygen by the brain in the early years of life are also accompanied by changes both in the level of cerebral blood flow and its distribution. This has been studied particularly in white matter in puppies (Kennedy, Grave,

Normal values for cerebral metabolism in man

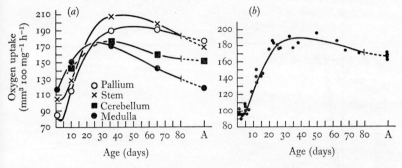

Fig. 11.2. The relation between oxygen consumption (mm^3 100 mg^{-1} h^{-1}) and the age of rats in days. 'A' is adult. In (a), the points are for identification of the curves representing parts of the brain as shown by the symbols. In (b), the oxygen consumption of whole brain is shown and each point represents an observation. (From Tyler & Van Harreveld (1942), *Am. J. Physiol.* **136**, 600–3, Fig. 1.)

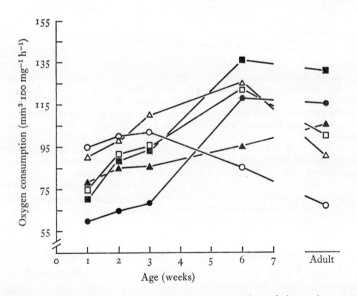

Fig. 11.3. The relation between oxygen consumption of the various parts of the dog brain at 1, 2, 3 and 6 weeks and in adulthood. In general, the cerebral metabolism increases during the first six weeks and then reverses slightly. Symbols: ● cortex; □ thalamus; ▲ cerebellum; ○ medulla; △ midbrain; ■ caudate nucleus. (From Himwich & Fazekas (1941), *Am. J. Physiol.* **132**, 454–9, Fig. 1.)

11. Cerebral blood flow and metabolism

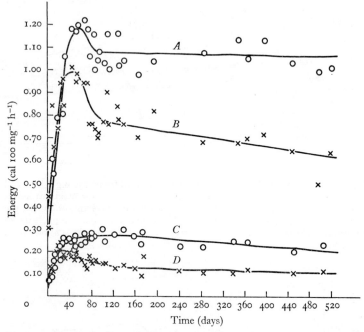

Fig. 11.4. In these curves, drawn by inspection, it may be seen that starting from a low level at birth the aerobic energy (A) rises to a maximum at about 6–8 weeks in the cortex, then falls rapidly until about the thirteenth week, after which the decrease is not significant. In the brain stem (B) the maximum is reached sometime between the fourth and seventh weeks. The rate then decreases sharply until the fifteenth week, and then more gradually to old age. In outline, though not in detail, the glycolysis curve matches the oxidative curve. In the cerebral cortex (C) glycolysis accelerates greatly until about the fifth week and then more slowly for about three months. After this time it appears to decline slightly. The rate of glycolysis in the brain stem (D) rises to a maximum at about three weeks and then decreases, precipitously at first, and then not so noticeably to old age. Each value presented in the graph is an average of a set of observations. (From Chesler & Himwich (1944), *Am. J. Physiol.* **141**, 513–17, Fig. 1.)

Jehle & Sokoloff, 1969). Using the labelled antipyrine method, the workers found that values for blood flow (all expressed as ml $100 \text{ g}^{-1} \text{ min}^{-1}$) in spinal cord and brain stem white matter were high, i.e. at birth 35 to 70 and at two weeks, 65 to 92, when flow was nearly as high as in the cortex. Cerebral white matter blood flow was lower 5 to 12 and reached its peak at seven weeks, 14 to 27. These authors also showed that the changes in blood flow could not be accounted for by changes in cell density:

it is possible that they are related to different rates of metabolism associated with myelinization but no data are available on this point.

A further feature of the changes in oxygen consumption with age is the alteration in ratio between energy supplied to the brain as a result of glycolysis and oxidative phosphorylation. This is shown for rat brain in Fig. 11.4, for cortex, curves A and C and for brain stem, exclusive of corpus striatium, curves B and D. The patterns of change with age are different and show that aerobic energy is low at birth and rises to a maximum in 6–8 weeks in the cortex, 4–7 weeks in the brain stem and then falls. The rate of glycolysis reaches a maximum in 3–5 weeks. If the aerobic energy is equivalent to the caloric requirement and the anaerobic energy that elaborated during hypoxia, then the deficit between curves A and C for the cortex and B and D for the brain stem represents the deficit which would arise with severe hypoxia. Clearly, this difference is smallest at birth; at 5 days the deficit for cortex is 0.29 cal 100 mg^{-1} h^{-1}, for brain stem 0.37. At 50 days the respective deficits are 0.92 and 0.81 cal 100 mg^{-1} h^{-1} and at 365 days 0.85 and 0.57 cal 100 mg^{-1} h^{-1}. These data have important physiological consequences for they imply that the infant accumulates its deficit at the slowest rate, while after seven weeks, this rate is greatest. This coincides with the period of maximum sensitivity to severe hypoxia as is shown for the rat in Fig. 11.5.

Cerebral metabolism in childhood. The first attempt to measure cerebral blood flow in the human child was by Baird & Garfunkel (1953). Children with a variety of cerebral diseases were studied and they were all sedated. The results were published in the following year (Garfunkel, Baird & Ziegler, 1954) and show a wide range for cerebral blood flow and metabolism. In general, there was a relation between both blood flow and metabolism and the severity of the disease. Thus in the less severely handicapped, average blood flow was 60 ml 100 g^{-1} min^{-1} and average metabolic rate was 3.1 ml 100 g^{-1} min^{-1}: the corresponding average figures for the severely handicapped children were 35 and 1.5. This study can be criticized on the grounds that the group consists of children with diseases which affect principally brain cells, or glia or the flow of c.s.f. and further the levels of sedation varied

11. Cerebral blood flow and metabolism

considerably. Nevertheless, there appeared to be a close relation between blood flow and the level of metabolism and the degree of impairment and presumed level of cerebral functional activity.

In a further study, normal children were studied: that is children who had had episodes of convulsions some time previously but who at the time of the study were clinically normal and who had normal EEG patterns (Kennedy & Sokoloff, 1957). In this group, both blood flow and metabolism were substantially above

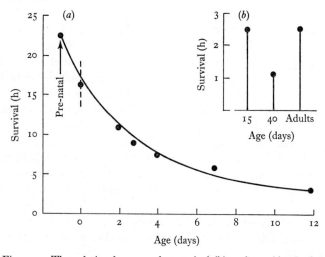

Fig. 11.5. The relation between the survival (h) and age (days) of rats kept in an atmosphere reduced to 320 mmHg. In (a), it will be noted that the greatest tolerance to hypoxia is to be seen prenatally and after the birth this tolerance is lost rapidly until the third week, when the mature level is attained. In (b), the range of age is extended and shows that decrease in resistance to hypoxia continues until the fortieth day when a gradual rise to the adult level occurs. (From Himwich (1951), *Brain metabolism and cerebral disorders*, Baltimore: Williams & Wilkins, Fig. 19.)

adult levels, the average values being 106.4 S.E. ± 3.3 and 5.17 S.E. ± 0.23 ml $100 \text{ g}^{-1} \text{ min}^{-1}$ respectively. The children were 3 to 10 years of age and although the numbers were few ($n = 11$) no correlation was found between either blood flow or metabolism and age over this range. For ethical reasons, it is unlikely that values for cerebral metabolic rate in completely normal children will be available until some more acceptable method of measurement is developed. But if the values obtained by Kennedy &

Normal values for cerebral metabolism in man

Sokoloff (1957) do not overestimate the true value, then it can be calculated that within this age range, the brain consumes approximately half the total body oxygen uptake.

Changes with aging. Table 11.4 includes the results of studies in which cerebral blood flow and metabolism have been measured at various ages throughout life. The results are also plotted in Fig. 11.6 which confirms that both blood flow and metabolic

Table 11.4. *Relation of cerebral blood flow and metabolic rate and age in man*

Mean age (years)	No.	Cerebral blood flow (ml 100 g^{-1} min^{-1})	Cerebral metabolic rate	References
5	6	104	5.1	Kennedy, Sokoloff & Anderson (1954)
10	7	90	5.3	Kennedy: Unpublished data
13	7	68	4.0	Kennedy: Unpublished data
19	4	60	3.70	Sokoloff, Wechsler, Mangold, Ralls & Kety (1953); Mangold, Sokoloff, Therman, Conner, Kleinerman & Kety (1955)
23	7	52	3.12	Sokoloff, Wechsler, Mangold, Ralls & Kety (1953); Mangold, Sokoloff, Therman, Conner, Kleinerman & Kety (1955)
25	14	54	3.3	Kety & Schmidt (1948a)
25	19	65	3.8	Scheinberg & Stead (1949)
29	12	62	4.0	Schieve & Wilson (1953b)
30	12	53	3.4	Shenkin, Novak, Goluboff, Soffe & Bortin (1953)
34	9	54	3.5	Fazekas, Alman & Bessman (1952)
40	10	57	3.5	Schieve & Wilson (1953b)
50	16	59	3.8	Scheinberg, Blackburn, Rich & Saslaw (1953)
63	16	51	3.4	Scheinberg, Blackburn, Rich & Saslaw (1953)
64	7	55	3.7	Schieve & Wilson (1953a)
68	15	43	2.4	Fazekas, Alman & Bessman (1952)
93	18	39	2.3	Fazekas, Kleh & Witkin (1953)

11. Cerebral blood flow and metabolism

rate fall from the high values in childhood, rapidly at first and then more slowly. It is possible that with further data, particularly in the younger age groups, the curve would not be found to be less smooth and would show a steeper fall from childhood to adult levels.

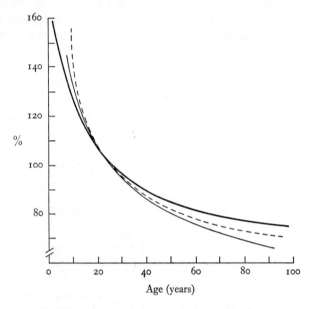

Fig. 11.6. The relation between cerebral blood flow, cerebral oxygen consumption and the density of cortical neurones in four parts of the cortex and age. All values are expressed as per cent of the respective mean value at the age of 25 years. (From Kety (1956), *J. Chron. Dis.* **3**, 478–86, Fig. 4.)

The curve could include an age dependent artifact. If, with the aging process, jugular venous sampling became less reliable, because of increasing contamination from extracerebral sources, this would be expected to give increasingly low values for cerebral blood flow and also a greater scatter of results. In the adult, this contamination has been found to be of the order of 3 per cent (Shenkin, Harmel & Kety, 1948): so it could scarcely account for the fall observed during adolescence. Further in two studies in aged subjects (Shenkin, Novak, Goluboff, Soffe & Bortin, 1953; Freyhan, Woodford & Kety, 1951), there was no evidence of a degree of contamination significantly greater than in young adults. These points, though circumstantial, argue against the existence of an artifact which systematically increases with age.

At the moment the precise reason why blood flow and metabolic rate fall is not fully understood. The nitrous oxide technique yields results in terms of blood flow per unit weight or volume of brain and this does not distinguish between different parts of the brain or tissues which may have very different rates of perfusion and metabolism. The changes observed in cerebral metabolism could therefore reflect a general reduction in the number of actively metabolizing units or areas or a reduction in the ratio of their mass to that of the brain as a whole. In Fig. 11.6 the reduction in mean cerebral blood flow and metabolic rate have been related to the reduction in neuronal mass, i.e. the neuronal density measured in discrete parts of the central nervous system (Brody, 1955). The striking similarity between the curves suggests that this is more than just a chance relationship.

The close relationship between changes in cerebral blood flow and metabolism over this range of age can be variously explained. On the one hand, it is possible to argue that this relationship is merely one example of blood flow being adjusted to metabolic requirement through some mechanism such as the changes in $pH_{e.c.f.}$ discussed in Chapter 8. We might therefore suppose that neurones have a definite life span and during the latter part of this, activity slowly falls and in consequence the level of metabolism. Alternatively, it could be argued that the rate of cerebral metabolism is itself limited and regulated by the rate of cerebral perfusion. This is likely to be the case in childhood where presumably there must be some mechanism to limit the level of growth and metabolism: it could equally apply in old age when cerebral blood flow diminishes partly as a result of degenerative changes in the vessel walls and partly as a result of atherosclerotic and other forms of vascular disease. At the moment, there is no evidence in the cerebral or in other vascular beds that would distinguish between these possibilities.

Cerebral metabolism and changes in blood gas tensions

Hypoxia. The circulatory changes in cerebral vessels which occur in response to arterial hypoxia have been considered in detail in Chapter 6. In essence, they consist of an increase in blood flow,

11. Cerebral blood flow and metabolism

vasodilatation and in the long term (possibly in the short term), an increase in the size of the capillary bed by the flow of blood through additional capillaries. These changes ensure that despite a reduction in the oxygen content of arterial blood, the oxygen supply to the brain is maintained while the capillary changes ensure that the fall in P_{a,O_2} is offset by a reduction in the length of the diffusion pathway for oxygen from capillaries to mitochondria. These changes are enchanced during asphyxia when there is a concomitant rise in P_{a,CO_2} and are limited by hypocapnia. The brain is thus protected to a certain extent against the effects of hypoxia. What has to be considered in this section are the limits of circulatory homeostasis, the effects on cerebral metabolism of more severe hypoxia and the means by which cerebral tissue hypoxia can be recognized *in vivo*.

In early experiments, the effect of arterial hypoxia was tested in man during the inhalation of 8 per cent oxygen at constant carbon dioxide (Turner, Lambertson, Owen, Wendel & Chiodi, 1957), and of 10 per cent oxygen without carbon dioxide being controlled (Kety & Schmidt, 1948b), and in neither of these studies was any alteration of total cerebral oxygen consumption observed. It does not follow from this that under these conditions, there was no tissue hypoxia because (*a*) there could have been small and local changes in oxygen consumption (Ingvar, 1961; Gleichman, Ingvar, Lübbers, Siesjö & Thews, 1962) which were concealed in the error of measurement, or (*b*) there could have been changes in carbohydrate metabolism without any demonstrable change in oxygen uptake.

There is in fact considerable evidence available from studies *in vitro* (for references see McIlwain, 1966) and from studies *in vivo* in particular, that with a comparable reduction of P_{a,O_2} to that of Turner *et al.* (1957) and Kety & Schmidt (1948b), marked alterations occur in cerebral carbohydrate metabolism and in energy production (McGinty, 1929; Stone, Marshall & Nims, 1941; Gurdjian, Webster & Stone, 1944, 1949; Olsen & Klein, 1947a, b; Biddulph, Van Fossan, Criscuolo & Clark, 1958; Geiger, 1958; Chance, Cohen, Jobsis & Schoener, 1962; Chance & Schoener, 1962; Lolley & Sampson, 1962; Lowry, Passoneau, Hasselberger & Schultz, 1964). When the supply of oxygen to the tissue fails, the acceptance of electrons, removed by dehydrogenation and transported in the electron transport chain is slowed. As a consequence, the various oxido-reduction systems shift to a more reduced state and protons accumulate. As part of this redox shift, NADH increases and, with it, the concentration of lactate.

$$\text{Pyruvate}^- + \text{NADH} + \text{H}^+ = \text{Lactate}^- + \text{NAD}^+.$$

The accumulation of H^+ in the cell is enhanced under conditions of hypoxia (a) by the production of two protons for each molecule of glucose broken down, and (b) during ATP hydrolysis as the ATP stores become exhausted. The intracellular acidosis is limited on the other hand by the breakdown of phosphocreatine which causes the pK' of the phosphate group to shift to the alkaline side and by the efflux of lactic acid from cell to extracellular fluid.

The finding of an increased intracellular lactate concentration has been confirmed by many of the workers cited above at levels of arterial hypoxia comparable to those induced by Kety & Schmidt (1948b), i.e. at P_{a,O_2} c. 35 mmHg and P_{v,O_2} 25 mmHg. With more severe hypoxia, P_{a,O_2} 25–30 mmHg, P_{v,O_2} 13–17 mmHg, Gurdjian, Webster & Stone (1944) observed a breakdown of phosphocreatine and this has been confirmed in experiments in which graded doses of cyanide were injected intravenously (Olsen & Klein, 1947a) or in which rabbits were ventilated with an N_2–CO_2 mixture (Thorn, Scholl, Pfleiderer & Mueldener, 1958). In these last experiments, there was also a progressive reduction in ATP and rise in ATP concentrations. The reduction of NAD in the cortex during brief periods of hypoxia has also been recorded *in vivo* (Chance, Cohen, Jobsis & Schoener, 1962) and this is illustrated in Fig. 11.7. The potential value of this technique is that the redox state can be correlated with tissue P_{O_2} measured with an oxygen micro-electrode and the kinetics of NAD reduction studied with some precision. However, the technical difficulties are considerable because of the presence of haemoglobin whose state of oxygenation is also changing.

The results which have been quoted so far have been obtained in experiments in which tissue analysis has been carried out post-mortem after rapid freezing. They are therefore of limited help in establishing a relation between an index of tissue hypoxia, the degree of arterial hypoxia and the circulatory response. A number of workers have sought to make use of the production

11. Cerebral blood flow and metabolism

of excess lactate for this purpose, or, as a refinement, the lactate–pyruvate ratio since this is closely coupled to the redox state of NAD. Hohorst, Betz & Weidner (1968), for example, have shown that with moderate hypoxia (P_{a,O_2} of the cortex = 10 mmHg), lactate in brain tissue increased but there was no correlation between this and the EEG pattern. The latter only changed consistently with more severe hypoxia (cortical P_{O_2} 0–2 mmHg) and was most closely correlated with the breakdown and resynthesis of phosphocreatine.

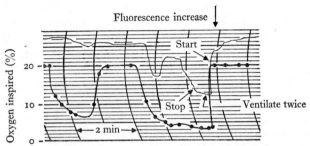

Fig. 11.7. Correlation of the percentage of oxygen in inspired air with the increase in fluorescence intensity as measured on the brain cortex of a rat under urethane anaesthesia. The record is read from right to left. $t = 2$ min. (From Chance et al. (1962), Science **137**, 499–507, Fig. 7.)

Several workers have emphasized the unreliability of lactate as an index of brain tissue hypoxia. Olson (1963) quoted several examples where excess lactate was found without hypoxia. Tobin (1964) showed that excess lactate could be produced by changes in pH rather than in oxygenation. Huckabee (1958) found excess lactate in arterial blood in dogs inhaling 7–8 per cent oxygen but whose P_{a,CO_2} was uncontrolled. Cain (1965) however did not detect excess lactic acid in arterial blood in dogs until P_{a,O_2} was reduced to less than 25 mmHg.

Some possible reasons for this variation emerge from recent studies on the changes in c.s.f. lactate during hypoxia. Some typical results are shown in Fig. 11.8. Rats were asphyxiated for 1–4 min and the hypoxia was severe enough to cause the breakdown of all the tissue phosphocreatine and half the ATP. Intracellular lactate rose to 25 m-mol l^{-1}, intracellular bicarbonate was reduced but even after a 10 min recovery period, c.s.f. lactate had barely altered. The most probable reason for this is the

Cerebral metabolism and hypoxia

relatively slow turnover of c.s.f. and a further consequence of this, together with the slow clearance of lactate is that following a brief period of asphyxia or cerebral ischaemia, c.s.f lactate remains high for a period of 1 hour or more. That c.s.f. lactate poorly reflects intracellular changes has been shown by Zwetnow (1968). Cerebral ischaemia was induced in dogs by raising intracranial pressure to 30–40 mmHg for 15 minutes. In brain tissue, the lactate–pyruvate ratio increased from a central level of 17 to

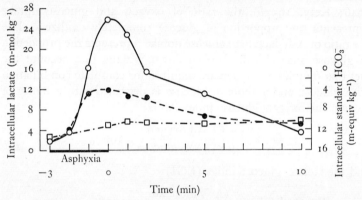

Fig. 11.8. The effect of three minutes asphyxia caused by stopping the respiratory pump upon the concentrations of intracellular lactate (○), intracellular standard bicarbonate (●) and c.s.f. lactate (□). Rats, anaesthetized with nitrous oxide. (From Kaasik *et al.* (1968), *Scand. J. clin. Lab. Invest.* Suppl. 102 IIIc, Fig. 1.)

30 in 40 minutes and lactate concentration increased over the same period and then declined to control levels after a further 60 min. C.s.f. lactate and lactate–pyruvate ratio increased to a lesser degree and considerably more slowly and were still high after 100 min recovery period. Although there is an obvious discrepancy between intracellular and c.s.f. lactate concentrations, it is of interest that the reactive hyperaemia which follows a brief period of cerebral ischaemia has approximately the same time course as the changes in c.s.f. lactate or $pH_{e.c.f.}$ measured on the surface of the brain (Cotev, Cullen & Severinghaus, 1968). This observation is consistent with the view that cerebral blood flow correlates closely with changes in e.c.f. or c.s.f. pH and may be regulated by them.

An alternative method of determining cerebral tissue hypoxia

11. *Cerebral blood flow and metabolism*

has been proposed by Cohen, Alexander, Smith, Reivich & Wollman (1967). They have found that the changes in lactate concentration in jugular venous blood in man more faithfully reflects changes in intracellular lactate than does c.s.f. and further, that by measuring the A–V lactate difference it is possible to estimate what proportion of glucose has been broken down by cerebral tissue to lactate. On the assumption that the cerebral glucose uptake represents the total amount degraded both oxidatively and anaerobically to lactate (Himwich & Himwich, 1946; Kety, 1957b), the ratio of oxygen and glucose uptake represents the proportion of glucose completely oxidized while the ratio of A–V lactate to glucose uptake represents the proportion of glucose anaerobically broken down. Taking into account the fact that 6 moles of oxygen are required for complete combination with glucose and 1 mole of glucose is degraded into 2 moles of lactate, the oxygen–glucose index (OGI)

$$\text{OGI} = \frac{\text{(A–V) oxygen}}{6 \times \text{(A–V) glucose}} \times 100\%$$

and the lactate–glucose index (LGI)

$$\text{LGI} = \frac{\text{(A–V) Lactate}}{2 \times \text{(A–V) glucose}} \times 100\%.$$

Using these estimates of oxidative and anaerobic metabolism for glucose in brain, Cohen *et al.* (1967) found that during arterial hypoxia in man (P_{a,O_2} reduced from 89 to 34.5 mmHg, P_{a,CO_2} held constant at 39–40 mmHg), the oxygen–glucose index was reduced from 92 to 75.7 per cent while the lactate–glucose index increased from 4 to 18.8 per cent.

This approach depends on a number of assumptions of which the more important are how rapidly and completely lactate in cells and venous blood equilibrate, whether oxygen or glucose are limited by diffusion and whether the calculations take into account metabolic interconversions between carbohydrates, lipids and amino-acids in the brain (Sacks, 1958, 1965). These points have been considered in detail by Alexander, Cohen, Wollman, Smith, Reivich & Vander Molen (1965) and although some uncertainties remain, it is probable that these estimates are valid in steady state conditions. They are of particular interest in showing that under conditions of arterial hypoxia where no reduction in the

Cerebral metabolism and carbon dioxide

metabolic rate for oxygen can be detected, i.e. at a P_{a,O_2} c. 35 mmHg and P_{v,O_2} 22–27 mmHg, and despite the evidence of tissue hypoxia, the energy production of the brain appears to remain unchanged.

Changes with carbon dioxide. The effect of raising alveolar and arterial P_{CO_2} upon cerebral metabolism has not been studied in any detail. In patients with chronic lung disease with retention of carbon dioxide, i.e. P_{a,CO_2} up to 60 mmHg and arterial pH reduced to 7.29 units, no significant changes in cerebral metabolic rate were observed (Scheinberg, Blackburn, Saslaw, Rich & Baum, 1953). Similarly, Hohorst et al. (1968) were unable to demonstrate any change in the levels of brain lactate or energy rich phosphates in cats given high carbon dioxide levels to breathe.

Of greater physiological interest are the changes associated with hypocapnia since a reduction of P_{a,CO_2} is observed at altitude and during the administration of anaesthesia with controlled ventilation. The changes in cerebral blood flow with hypocapnia have been documented and discussed in Chapter 7. Here it should be emphasized that although it is clear that a fall in P_{a,CO_2} is accompanied by a fall in mean cerebral blood flow, it is probable that this reduction in flow is not uniform throughout the brain. The circulatory changes are opposite to those seen in arterial hypoxia. There is a fall in oxygen supply which is proportional to the fall in blood flow, a constriction of arteries and probably a reduction in capillary blood flow which will affect the diffusion pathway for oxygen. Some degree of cerebral tissue hypoxia would therefore be expected in severe hypocapnia and changes in cerebral function, indicated by altered EEG pattern and in the response to the flicker-fusion test have been reported (Geddes & Gray, 1959; Hughes, King, Cutler & Markello, 1962; Allen & Morris, 1962).

The effects of hypocapnia in man have been studied as part of an investigation into the effects of various kinds of anaesthesia upon cerebral metabolic rate. The most clear-cut changes have been obtained by Alexander et al. (1965) who studied the effect of hypocapnia on the energy production in brain in patients anaesthetized with nitrous oxide. Using the estimates of oxidative and anaerobic metabolism of glucose by the brain, described in the last section, these workers found that when P_{a,CO_2} was reduced from 41 to 18 mmHg, there was no obvious change in cerebral

11. Cerebral blood flow and metabolism

oxygen consumption. However, the index of aerobic metabolism fell from 98 to 87 per cent while the index of anaerobic metabolism approximately doubled, 7 to 15 per cent. Estimates of 'excess lactate' as defined by Huckabee (1961) showed that this was unchanged while the EEG changes were variable.

These changes in the proportion of glucose degraded aerobically and anaerobically with hypocapnia are compared with the equivalent changes during arterial hypoxia, Table 11.5 and this indicates that although the rate of production of lactate is approximately the same in hypoxia and severe hypocapnia, the reduction in aerobic metabolism of glucose was most marked in severe hypocapnia.

Table 11.5. *Comparison of the effects of arterial hypoxia and hypocapnia upon various indices of cerebral energy production**

	Arterial hypoxia P_{a,O_2} 34.6 mmHg P_{a,CO_2} 39.1 mmHg	Moderate hypocapnia P_{a,O_2} 157 mmHg P_{a,CO_2} 19.4 mmHg	Severe hypocapnia P_{a,O_2} 187 mmHg P_{a,CO_2} 10.1 mmHg
Cerebral metabolic rate:			
for glucose (mg 100 g^{-1} min^{-1})	+1.25	+0.42	+1.04
for lactate (μmol/100 g/min)	+8.42	+0.63	+7.65
Aerobic index (%)	−16.1	−14.9	−38.8
Anaerobic index (%)	+14.4	−3.30	+17.2

* Expressed as the change from control values in unanaesthetized man breathing air.

Corroborative evidence for tissue hypoxia during severe hypocapnia – < 20 mmHg P_{a,CO_2}, has been given by Meyer & Gotoh (1960) who demonstrated marked reductions in cortical P_{O_2} and by Granholm, Lukjanova & Siesjö (1968) who showed a progressive rise in lactate concentrations in brain tissue and c.s.f., in the lactate–pyruvate ratio and in tissue NADH levels as measured by microfluorimetry.

How is this tissue hypoxia caused? Together with the circulatory changes mentioned previously, the shift of the oxyhaemoglobin dissociation curve to the left with a rise in arterial pH and a possible direct effect of hypocapnia or alkalosis upon cerebral metabolism must be considered. In some recent experiments, Reivich, Dickson, Clark, Hedden & Lambertsen (1968) have shown that the hypocapnia by itself is not responsible for the tissue hypoxia. However when the subjects inhaled 100 per cent oxygen at 3.5 atmospheres, the evidence of anaerobic glycolosis disappeared although cerebral blood flow remained at low levels. These findings support the view that the reduction in blood flow during hypocapnia causes impaired diffusion of oxygen from capillaries to tissues but that this can be overcome by creating a very large gradient for oxygen diffusion.

Cerebral blood flow, metabolism and mental activity

The idea that an increase in mental activity should be accompanied by an increase in blood flow and metabolic rate in the same way as in the heart or salivary glands is not new. A number of early studies purported to show that a rise in mental activity was sufficient to increase total body oxygen consumption (Becker & Olsen, 1914; Day, 1923; Gillespie, 1924; Liebermann, 1926) but in none of these studies is it clear that the effects of muscular tension which accompanies sustained mental effort had been excluded.

Fulton (1928) gives an account of the responses to an increase in illumination of blood flowing through an angioma of the occipital cortex. The patient had noted an increased noise in the head whenever he used his eyes and Fulton was able to demonstrate an undoubted increase in the frequency and amplitude of vibrations from the tumour when the patient read newsprint after some minutes of dark adaptation. Various safeguards were taken to ensure that there was no accompanying muscular movements and it was also established that other types of sensory stimulation were without effect. Cobb & Talbot (1927) approached the matter another way by estimating the capillary density in the motor cortex, in the olfactory bulb and visual area in animals which immediately before death were subjected to olfactory stimulation. In unstimulated animals, there was an average of 110 μm of open capillaries per 100,000 μm^3 of brain substance, in animals stimulated with ether the average value was 149 μm and with ammonia 171 μm. This index of increase in the capillary bed and therefore of overall blood flow provides some evidence that blood flow and presumed neuronal activity increase together. Experiments by

11. Cerebral blood flow and metabolism

Schmidt & Hendrix (1938) and Gerard (1938) are discussed in a later section.

Lennox (1931b) approached the problem by measuring the oxygen content of jugular venous blood in subjects before, during and after the performance of mental arithmetic and found that in ten out of the fifteen subjects there was a rise in oxygen content. These results give little direct information as to whether blood flow actually increased since there may have been changes in cerebral metabolism and in blood gas tensions associated with altered emotions.

Further studies to clarify this matter followed the introduction of the nitrous oxide method of measuring cerebral blood flow and its application to man. Sokoloff, Mangold, Wechsler, Kennedy & Kety (1955) measured blood flow and calculated cerebral metabolism in 13 young normal subjects before and during the performance of mental arithmetic. Despite a significant rise in arterial pressure and pulse rate, no changes in oxygen consumption or blood flow were observed and no changes in blood gas tensions were observed which might have obscured changes in flow. Mental activity of this sort however caused a shift in the EEG pattern to a lower voltage, higher frequency and asynchronous activity.

In discussing these results, Sokoloff et al. (1955) pointed out that areas in which there was an increase in neuronal activity could be counterbalanced by areas of reduced activity, and if there was a redistribution of blood flow in relation to these areas, the overall changes in mean blood flow might be negligible or too small to be detected by the nitrous oxide method. To this might be added the fact that the subjects may have been aroused or unduly apprehensive during the test since the average value for cerebral blood flow was 69.6 ml 100 g^{-1} min^{-1} and for oxygen consumption 3.99 ml 100 g^{-1} min^{-1} and these should be compared with the average normal values of Kety & Schmidt (1948a), 54 and 3.3 ml 100 g^{-1} min^{-1} respectively.

More recently an attempt to measure the local changes in cerebral blood flow during mental activity has been made (Ingvar & Risberg, 1967). They used the ^{133}Xe desaturation method (Lassen, Høedt-Rasmussen, Sorensen, Skinhøj, Cronquist, Bodforss & Ingvar, 1963; Høedt-Rasmussen, Sveinsdottir & Lassen, 1966) in order to determine blood flow through grey and white matter and used multiple detectors over one hemisphere. During mental activity, changes in blood flow were localized to the cortex where an average 8 per cent increase was observed. But the increase in flow was considerably greater in certain areas notably the suprasylvian area while in other areas flow actually fell.

The interpretation of these results is difficult for it is probable that with increased mental performance, two separate processes exist. First, there is an element of arousal and the maintenance of vigilance and it is this which is most probably associated with blocking of the alpha rhythm of the EEG (Berger, 1929). In addition there may be more peripheral and autonomic effects such as an increase in muscle tone (Davis, 1938; Von Euler & Soderberg, 1956) and reactions of tension and anxiety which may be causing an afferent inflow, cause localized changes in metabolism and flow. A second and more specific change in neuronal activity as a consequence of mental arithmetic and other test procedures is possible but at the moment a matter of speculation.

This study has been taken one stage further by Risberg & Ingvar (1968) who studied changes in cerebral blood volume in seven healthy young subjects during a number of mentation procedures using radioactive iodinated serum albumen (RISA). Although there has been no satisfactory correlation between changes in cerebral blood volume and blood flow, it is of interest that the changes in cerebral blood volume, expressed semi-quantitatively as per cent of control, during mental activity were similar in location to those obtained by the same workers using ^{133}Xe (Ingvar & Risberg, 1967). Thus it was confirmed that in frontal and temporal areas, blood volume increased whereas there was an insignificant change or actual fall in blood volume in parietal or sylvian areas. The method does not give any indication of how the volume increase takes place, whether by vasodilatation or by the opening up of new capillaries. Some evidence is available that the first alternative is more likely (Opitz, 1951) but the second remains to be excluded. A feature in this second method which is of particular advantage in this type of experiment is that there is no need for intracarotid injections and the measurements are continuous. Thus there is likely to be much less cause for apprehension on the part of the subject than with the ^{133}Xe method: accordingly the results obtained with RISA may represent more specific vascular changes associated with mental procedures and less of the non-specific consequences of apprehension, etc.

11. *Cerebral blood flow and metabolism*

The cerebral circulation and sleep

Sleep has been attributed to hypoxia, cerebral ischaemia, or to narcosis on the basis of some form of metabolic alteration. These hypotheses have been based either on guesswork or the results of early experiments in which some index of cerebral blood flow was measured or both. Thus Mosso (1881) measured changes in brain volume using trephine holes and the skull as a plethysmograph during wakefulness and sleep. His results were contradictory but on the majority of occasions, he found that sleep was associated with a reduction in brain volume. Tarchanoff (1894) observed blanching of pial vessels in puppies when sleep occurred and this would confirm Mosso's conclusion that blood flow to the brain fell. Other workers however were unable to confirm these results and instead found evidence of cerebral engorgement (Czerny, 1896; Shepard, 1914; Stevenson, Christiensen & Wortis, 1929; Vujic, 1933; Nygard, 1937). None of these studies yielded direct information on cerebral blood flow since, as has been pointed out before (p. 313) there is not necessarily a correlation between cerebral volume and blood flow. Using better methods of recording, Kleitman (1939) observed a small reduction of blood flow on passing from wakefulness to sleep: Gibbs, Gibbs & Lennox (1935c) however observed no difference. In the latter study, the thermoelectric method which was used would have registered gross effects but would also have been susceptible to changes in cerebral metabolism and these were not measured.

The first quantitative measurement of cerebral blood flow during states of wakefulness and sleep was carried out by Mangold, Sokoloff, Therman, Connor, Kleinerman & Kety (1955). The method used was that described by Kety & Schmidt (1948a) and the study was carried out on young healthy men. They found that during sleep, blood pressure fell, cerebral blood flow increased and a consequently cerebrovascular resistance fell. This rise in blood flow could not be accounted for by changes in cerebral metabolism, blood gas tensions or pH or in arterial haematocrit for all these were unchanged from the last measurement taken before sleep. These results are of interest for a number of reasons. First they dispose of hypotheses based on the idea that sleep is related to cerebral hypoxia or ischaemia or that it is due to alterations of metabolism or changes in blood gas tensions or pH.

Cerebral circulation and sleep

Secondly, the fact that blood flow rises in the face of a rise in intracranial pressure (Stevenson, Christiensen & Wortis, 1929; Vujic, 1933) and a fall in blood pressure (Brooks & Carroll, 1912; Müller, 1921; Landis, 1925) indicates a genuine reduction in vascular resistance. What the cause of this vascular relaxation is has not been determined. Mangold et al. (1955) rejected the possibility that it was due to neural influences, e.g. a fall in sympathetic vasoconstrictor activity because it had been shown in a group of patients that bilateral stellate block was without effect upon cerebral blood flow (Harmel et al. 1949). The difficulties of interpreting the data from this latter series have been discussed elsewhere (p. 265) and in view of more recent evidence that cerebral blood flow can be affected by extrinsic neural activity (James, Millar & Purves, 1969), the mechanism of the rise in blood flow must remain open at the moment.

A third point of interest arising from the results of Mangold et al. (1955) is that it provides an example of the apparent dissociation of changes in cerebral blood flow and metabolism. Further, the maintenance of an unchanged cerebral metabolism during sleep distinguishes it from the state of anaesthesia or coma in which cerebral metabolism is markedly depressed (Wechsler, Dripps & Kety, 1951; Kety, Woodford, Harmel, Freyhan, Appel & Schmidt, 1948; Kety, Polis, Nadler & Schmidt, 1948). Nor can the state of cerebral function activity in the mature foetus be defined as sleep since, with occlusion of the umbilical cord and the onset of respiration, there is a large increase in cerebral metabolic rate (Purves & James, 1969).

More recent experiments have taken into account the different types of sleep particularly the phenomenon of paradoxical or rapid eye movement (S–REM) sleep (Aserinsky & Kleitman, 1953). During S–REM, there is evidence of marked central nervous system activity with increased unit activity in the occipital cortex of the cat (Evarts, 1962), increased frequency of pyramidal tract neurons, in contrast to that seen during sleep with electroencephalographic slow waves (S–SW) (Evarts, 1964), and an increase in spontaneous unit discharge in the mesencephalic reticular formation (Huttenlocker, 1961), in the lateral geniculate body, and vestibular nuclei (Bizzi, Pompeiano & Somegzi, 1964). The idea that blood flow might change in the same direction as increased functional activity in specific parts of the brain has

11. Cerebral blood flow and metabolism

been tested in two ways. First, a relative increase in blood flow in the cortex and rhombencephalic reticular formation measured with implanted thermocouples was found by Kanzow, Krause & Kuhnel (1962) during S-REM but a relative decrease in flow in the mesencephalic reticular formation was noted using the same methods (Baust, 1967). Kawamura & Sawyer (1965) measured increases in temperature of the fore-brain and hypothalamus during S-REM in cats. These methods, of course, give only the most indirect indication of the changes in blood flow and are always likely to confuse because of local changes in heat production as a result of altered metabolic rate.

A second method employed by Reivich, Isaacs, Evarts & Kety (1968) and which was a modification (Reivich, Jehle, Sokoloff & Kety, 1969) of the autoradiographic technique described by Landau, Freygang, Rowland, Sokoloff & Kety (1955) gives a more direct measure of local changes in blood flow in the brain and is unaffected by local changes in temperature. The method has been described in Chapter 5: briefly it consists of the intravenous injection of [^{14}C]antipyrine at a constant rate over 1 min with frequent measurement of the arterial concentration. Immediately after injection the cat is killed, the brain is removed, frozen and sectioned. From the calibration of optical density and regional [^{14}C]antipyrine concentration, the brain–blood partition coefficient for antipyrine and arterial concentration, regional blood flow is calculated. These authors compared values for blood flow at various sites in the brain in awake cats, and during S-SW and S-REM. These changes are shown in diagrammatic form in Fig. 11.9 and from this it is obvious that not only is blood flow greater in S-REM than S-SW, but that in most parts of the brain, blood flow during S-REM is substantially greater than in awake cats. If allowance is made for weight of the structures studied, then the blood flow would be approximately 15 per cent greater than control during S-SW, and 80 per cent greater than control during S-REM. This rise in blood flow during S-SW is of the same order as the 10 per cent increase measured by Mangold *et al.* (1955). Whether there are corresponding changes in cerebral metabolism have not yet been determined. Brebbia & Altshuler (1965) found that total body oxygen utilization in man was greatest during S-REM and least during stages III and IV of S-SW. Hyden & Lange (1965) observed a two- to threefold increase in succinate oxidase activity in cells of the reticular formation during S–SW. Mangold *et al.* (1955) found no significant change in cerebral metabolic activity.

These more recent series of experiments emphasize the fact that the mechanism of the increase in flow remains unexplained.

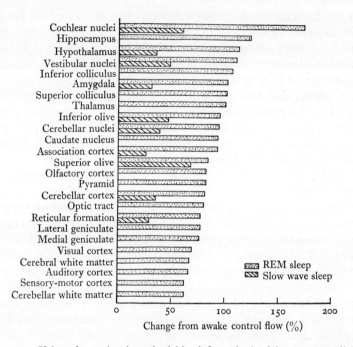

Fig. 11.9. Values for regional cerebral blood flow obtained by an autoradiographic technique in the cat during slow wave (S–SW) and rapid eye movement (S–REM) sleep expressed as a percentage change from blood flow when the cat was awake. (From Rievich, Isaacs, Evarts & Kety (1968), *J. Neurochem.* **15**, 301–6, Fig. 3.)

Certainly Reivich, Isaacs, Evarts & Kety (1968) found no change in P_{a,CO_2}, P_{a,O_2}, pH or haematocrit. From other evidence, (Snyder, Hobson, Morrison & Goldfrank, 1964), no changes in blood pressure which have been observed during S–SW or S–REM could account for the changes in blood flow. The possibility that the reduction in cerebral vascular resistance could be neurogenic, e.g. a fall in sympathetic vasoconstrictor activity, was rejected by Reivich, Isaacs, Evarts & Kety (1968) on the grounds that such

11. Cerebral blood flow and metabolism

a tonic influence had not been demonstrated. The observations of James, Millar & Purves (1969) that cerebral vascular responses can be modified by activity mediated by cervical sympathetic nerves suggest that this question remains open.

Changes during hypothermia

It has been known for many years that in adult man (but not in the newborn) a fall in body temperature is accompanied by a fall in total body oxygen consumption (Penrod, 1949; Bigelow, Lindsay, Harrison, Gordon & Greenwood, 1950). This fact was made use of in a variety of cardiovascular surgical operations on the assumption that the brain and spinal cord would be less sensitive to periods of reduced blood flow. At the time when such operations were introduced, there was virtually no evidence that the cerebral metabolic rate participated in the general reduction of metabolism with hypothermia.

Field, Fuhrman & Martin (1944) had shown that the oxygen consumption of brain *in vitro* varied as a constant function of the temperature. The first determination *in vivo* of cerebral oxygen consumption and blood flow was by Rosomoff & Holaday (1954) who showed that both variables diminished with body temperature. The fall was approximately linear being 3.1 ml $100\ g^{-1}\ min^{-1}\ degC^{-1}$ and 0.47 ml $100\ g^{-1}\ min^{-1}\ degC^{-1}$ for blood flow and oxygen consumption respectively. The absolute figures are not likely to be accurate since these authors attempted to isolate the dog's brain and measure blood flow through a rotameter placed in series with the carotid arteries. Oxygen consumption was calculated from blood samples from the sagittal sinus and these manoeuvres resulted in values for blood flow which are rather low and for oxygen consumption which are distinctly high, suggesting that venous blood was contaminated from non-cerebral sources. Further, no account was taken of the changes in respiration which accompany hypothermia and the consequent changes in P_{a,CO_2}. Nevertheless, this study clearly confirmed the relation between cerebral blood flow, cerebral oxygen consumption and body temperature and further useful application of the method of cooling the body was suggested by Rosomoff (1956) who showed that the effects of occlusion of the middle cerebral artery in the dog, which in the normothermic animal included substantial infarction of the brain (Harvey & Rasmussen, 1951; Hain,

Changes during hypothermia

Westhaysen & Swank, 1952), could be greatly mitigated in the hypothermic animal.

A more recent study by Forrester, McDowall, Harper & Nisbet (1964) has confirmed these points. As body temperature was reduced from 38 °C to 30 °C, cortical blood flow fell by 37 per cent of control and from the associated fall in arterial pressure, it was calculated that cerebral vascular resistance rose by +33 per cent of control. These changes occurred at constant P_{a,CO_2} and it was further shown that at the lower temperature, cerebral vessels maintained their ability to respond to changes in carbon dioxide. The mechanism whereby blood flow is reduced under these circumstances so that the cerebral A–V oxygen difference remains virtually constant is unknown. Rosomoff & Gilbert (1955) showed that as temperature was lowered, c.s.f. pressure also fell, on average by 50 per cent of control and this would certainly have the effect of diminishing capillary and post-capillary resistance to blood flow. On the other hand, the transmural pressure gradient would increase and provided that cerebral vessels maintain their ability to respond to such changes, this would lead to precapillary constriction. It may be that the final level of blood flow reflects the resultant of these effects upon cerebral vessels in addition to changes in the chemical environment of the vessels as cerebral metabolic rate falls.

Changes with the administration of analeptic and anaesthetic drugs

Analeptic drugs. It was originally thought that epileptic convulsions were caused by cerebral ischaemia (for references see Hill, 1896; Gibbs, 1933a). This was not an unreasonable idea since convulsions are a terminal feature of prolonged cerebral ischaemia. In early experiments, however, the majority of results suggested that on the contrary, cerebral blood flow increased both during spontaneous convulsions (Grant, Spitz, Shenkin, Schmidt & Kety, 1947) and those induced by the local application of strychnine (Florey, 1925) or injection of oil of absinthe, camphor and picrotoxin (Gibbs, 1933a). Koopmans (1939), however, found that the intravenous injection of strychnine caused a reduction of brain volume used as an index of cerebral blood flow. Other workers (e.g. Finesinger & Cobb, 1933) studying the effect of analeptic drugs applied locally upon the calibre of pial vessels

11. Cerebral blood flow and metabolism

obtained variable results and were not able to say whether the drugs affected the vessels primarily or whether the vessels were affected secondarily by the (presumed) changes in cerebral activity, systemic arterial pressure, respiration and blood gas tensions. In some later experiments in which the last factors were controlled, it was clearly established that if strychnine was given in doses sufficient to cause convulsions in cats and monkeys, both cerebral blood flow and cerebral oxygen uptake increased (Geiger & Magnes, 1947; Dumke & Schmidt, 1943; Schmidt, Kety & Pennes, 1945).

The time course of these changes is of interest. Davies & Remond (1946) and Davies, Grenell & Bronk (1948) devised a system whereby the P_{O_2} of the cortex could be measured polargraphically and the local oxygen consumption computed from the rate at which P_{O_2} fell when the electrode was pressed on to the tissue and caused ischaemia. These workers observed that when convulsions were caused by electrical stimulation, local oxygen consumption doubled and then fell to some two thirds of the resting rate. At this point, venous blood draining the area was observed to redden. The oxygen consumption then rose to resting levels.

These results confirmed predictions made by Davis, McCullogh & Roseman (1944) that convulsions were associated with a rise in oxygen consumption but that the rise in blood flow was either inadequate or delayed since local tissue P_{O_2} fell on the surface of the brain during convulsions to levels even lower than those seen when the animal was given nitrogen to inhale. If this view of the mechanism of the fall in P_{O_2} is correct, then it suggests that the rise in oxygen consumption must occur very soon after the stimulus is applied for Davis et al. (1944) observed that the changes in P_{O_2} occurred before any change was observed in the EEG pattern.

The pattern of change of P_{O_2} on the surface of the brain during the administration of analeptic drugs (pentamethylene-tetrazol) has also been studied by Ingvar, Lübbers & Siesjö (1962). They confirmed that P_{O_2} fell during a convulsion but that if the EEG pattern was merely activated or desynchronized by electrical stimulation of the brain stem, P_{O_2} remained either constant or increased. In both circumstances, there was indirect evidence of an increase in blood flow (Ingvar, 1958) and, in a proportion

Effects of analeptic drugs

of experiments, a pressor response. It should be noted that these experiments were carried out in decerebrate animals (encephalé or cerveaux isolés) in contrast to the earlier experiments which have been quoted and it has also been shown that in the non-anaesthetized animal, EEG activation is invariably accompanied by a rise in cerebral blood flow (Kanzow, Held & Richtering, 1960).

Recently, however, it has been shown that although the calibre of pial vessels increases with the local application of 0.5 per cent strychnine (Mchedlishvili, Baramidze & Nikolaishvili, 1967), the small cortical vessels of resting diameter 10–34 μm diminish in calibre (Mchedlishvili & Baramidze, 1967). These changes progress with time, i.e. for some minutes after application of the drug and they are similar to those seen after brief periods of cerebral circulatory arrest. From these results, Mchedlishvili and his colleagues have proposed that, in general, the pial rather than the smaller cortical vessels regulate the amount of blood which flows to the brain and that, in particular, cortical circulation is deficient during convulsions. For reasons which have been fully discussed in previous chapters, it is easy to draw erroneous conclusions about changes in blood flow from changes in the calibre of pial or other vessels particularly when, as in these experiments, the calibre of vessels altered in opposite directions. However, if these observations are generally confirmed, i.e. that convulsive activity is accompanied by a reduction in calibre of some vessels, then the results obtained by other workers which have shown a rise in blood flow can only be explained by postulating that vasodilatation occurs in another part of the vascular tree. The mechanisms whereby such vasodilatation occurs have been discussed in Chapters 9 and 10. Some support for the notion that a fall in $pH_{e.c.f.}$ is involved is given by the observation that during convulsions, cortical pH falls (Jasper & Erikson, 1941) and lactic acid accumulates (Stone, Webster & Gurdjian, 1945).

There is some evidence from other series that the primary action of analeptic drugs is to cause vasoconstriction. The xanthine group, caffeine (1,3,7-trimethylxanthine), theophylline (1,3-dimethylxanthine), aminoplylline (theophylline-ethylenediamine) have been extensively tested. In early experiments, the effect of giving these drugs in convulsive or subconvulsive doses was so variable that no conclusions concerning their effect upon cerebral

11. Cerebral blood flow and metabolism

blood vessels and flow could be drawn (Wolff, 1936; Schmidt, 1950; Sokoloff, 1959). The most consistent results have been obtained in man where the intravenous injection of aminophylline in therapeutic but non-convulsive doses (0.5 g) has given rise to a fall in cerebral blood flow of up to 25 per cent of control and very large falls in venous oxygen content without evidence of any change in cerebral oxygen uptake (Wechsler, Kleiss & Kety, 1950). Similarly, the injection of sodium benzoate caused a fall in cerebral blood flow (Gibbs, Gibbs & Lennox, 1935 a). In neither of these series could the changes in carbon dioxide which accompanied the injection have accounted for the changes in blood flow.

In man, therefore, the xanthine group of drugs appear to have a predominantly vasoconstrictor action and this can be seen only when the dosage is insufficient to cause an increase in cerebral metabolic rate. Their therapeutic effectiveness could therefore have nothing to do with their alleged dilator action upon cerebral vessels but could be due to the rise in cerebral tissue [H^+] as a consequence of a fall in cerebral blood flow or to a direct effect of these drugs on specific medullary neurones.

Anaesthetic drugs. It has been found equally difficult to distinguish between a direct effect of anaesthetic agents upon cerebral blood vessels and secondary effects related to changes in cerebral metabolism, arterial pressure, alveolar ventilation and blood gas tensions which accompany general anaesthesia. The effects of changes in body temperature with anaesthesia upon cerebral vessels and metabolism are important and often overlooked. There is a further difficulty which arises from the fact that it is not easy to compare values in the same animal unanaesthetized and anaesthetized. In consequence, the effect of inhalational anaesthetics have to be compared with one another: that is, what additional change in blood flow occurs when, for example, halothane is added to nitrous oxide and oxygen. This difficulty has been circumvented to a certain extent in man by the use of normal volunteers: in most series, however, data has been obtained from patients with a variety of disorders. For all these reasons, the vascular response of cerebral vessels to anaesthetic agents is only partially known.

Diethyl ether. Early experiments suggested that ether dilates

Effects of anaesthetics

cerebral vessels. Using the cranial window technique, Finesinger & Cobb (1935) showed that cerebral vessels constricted as the effects of ether wore off. Using a thermocouple as an index of cerebral blood flow, Schmidt & Hendrix (1938) demonstrated a rise in flow with ether in the parietal cortex, in the hypothalamus (Schmidt, 1934) and in the medulla (Schmidt & Pierson, 1934). In none of these experiments were any measurements made of carbon dioxide in blood: we are merely told that there was no evidence of respiratory depression. It is therefore by no means certain whether the changes in vessel diameter or blood flow could not have been accounted for by changes in blood gas tensions with anaesthesia. The evidence with regard to the action of ether has been further confused by a number of indirect studies, e.g. measurement of carotid artery flow (Bennett, Bassett & Beecher, 1944); measurements of c.s.f. pressure (Stephen, Woodhall, Golden, Martin & Nowill, 1954), intracranial pressure (Koopmans, 1939) or intracranial volume (White, Verlot, Silverstone & Beecher, 1942). There have been no satisfactory qualitative measurements of the effects of ether upon cerebral blood flow.

Ether is reported to reduce cerebral metabolism *in vivo* (Kety, unpublished data quoted by Sokoloff, 1959) and *in vitro* (Butler, 1950; Hunter & Lowry, 1956) although the concentrations used were considerably greater than those found in anaesthetized subjects.

Chloroform. Finesinger & Cobb (1935) observed that chloroform caused dilatation of pial vessels. It is not certain how specific this response was since P_{a, CO_2} was not measured and the animals had very low blood pressures: both a rise in carbon dioxide with anaesthesia and hypotension of the order recorded could give rise to cerebral vasodilatation. McDowall (1965) reported that cortical blood flow in dogs increased significantly at constant P_{a, CO_2} and without a fall in arterial pressure which would be likely to affect blood flow. There was no obvious reduction in the level of oxygen uptake by the brain.

Other inhalational anaesthetics. The effect of *nitrous oxide* given in anaesthetic doses (usually with high oxygen) upon cerebral blood flow has not been precisely determined. McDowall, Harper & Jacobsen (1964) found that in dogs under nitrous oxide anaesthesia the average value for cortical blood flow was 76 ml 100 g^{-1} min^{-1}. These values fall within the range found in conscious

11. Cerebral blood flow and metabolism

human subjects in whom blood flow was measured using the same technique (Ingvar & Lassen, 1962) and are substantially lower than those found in dogs under pentobarbitone anaesthesia, 99 ml 100 g^{-1} min^{-1} (Harper, Glass & Glover, 1961). The variation in results may reflect no more than differences in technique and in interpreting the radioactive indicator washout curves. Thus, the evidence, admittedly sparse, suggests that nitrous oxide has little effect upon overall cortical flow or metabolism. It is of interest, however, that the differences in blood flow between various areas of the cortex observed by Landau et al. (1955) in conscious cats were largely eliminated under nitrous oxide anaesthesia (Harper et al. 1961) as they were with barbiturates. Cortical blood flow was therefore more uniform. Some evidence that nitrous oxide may have a direct action upon vascular smooth muscle has been given by Price & Price (1962) who showed that nitrous oxide augmented the response of aortic strips to noradrenaline by 25 per cent: this observation may or may not be relevant to cerebral vessels.

Halothane. This, of all the inhalational anaesthetics, has the most profound effect upon cerebral blood flow and metabolism. McDowall, Harper & Jacobson (1963) found that with inhaled concentrations of 0.5 per cent in order to minimize hypotension and with P_{a,CO_2} controlled, halothane caused a 46 per cent reduction in cortical blood flow and 49 per cent reduction in cerebral oxygen uptake in the dog, these values being relative to those obtained with nitrous oxide–oxygen anaesthesia. Precisely opposite results were obtained by Wollman, Alexander, Cohen, Chase, Melman & Behar (1964) and by Cohen, Wollman, Alexander, Chase & Behar (1964) in man anaesthetized with rather larger concentrations of halothane, 1.2 per cent. In these series, mean cerebral blood flow increased by 15 per cent of control while oxygen uptake fell by between 9 and 21 per cent although, as admitted by the authors, an unknown proportion of this fall could have been due to the associated fall in temperature.

These differences have been resolved in a subsequent paper by McDowall (1967) in which it was shown that halothane does cause a rise in cerebral blood and since arterial pressure falls, a striking fall in cerebral vascular resistance which is proportional to the concentration of inhaled halothane. Even greater increases in cerebral blood flow with halothane have been reported (McHenry,

Slocum, Bivens, Mayes & Hayes, 1965) but it is probable that part of this increase may have been due to a rise in P_{a,CO_2}.

The mechanism of action of halothane remains uncertain. The rise in blood flow is unlikely to be due to the release of catechol amines from the adrenal medulla (Price, Linde, Jones, Black & Price, 1959; Hamelberg, Sprouse, Mahaffey & Richardson, 1960). In other vascular beds, halothane has been found to be dilator (hind limb preparation in the dog, Burn & Epstein, 1959) or constrictor (pulmonary circulation, Wyant, Merriman, Kilduff & Thomas, 1958). Price & Price (1962) found that halothane differed from all other anaesthetics tested in its action upon aortic strips for it invariably reduced the response of these strips to noradrenaline.

This action was noticeably opposite to that of *cyclopropane* which substantially augmented the response of vascular smooth muscle to noradrenaline. The precise way in which halothane and cyclopropane evoke their vascular responses is unknown. Price & Price (1962) favoured the idea of a direct action of these drugs upon smooth muscle cells since, with respect to cyclopropane, the alternative possibilities, i.e. a reduction in the metabolism of noradrenaline or 'sensitization' of receptors to noradrenaline were excluded experimentally.

No direct studies on the effects of cyclopropane upon cerebral vessels appear to have been carried out. The effects of *trichloroethylene* have been tested by McDowall, Harper & Jacobson (1964) in dogs. They reported that this drug did not affect cerebral blood flow, that the response of cerebral vessels to carbon dioxide was unaffected but that cortical oxygen uptake was reduced by approximately 20 per cent of control. These workers have placed reliance on the method of collecting blood from the sagittal sinus in the dog, developed by Gleichmann, Ingvar, Lassen, Lübbers, Siesjö & Thews (1962). As the latter authors point out, it is difficult if not impossible to show with certainty that this blood drains the cortex only or that it is representative of the area of cortex, some 1.5 cm in diameter from which blood flow measurements are made using ^{85}Kr. For these reasons, it is doubtful whether the measurements of local oxygen consumption have much meaning. This is especially true during the testing of anaesthetic drugs which apparently cause subtle shifts of blood flow in various areas of the cortex (McDowall, 1967).

11. Cerebral blood flow and metabolism

Barbiturates. The early experiments in which the effect of barbiturates was tested either by their action upon pial vessels (Finesinger & Cobb, 1935: Sohler, Lothrop & Forbes, 1941) or upon blood flow measured with thermocouples (Field, Grayson & Rogers, 1951) have been reviewed by Sokoloff (1959). The effects were extremely variable and this variation in response was also observed in studies in which blood flow was measured quantitatively. All of these studies have confirmed the original observation that barbiturates given in anaesthetic doses cause a fall in cerebral oxygen uptake (Schmidt, Kety & Pennes, 1945; Homberger, Himwich, Etstein, York, Maresca & Himwich, 1946). But the response is a graded one, for if barbiturates are given in sedative doses, neither cerebral metabolic rate nor blood flow is affected (Kety, Woodford, Harmel, Freyhan, Appel & Schmidt, 1948; McCall & Taylor, 1952).

In some series of experiments, barbiturates have been reported as causing a rise in blood flow (Wechsler, Dripps & Kety, 1951); no change (Fazekas & Bessman, 1953; Wilson, Odom & Schieve, 1953) or a fall in blood flow (Himwich, Homberger, Maresca & Himwich, 1947; McCall & Taylor, 1952; Schieve & Wilson, 1953*b*; Pierce, Lambertsen, Deutch, Chase, Linde, Dripps & Price, 1962). Most of these discrepant results can be resolved by the fact that in few studies were important variables such as P_{a,CO_2} and arterial pressure controlled and the dosage of drugs administered varied considerably. The last series of experiments is of particular interest since it was shown that thiopentone sodium given intravenously 10–55 mg kg^{-1} gave rise to blood concentrations of approximately 25 mgl^{-1}, that with this dosage and at constant P_{a,CO_2} (but probably not at constant body temperature) cerebral blood flow and oxygen uptake approximately halved and that if the subjects were subsequently hyperventilated so that P_{a,CO_2} was reduced to 17.5 mmHg, there was a further fall in cerebral blood flow. Despite the fact that a fall in body temperature of the subjects may have contributed to a fall in blood flow and oxygen uptake, these results provide good evidence that barbiturates cause a fall in blood flow when there is a concomitant fall in oxygen uptake and that cerebral vessels still respond to carbon dioxide. This is not consistent with the earlier findings of Schieve & Wilson (1953*b*) who reported a loss of sensitivity of cerebral blood vessels to carbon dioxide under

Effects of anaesthesia

barbiturate anaesthesia. Their studies, however, were based on determinations of blood flow from A–V oxygen differences and in the presence of changes in oxygen consumption, this method is quite unreliable.

The effects of barbiturate anaesthesia do not appear to be uniform throughout the brain. Landau *et al.* (1955) and Sokoloff (1957) reported that the reduction in blood flow was virtually limited to grey matter of the cortex and the reduction in blood flow was greatest in these areas of the brain in which blood flow in the conscious animal (the cat) was greatest, viz. the auditory, visual and sensory areas. These observations indicate that local blood flow could be reduced before changes in mean cerebral blood flow can be detected.

The effect of barbiturates upon aortic strips has been tested in the rabbit by Price & Price (1962). They observed that thiopentone caused a contraction of vascular smooth muscle equivalent to between 5 and 15 per cent of the response to 0.01 to 0.08 μg noradrenaline and that it powerfully augmented the smooth muscle response to noradrenaline. If these findings can safely be applied to cerebral blood vessels, this would suggest that barbiturates have a direct constrictor action upon vascular smooth muscle and that this is independent of the secondary changes which follow alterations in cerebral metabolic rate and systemic changes in pressure and blood gas tensions. The way in which barbiturates and inhalational anaesthetics affect neural and other excitable tissue is uncertain. The gross effects of anaesthetics upon cerebral vessels *in vivo* and *in vitro* suggest that they may have a direct action upon vascular smooth muscle which is independent of, but may be reversed by, other local and systemic changes which accompany anaesthesia. Interest has recently been concentrated upon lipid fractions associated with cell membranes since this is consistent with the known solubility of most anaesthetic agents in fat. Johnson & Millar (1970) have shown an increased permeability to K^+ and Rb^+ in liposomes when exposed to anaesthetic agents and Metcalfe & Burgen (1968) and Metcalfe, Seeman & Burgen (1968) have shown by studying the nuclear magnetic resonance of anaesthetics an increased freedom of movement of membranes of red cells, synaptosomes and myelinated nerves. It is possible that the primary action of anaesthetic agents in these and other sensitive tissues (including vascular smooth muscle)

11. Cerebral blood flow and metabolism

is to cause an increase in the lipid part of the membrane and changes in associated proteins (Johnson & Miller, 1970). It will be interesting to see how far a general theory of anaesthetic action can account for the changes in both cerebral oxygen uptake and blood flow discussed in this section.

Cerebral blood flow and the EEG

Because regional oxygen consumption in the brain is technically difficult to measure, an alternative approach which has been taken is to correlate changes of cerebral blood flow with EEG appearances,

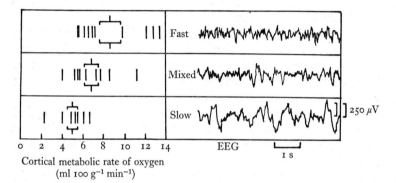

Fig. 11.10. Correlation between regional cortical oxygen uptake and EEG in the unanaesthetized dog. The EEG pattern prevailing in all leads at the time of the metabolic determinations was classified as I, *Fast* when dominated by frequencies of 5–12 Hz with absent slow waves; as II, *Mixed* when fast and slow waves were present together and as III, *Slow* when high voltage slow waves of 0.5–4 Hz dominated the pattern. On the left, each vertical bar represents one local oxygen consumption determination in the post-sigmoid gyrus. The mean and standard error of the mean are also included for each class. On the right, the corresponding EEG patterns are shown. (From Gleichman, Ingvar, Lassen, Lübbers, Siesjö & Thews (1962), *Acta Physiol., Scand.* **55**, 82–94, Fig. 2.)

using the latter as an index of functional activity of the cortex. Gleichman, Ingvar, Lassen, Lübbers, Siesjö & Thews (1962) have provided some evidence that a correlation between cortical oxygen consumption and EEG pattern exists although the method of measurement of oxygen consumption which they used is based on a number of assumptions not all of which can be validated at the moment. The correlation which they found is shown in Fig. 11.10 and consists of the finding of a predominantly fast

(5–12 Hz) pattern with high rates of oxygen consumption and predominantly slow high voltage waves (0.5–4 Hz) with low oxygen consumption. Such a correlation is consistent with the findings in man. Thus in states of coma when total cerebral oxygen consumption is reduced from 3.3 to 2.0 ml 100 g^{-1} min^{-1} (Lassen, 1959) the EEG shows a slow wave pattern (0.5–3.0 Hz) with high voltage delta waves. The coexistence of fast or mixed EEG pattern with the higher rates of cerebral metabolism is consistent with observations of Homberger, Himwich, Etstein, York, Maresca & Himwich (1946) in the dog. Further there is some evidence in the literature that with the fast EEG pattern occurring with arousal from painful stimuli or anxiety, cerebral oxygen consumption is substantially higher than at rest (Schmidt et al. 1945; Sokoloff, 1959; Kety, 1960; Lassen, Feinberg & Lane, 1960). Gleichman, Ingvar, Lassen, Lübbers, Siesjö & Thews (1962) for example compute that under such conditions, oxygen consumption of the sensory–motor cortex may be almost twice that found under more normal conditions in the lightly anaesthetized dog.

An alternative approach has been to estimate changes in local cerebral oxygen consumption by measuring local blood flow by means of a heated thermocouple and local oxygen tension by means of a Clark membrane covered oxygen electrode. The latter is placed on the cortical surface and gives a mean oxygen tension value of a volume of tissue which depends, among other things, on the diameter of the platinum cathode. Some evidence that P_{O_2} measured at the cortical surface correlates with mean tissue P_{O_2} calculated independently from P_{a,O_2} and local oxygen consumption has been provided by Gleichman, Ingvar, Lübbers, Seisjö & Thews (1962). Early qualitative studies using surface oxygen electrodes were carried out by Davis, McCulloch & Roseman (1944), Davies & Rémond (1946), Mochizucki (1951), Meyer & Denny-Brown (1955), Ingvar, Lübbers & Siesjö (1960), Meyer & Gotoh (1961). In only a few of these studies were simultaneous measurements made of local cerebral blood flow and even in a number of these, the type of thermocouple used made it likely that factors other than changes in flow were being measured. In consequence, a number of these papers include unwarranted assertions about changes in 'oxygen consumption' or 'available oxygen'. A more satisfactory study using these methods is by Ingvar, Lübbers & Siesjö (1962) in which EEG

11. Cerebral blood flow and metabolism

patterns were correlated with cortical oxygen tensions during normal conditions and when the mesencephalon was stimulated. Under normal conditions they found that slowly changing EEG patterns were accompanied by negligible changes in tissue P_{O_2} and this was attributed by these workers to the compensatory changes in blood flow which accompanied small presumed changes in local oxygen consumption. But since neither oxygen consumption was measured nor changes in blood flow observed, this conclusion is hardly supported by the data. On the other hand, relatively large changes in oxygen tension and regional blood flow were observed when the mesencephalon was stimulated or when epileptic convulsions were caused by metrazol injections. Epileptic discharges were accompanied by an early rise in blood flow followed by a fall in P_{O_2} and these changes were in turn followed by a rise in P_{O_2} and further increase in blood flow. These changes were interpreted by the authors and, in another context by Ingvar (1958) as being due to an increase in cortical oxygen consumption (marked by the fall in P_{O_2}), the subsequent rise in P_{O_2} being due to the further increase in blood flow and in oxygen supply. That the changes in P_{O_2} do probably represent alterations of tissue metabolism is supported by the observation that cortical P_{CO_2} rises during the activation period in the EEG and that this rise occurs despite the changes in blood flow described above (Seisjö, 1961; Ingvar, Seisjö & Hertz, 1959). One point of interest in these experiments is the initial dissociation between blood flow and P_{O_2}. If, as the authors conclude, the late and further rise in blood flow is secondary to the changes in metabolism, this will not explain the rise in blood flow which occurs initially before any changes in P_{O_2} are seen. This early increase in blood flow could well represent a reflex response since it is known that electrical stimulation of the fore-brain and mid-brain is accompanied by various autonomic responses (Hodes & Magoun, 1942). There is a considerable literature on the relation between cerebral blood flow and the EEG. The more important points which have emerged deserve to be emphasized in the present context. The effect of extreme changes in blood flow upon the EEG has been studied in detail by Baldwin (1960) in the sheep and calf (Baldwin & Bell, 1963b) and this has also yielded valuable information on the functional blood supply to the brain in these species. Partial cerebral ischaemia induced by occlusion of one carotid or both

Cerebral blood flow and the EEG

vertebral arteries had no effect upon the EEG. More severe ischaemia caused by bilateral occlusion of the arteries caused the EEG to display slow rhythmic activity and complete cerebral ischaemia caused abolition of EEG activity after about 8 s. These changes were broadly consistent with those found in other species, in the cat (Sugar & Gerard, 1938), in the dog (ten Cate & Horsten, 1952), in the rabbit (Leão & Morison, 1945; Van Harreveld & Stamm, 1952), in the monkey (Meyer, Feng & Denny-Brown, 1954). The time to onset of slow wave EEG pattern and final abolition of activity was also found to be a function of age, the younger the animal the longer the period of resistance (Baldwin & Bell, 1963b) and this finding is consistent with the other evidence discussed earlier in this chapter that the newborn is considerably more resistant to hypoxia than the adult.

The relation between cerebral blood flow, metabolism and EEG has been studied in man by Obrist, Sokoloff, Lassen, Lane, Butler & Feinberg (1963). They compared these variables in three groups of subjects, age controls (mean 71.0 years), and patients with various psychiatric disturbances and forms of mental deterioration. The control subjects had values for cerebral blood flow and oxygen consumption which were not significantly different from young normal adults, while the psychiatric patients had, by comparison, significant reduction in both blood flow and metabolic rate.

In the control group, there was no correlation between EEG characteristics and either blood flow or metabolism, but in the groups of patients, correlations between blood flow and oxygen uptake and the peak EEG frequency and per cent slow wave activity were found. The most obvious of these were between cerebral oxygen uptake and slow wave activity which indicated an association between depressed metabolic rate and the occurrence of waves 1–7 Hz. A further point of interest which emerged from this study was the observation that in patients in whom bilateral studies were carried out using ^{85}Kr, a higher correlation existed between EEG and circulating variables on the left, as opposed to the right, side. This finding raises the question as to whether significant differences in cerebral metabolic function are to be found between sides. Lassen, Feinberg & Lane (1960), Feinberg, Lane & Lassen (1960) reported higher correlation between intelligence test scores and cerebral oxygen metabolic rates on

11. Cerebral blood flow and metabolism

the left compared with the right with parallel EEG findings. In the study of Obrist *et al.* (1963), cerebral oxygen consumption calculated from left jugular venous blood was 15 per cent lower than on the right and the variance was twice as great. Previous studies in a younger age group have failed to show such consistent difference between left and right sides (Scheinberg, 1950; Wechsler *et al.* 1951; Munck & Lassen, 1957), and it is possible that mental deterioration is associated with a relative depression of metabolism on the left side. This would be consistent with the observation that EEG abnormalities, in particular the presence of increased slow wave activity, are prevalent over the left hemisphere in elderly people (Silverman, Busse, Barnes, 1955; Bruens, Gastaut & Giove, 1960).

Cerebral blood flow or an index of it has been related with EEG changes in two other sets of circumstances. In the first, pial artery calibre was measured with EEG electrodes placed within 5 mm (Darrow & Graff, 1945). Manoeuvres which caused pial artery dilatation, e.g. the inhalation of carbon dioxide, were associated with an increase in frequency and potential of fast waves and a reduction in slow wave activity: manoeuvres which caused vasoconstriction also caused an increase in slow wave activity and a reduction in fast wave activity. This type of study was extended to examine the relation between parasympathetic activity and the EEG (Darrow, Green, Davis & Garol, 1944). Previous work by Darrow & Pathman (1943, 1944) had shown that the tachycardia which accompanied hyperventilation and fall in carbon dioxide tension was largely due to inhibition of vagal efferent activity to the heart. It had also previously been shown that stimulation of the vagi caused an increase in fast wave EEG. Bailey & Bremer (1938) and Darrow *et al.* (1944) used this to test a possible mode of action of the vasodilator pathway, described by Cobb & Finesinger (1932), Putnam & Ask-Upmark (1934), carried by the VIIth facial nerve to cerebral blood vessels. When the VIIth cranial nerves were intact, 2 min periods of hyperventilation were without effect upon the EEG but when the VIIth nerves had been cut, and the animals again hyperventilated, the EEG developed the slow wave activity which was typical of hypoxia and depression of tissue metabolism. These slow waves were reduced or abolished when the facial nerves were stimulated: when the nerves were intact, physostigmine prevented the effects

Cerebral blood flow and the EEG

of hyperventilation and they were exaggerated with the administration of atropine.

These results are of interest in providing further evidence for an important cholinergic pathway to blood vessels of the brain and a reasonable interpretation of the results would be that the vasodilatation limited the constriction which normally occurs with hyperventilation and hypocapnia. When the vasodilatation pathway is interrupted either by division of the facial nerve or by atropine, unopposed vasoconstriction causes marked tissue hypoxia with accompanying EEG changes. It is also probable that carbon dioxide and acetylcholine (ACh) interact since it was known that high carbon dioxide reduced the slow spiked EEG activity while Plattner, Galehr & Kodera (1928) and Gesell, Brassfield & Hamilton (1942) have shown *in vitro* and *in vivo* respectively that changes in pH affect cholinesterase activity and the effectiveness of ACh. This pathway involving vasodilator fibres to cerebral vessels and consequent changes in blood flow may explain the changes in EEG activity which occur in a variety of emotional states involving the autonomic nervous system and changes in blood gas tensions which have been extensively documented (Darrow & Gellhorn, 1939; Darrow, 1943; Darrow & Solomon, 1940; Solomon, Darrow & Blaurock, 1939).

A second and more direct approach was used by Ingvar & Soderberg (1958) to study changes in cortical blood flow and EEG when the brain stem was stimulated. As would be expected, stimulation of the brain stem in different areas evoked different EEG patterns and vascular responses. Four types of EEG response were obtained: (1) the arousal reaction; (2) the 'flattening' reaction with reduction of amplitude and with only occasional slow waves; (3) sleep spindle type of discharge; or (4) a response characterized by generalized high voltage slow waves and which was commonly seen with evidence of circulatory failure and cerebral oedema. With only the first of these responses was there clear evidence of a rise in blood flow and, since no changes in perfusion pressure were observed, this indicated that vasodilation had occurred. Ingvar & Soderberg (1958) interpreted these changes largely in terms of the alterations in oxygen consumption that are known to occur. As has been mentioned previously in this section, the alternative possibility, namely, that activity in vasodilator nerves to cerebral vessels increases should also be considered.

12 SOME ASPECTS OF THE PHARMACOLOGY OF CEREBRAL VASCULAR SMOOTH MUSCLE

The experimental methods used in the past to study the responses of cerebral blood vessels to the administration of various drugs have been almost without exception so inadequate that the results have been indecisive and the literature correspondingly confused. Thus up until the last year or so, only one series of studies *in vitro* had been carried out (Cow, 1911), and it is uncertain even in this study whether the strips of carotid artery studied came from within or without the skull. Although the results of this study have often been quoted, they must be considered as preliminary and qualitative since they deal with responses elicited when drugs of unknown concentration were added to a perfusate of uncertain ionic composition.

The results of experiments carried out *in vivo* have, in general, been similarly unsatisfactory. If cerebral vascular responses have been elicited, it has been difficult if not impossible to interpret their significance. For example, the intravenous infusion of adrenaline has been shown to cause an increase, a fall or no change in the calibre of pial vessels and blood flow, and whether these changes can be ascribed to adrenaline is uncertain because of the variable and often large changes in systemic arterial pressure which accompany the infusion. Furthermore, adrenaline and many other drugs also affect the rate of cerebral metabolism, changes in which may also affect the level of cerebral blood flow through mechanisms discussed in Chapter 10. As has been shown in Chapter 11, in the context of anaesthetic and analeptic drugs, the problem of distinguishing between direct effects of drugs upon cerebral vessels and indirect effects mediated through changes in metabolism is particularly acute.

For all these reasons, we know very little about the precise way in which drugs affect cerebral vascular smooth muscle. The literature in this field up to 1959 has been comprehensively

The pharmacology of cerebral vascular smooth muscle

reviewed by Sokoloff. In the present chapter the more recent developments will be considered and particular attention paid to the action of drugs whose action may help in resolving the contraversial question of how or whether the cerebral vessels are innervated.

Action of catechol amines upon cerebral blood vessels

Adrenaline

The effect of either adrenaline or L-adrenaline upon cerebral blood flow or the calibre of pial vessels as measured in early experiments may be summarized as follows. When applied locally to individual vessels on the cerebral cortex, adrenaline was found to cause vasoconstriction in the majority of studies (e.g. Forbes & Wolff, 1928). Given by intravenous or intracarotid arterial injection, adrenaline has been found to cause either vasoconstriction or vasodilatation, the response depending on the dose administered and on the degree to which systemic arterial pressure rises. If arterial pressure is controlled, then adrenaline invariably causes vasoconstriction of the pial vessels of the cat and monkey (Finesinger & Putnam, 1933, Fog, 1939a). Some studies in which thermocouples have been used as an index of blood flow have suggested that the reduction in blood flow in response to adrenaline varies in different parts of the brain.

In man, adrenaline has been found to have little effect upon cerebral blood vessels. Thus Gibbs, Gibbs & Lennox (1935a) found that blood flow and arterial pressure rose in parallel while King, Sokoloff & Wechsler (1952) found that blood flow increased proportionately to cerebral oxygen uptake. It is probable that the increase in blood flow observed in these two studies cannot be accounted for by the rise in pressure since blood flow has been found to be independent of blood pressure for some way above the physiological range of pressure: on the other hand, a rise in pressure would be expected by itself to cause constriction of pial vessels (Fog, 1939b). It is therefore difficult to assess how far the pial vasoconstriction noted by early workers was due to the action of adrenaline and how far it was due to the accompanying rise in blood pressure. The increase in blood flow observed in man in response to the infusion of adrenaline is also difficult to assess because of the accompanying rise in cerebral oxygen uptake

12. *The pharmacology of cerebral vascular smooth muscle*

– that is, the rise in flow may have been secondary to the rise in e.c.f. [H^+] (see Chapter 10). If this was so, then any direct vasoconstricting effect of adrenaline upon cerebral vessels would have been modified or reversed.

In view of these uncertainties, it is remarkable that, until very recently, no-one has tested the effect of adrenaline upon cerebral vessels *in vitro*. The first study of this kind was undertaken by Bohr, Goulet & Taquini (1961) who tested the effects of various amines upon cerebral resistance vessels (200–300 μm in diameter) in the dog. In their hands, neither adrenaline nor noradrenaline was found to have any effect upon the contraction of the arterial strips. It is possible that this may have been due either to the method of storage of the material after death for 24 hours at 4 °C or to the cutting of these small arteries into strips which is likely to affect the contractile response of smooth muscle or to the method of mounting these strips which these authors describe and which is likely seriously to diminish the sensitivity of the response.

In a further study, the effects of adrenaline were tested upon 4 mm lengths of middle cerebral artery in the cat *in vitro* (Nielsen & Owman, 1971). The log dose–response curve is shown in Fig. 12.1 and confirms that under these conditions, adrenaline causes a contraction of cerebral vascular smooth muscle which, in quantitative terms is only exceeded by 5-hydroxytryptamine. These workers have also shown that the maximum response was only achieved gradually, that is after 2–3 min, Fig. 12.2. This type of response could be explained by supposing that adrenaline diffuses only slowly to the receptor sites. Alternatively, it could mean, according to the hypothesis proposed by Ariens & Simonis (1964) that the intrinsic activity of adrenaline is determined by the number of receptors occupied rather than the rate, until equilibrium is established.

Noradrenaline

In contrast to adrenaline, the actions of L-noradrenaline upon cerebral blood vessels have been found to be more clear cut and consistent. When given intravenously in pressor doses, noradrenaline causes a reduction in blood flow without obvious change in cerebral oxygen uptake (King, Sokoloff & Wechsler, 1952). As mentioned in the previous section, Bohr *et al.* (1961)

Action of catechol amines upon cerebral blood vessels

did not observe any direct action of noradrenaline upon cerebral vessels *in vitro*: on the other hand, Nielsen & Owman (1971) have confirmed that noradrenaline causes contraction of cerebral vascular smooth muscle and have shown that in equivalent dosage, its effect is substantially less than for adrenaline. The contraction

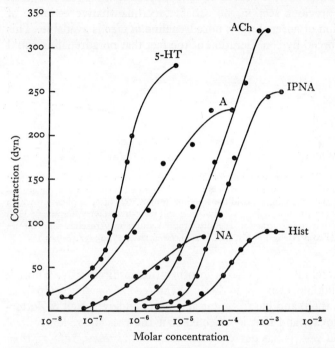

Fig. 12.1. Log dose–response curves showing contractile responses of lengths of isolated middle cerebral artery from three different cats after fractionated application of 5-hydroxytryptamine (5-HT), adrenaline (A), noradrenaline (NA), acetylcholine (ACh), isoprenaline (IPNA) and cumulative application of histamine (Hist). The catecholamine curves were obtained from one and the same artery. (From Nielsen & Owman (1971), *Brain Res.* **27**, 25–32, Fig. 2.)

of smooth muscle is also different in that it is rapid in onset, resembling the response to acetylcholine, Fig. 12.2, and this according to Ariens & Simonis (1964) would be consistent with the notion that the activity of noradrenaline is determined by the rate of occupation of receptors.

The fact that noradrenaline *in vitro* appears to be a less powerful vasoconstrictor of cerebral vascular smooth muscle than adrenaline

12. The pharmacology of cerebral vascular smooth muscle

appears at first sight to be inconsistent with the relative effectiveness of these drugs *in vivo*. Nielsen & Owman (1971) have suggested that one reason for this may be that *in vitro*, a proportion of the noradrenaline added to the perfusate is taken up by the adrenergic nerves present and less is available for action at the receptor sites. At the same time, it should be remembered that for reasons given in the previous section, no satisfactory quantitative estimate of the action of adrenaline or noradrenaline *in vivo* is available. This is reinforced by consideration of the fact that noradrenaline could

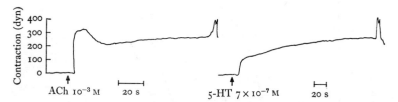

Fig. 12.2. Time–response curves illustrating the two modes of contraction observed in the various experiments. The illustrated responses are for acetylcholine (ACh) showing a rapid contraction immediately after application which then fades to an intermediate steady level; and for 5-hydroxytryptamine (5-HT) showing that the contraction is more gradual and slowly reaches a maximum plateau. (From Nielsen & Owman (1971), *Brain Res.* **27**, 25–32, Fig. 3.)

affect cerebral blood vessels in other than a direct way. Thus, it could initiate changes in both chemoreceptor activity (Joels & White, 1968) and baroreceptor activity (Landgren, Neil & Zotterman, 1952); and if it is objected that the case for reflex control of blood vessels has not yet been made out (Lassen, 1968a), then it is worth recalling that noradrenaline has been shown to cause a marked increase in alveolar ventilation in man (Whelan & Young, 1953; Cunningham, Hey & Lloyd, 1958), and in the cat (Joels & White, 1968), which would itself have the effect of reducing P_{a,CO_2}, causing cerebral vasoconstriction and a fall in cerebral blood flow. The situation *in vivo* is, as expected, always more complex than it at first appears and it is unlikely that we shall know the effect of noradrenaline or other amines upon cerebral vessels in quantitative terms until many of the earlier experiments are repeated under more carefully controlled conditions.

Action of catechol amines upon cerebral blood vessels

Isoprenaline

The pharmacological interest in isoprenaline lies in the fact that it appears to act upon vascular smooth muscle principally if not entirely through beta-adrenergic receptors. It has been shown by Furchgott (1967) that under defined conditions with respect to the ratio of antagonist and agonist, the potency series is isoprenaline > noradrenaline > adrenaline > phenylephrine whereas with respect to the alpha-receptors, isoprenaline is considerably less potent than the other amines.

The action of isoprenaline upon cerebral vessels *in vitro* has been tested by Nielsen & Owman (1971) and was found to cause a contraction of smooth muscle with a potency less than that of 5-HT, adrenaline and acetylcholine but greater than that of noradrenaline and histamine, Fig. 12.1. The time course of the response resembled that of adrenaline and 5-HT, Fig. 12.2.

The effect of isoprenaline upon cerebral blood vessels *in vivo* is very much more complex. Laubie & Drouillat (1967) have shown that the brain participates in the whole body increase in oxygen uptake which occurs with the intravenous infusion of isoprenaline. In fact, the changes in cerebral oxygen uptake are more striking than those of the whole body being +100 and +33 per cent of control respectively. These workers also reported that isoprenaline caused an increase in cerebral blood flow and a fall in calculated vascular resistance. The response in this respect was thus similar to that seen in the coronary circulation (Nakano, McGiff, Zeckert, Guggenheim & Wegria, 1961), in the splanchnic circulation (Lacroix, 1962) and in skeletal muscles (Green & Krepchar, 1959).

The methods used by these workers to measure cerebral blood flow and oxygen uptake must be considered as less than satisfactory since both measurements depend on the demonstration that the venous blood sampled (for nitrous oxide and oxygen contents, respectively) drains the brain only and is not contaminated by blood from extracerebral sources. In the dog, this is difficult if not impossible to achieve (see the discussion in Chapter 1) and the method used by these authors to ensure separation of venous drainage, namely, the ligation of the internal maxillary and posterior auricular veins is unlikely to have been effective when blood is sampled from the jugular vein. For these reasons, it is uncertain whether the values given by these authors refer only to brain or

12. The pharmacology of cerebral vascular smooth muscle

whether they include extracerebral vascular beds, notably in muscle.

More recently, however, these results have been largely confirmed by James, Xanalatos & Nashat (1970). They measured cortical blood flow in the dog using ^{85}Kr and sampled blood for measurements of cortical oxygen consumption from the sagittal vein. They confirmed that the intravenous infusion of isoprenaline causes an average increase in cerebral blood flow and oxygen consumption of 50 per cent and an increase in glucose uptake by 80 per cent of control. They also showed that these increases could be almost completely abolished by the inhalation of 5 per cent carbon dioxide and this would be consistent with the observation that the vascular effect of isoprenaline varies with the pH of arterial blood (Schroeder, Robison, Miller & Harrison, 1970).

These results are not easy to interpret. If we accept that isoprenaline *in vitro* causes constriction of cerebral vessels, the dilatation seen *in vivo* could be associated with the increased rate of cerebral metabolism. It may be pertinent that in these experiments, cerebral blood flow and metabolism increased by the same amounts and that these changes could not be accounted for by alterations in P_{a,CO_2} (since this was controlled) or by the relatively small changes in perfusion pressure. Clearly, further experiments are required to see how far these changes can be affected or modified after beta-receptor blockade.

The effect of adrenergic blocking agents

Early studies in which the effects of ergotamine tartrate and the dihydrogenated ergot compounds dihydroergocornine, dihydroergocristine and dihydroergocryptine upon cerebral vessels was studied have been reviewed by Sokoloff (1959). The results showed that if ergot compounds are given intravenously, there is a widespread vasodilatation and fall in arterial pressure. The effect upon cerebral vessels, measured by direct inspection of pial vessels, the use of thermocouples and the nitrous oxide method has been found to be very variable: the more consistent results have suggested that cerebral blood vessels dilate, blood flow remains unchanged and cerebral vascular resistance falls. As occurs often in this type of experiment, the results are difficult to interpret because of the associated changes in arterial pressure. At the same

Action of catechol amines upon cerebral blood vessels

time they provide a totally inadequate basis for any hypothesis concerning the aetiology or therapeutic value of ergot compounds in migraine and allied cerebral vascular derangements. The results neither confirm nor contradict the proposition that adrenergic vasomotor fibres are involved in the reflex maintenance of cerebral vascular tone: similarly, the results of experiments in which autonomic ganglionic transmission has been blocked with hexamethonium and other similar drugs are of limited value in determining whether these drugs have a direct effect upon cerebral vessels or whether the inhibition of post-ganglionic vasomotor activity affects cerebral blood vessels or flow. Such effects as have been observed can almost certainly be accounted for by the large, often very large, changes in systemic arterial pressure which occur. The effects of these blocking drugs may therefore be non-specific and indistinguishable from those observed during hypotension induced by differential spinal sympathetic block and high spinal anaesthesia.

Of more immediate interest are the results of experiments in which the action of drugs which block alpha- or beta-adrenergic receptors has been tested since these are likely to show in what respects, if at all, cerebral blood vessels differ from those in other vascular beds in their response to sympathomimetic drugs and in their pattern of adrenergic innervation.

One of the first attempts to study the mechanism of action of catechol amines upon cerebral vessels was undertaken by Rosenblum (1969b). His motive was to find a suitable way of reversing cerebral vasoconstriction which is a feature of a number of intracranial disasters, notably following haemorrhage. In order to produce a suitable experimental vasoconstriction, Rosenblum applied 0.5 ml of 0.2 M barium chloride locally to pial vessels in mice and measured the changes in pial arterial calibre. The interest of barium in this connection is that it appears to activate genuine contractile mechanisms in smooth muscle and that among the inhibitors of this action are the drugs which block alpha- and beta-adrenergic receptors. This has been shown in rat's aortic tissue (Rosenblum, 1969a) and it is possible that barium acts by causing the release of neurosecretory agents from their storage sites (Ambache, 1949; Sachs, Share, Osinchak & Carpi, 1967).

In the cerebral vascular bed, Rosenblum (1969b) showed that

12. *The pharmacology of cerebral vascular smooth muscle*

the contraction caused by the application of barium was abolished by dichloroisoproterenol or propranalol, both beta-blocking agents and that the probability of inhibition increased with the dose used. Phentolamine mesylate, an alpha-receptor blocking agent or lignocaine hydrochloride failed in the majority of studies to reverse the contraction. These results largely confirm those of an earlier study (Lende, 1960) in which it was shown that the vasoconstriction of pial vessels induced by mechanical irritation could be reversed by the local application of beta-blocking agents. Taken together, these studies indicate that these particular types of vasoconstriction are mediated principally by beta-receptors but they yield little direct information about the mechanism of vasoconstriction caused either by catechol amines or by stimulation of vasomotor nerves. Some further evidence that beta-receptors could be involved in other types of vasoconstriction was given by Falck, Nielsen & Owman (1968) when they showed that the vasoconstriction caused by hyperventilation causing a lowered P_{a,CO_2} could be reversed by the administration of D-N-isopropyl-p-nitrophenylethanolamine (D-INPEA) a beta-blocking agent. There have been no satisfactory studies *in vivo* in which the action of specific alpha-receptor blocking agents have been tested.

Nielsen & Owman (1971) have however studied the effects of both alpha- and beta-receptor blocking agents *in vitro*. The beta-blocking agent D-INPEA did not affect the contraction of cerebral arteries induced by noradrenaline, Fig. 12.3: it reduced the amplitude of the contraction caused by adrenaline and, even more markedly, the response to isoprenaline. Addition of the alpha-blocking agent phenoxybenzamine abolished the response to noradrenaline, reduced the response to adrenaline and only slightly affected the response to isoprenaline. If both alpha- and beta-receptor blocking agents were added together, the vascular responses to these three amines were abolished.

These findings are in general agreement with the system proposed by Ahlquist (1948) and reviewed in greater detail in the context of vascular smooth muscle (Ahlquist, 1965). It is also probable that the further differentiation of the beta-class of adrenergic receptors proposed by Furchgott (1967) will become yet further refined as sub-classes of beta-adrenergic receptors are recognized.

Action of catechol amines upon cerebral blood vessels

The significance of these results is that cerebral blood vessels (to be precise, the middle cerebral artery which has been most thoroughly studied) share the same system of adrenergic receptors which have been found in most other vascular beds. If nothing else, this provides further support for the existence of a functionally

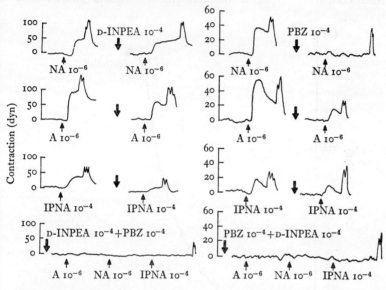

Fig. 12.3. Representative experiments with adrenergic blocking agents on two arteries (middle cerebral from the cat).

Left: the beta-blocking agent D-INPEA does not affect the noradrenaline (NA) contraction, it does reduce the adrenaline (A) contraction and the effect of isoprenaline (IPNA) is almost abolished.

Right: the alpha-receptor blocking agent phenoxybenzamine (PBZ) completely blocks the NA response, the response to A is more markedly affected than with the beta-blocker D-INPEA and there is only a slight reduction in the IPNA response.

Below: the catecholamine-induced contractions are abolished in the combined presence of both the alpha- and beta-blocking agents. Powers of ten are molar concentrations. (From Nielsen & Owman (1971), *Brain Res.* **27**, 33–42, Fig. 5.)

active adrenergic innervation of cerebral vessels though, as has been pointed out by Falck, Nielsen & Owman (1968), the distribution of adrenergic fibres as shown by fluorescent techniques is not even on cerebral vessels: it is therefore quite possible that the distribution of alpha-receptors and the ratio of these to beta-receptors varies from one cerebral artery to another.

12. The pharmacology of cerebral vascular smooth muscle

Tyramine

This is thought to exert its sympathomimetic effect by displacing the adrenergic transmitter (Burn & Rand, 1958; Langer, Draskoczy & Trendelenburg, 1967) but it is also probable that it exerts an independent effect upon vascular smooth muscle (Carlsson &

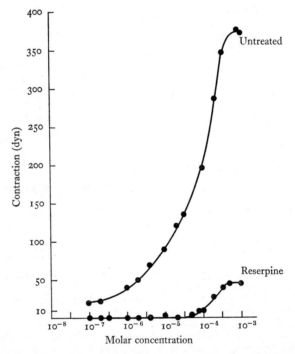

Fig. 12.4. Cumulative dose–response curves showing the tyramine-induced contractile response in a segment of the middle cerebral artery from an untreated animal and from an animal pretreated 21 h previously with 5 mg kg^{-1} reserpine intraperitoneally. The slight response remaining after reserpinization represents the direct, unspecific effect of tyramine (Nielsen, Owman & Sporrong, unpublished data).

Waldeck, 1963; Takenaka, 1963; Zaimis, 1968). Its action on cerebral vascular smooth muscle has been tested by Nielsen, Owman & Sporrong and this is illustrated in Fig. 12.4 which shows that tyramine causes contraction of smooth muscle *in vitro* and that this action is largely abolished if the animals had been

Action of catechol amines upon cerebral blood vessels

pretreated with reserpine or if the superior cervical ganglion had been removed one week before the experiment. These results indicate that the nerves accompanying cerebral vessels in the untreated animals can release sufficient transmitter to cause contraction of cerebral vascular smooth muscle. Whether they do so is another matter.

Mechanism of action of catechol amines upon cerebral vascular smooth muscle

It must be admitted that so far no studies in which the sucrose gap or intracellular recording techniques have been used have been carried out on cerebral vessels and, in consequence we do not know whether the correlation between electrical and mechanical events studied in other types of smooth muscle hold with respect to cerebral vessels.

In the smooth muscle of the guinea-pig ileum, adrenaline blocks the spontaneously occurring spike discharge when given in doses of 10^{-7} g ml^{-1} *in vitro* and causes hyperpolarization and relaxation of smooth muscle (Bülbring & Kuriyama, 1963). The associated rise in membrane potential is almost certainly due to an increase in the membrane permeability to potassium ions. In vascular smooth muscle, no such spontaneous action potentials have been found *in vitro* in the presence of physiological concentrations of potassium and sodium although action potentials can be induced by the addition of vasoconstrictor drugs (Keatinge, 1964). Further studies by Keatinge (1968) have shown that action potentials in vascular smooth muscle are carried by sodium ions so that, in this respect, vascular smooth muscle resembles striated muscle more closely than intestinal smooth muscle.

In vascular smooth muscle, adrenaline and noradrenaline cause depolarization and contraction (Keatinge, 1964) and *in vivo* these amines have been found to cause a rise in frequency of action potentials in rat mesenteric arteries when applied locally (Steadman, 1966). However, when injected intravenously, these drugs give rise to variable responses and this raises the point that apart from complicating reflex and chemical factors (see p. 338), the action of catechol amines *in vivo* and *in vitro* may be quite different. Thus the sustained depolarization with noradrenaline observed *in vitro* by Keatinge has not been confirmed *in vivo*; Speden (1967) has shown other differences, notably values for membrane and

12. The pharmacology of cerebral vascular smooth muscle

action potential and Su & Bevan (1966) have suggested that smooth muscle contraction due to electrical and other means may vary considerably in the two situations. It is also possible that, the evidence *in vitro* notwithstanding, cerebral vascular responses *in vivo* are more akin to those of the coronary circulation. Noradrenaline might then exert principally beta effects – a rise in blood flow and oxygen uptake – effects which would be abolished by propranolol (Barratt, 1965). Some preliminary evidence that this in fact occurs has been obtained by James (personal communication). In view of previous discrepancies, this whole question could well be re-examined.

The action of acetylcholine upon cerebral vessels

Interest in the action of acetylcholine upon cerebral blood vessels arises in part from the demonstration of a vasodilator pathway and the more recent evidence which suggests the existence of cholinergic fibres which accompany pial vessels (Lavrentieva, Mchedlishvili & Plechkova, 1968).

Early studies showed that acetylcholine or acetyl-β-methylcholine cause dilatation of pial vessels (Schmidt & Hendrix, 1938) and a rise in cerebral blood flow (Lubsen, 1941; Norcross, 1938; Dumke & Schmidt, 1943). The changes in blood flow occur despite the accompanying changes in systemic arterial pressure and persist after blood pressure has returned to normal levels. In this respect therefore, the cerebral vessels respond as do vessels in other vascular beds. The action of acetylcholine is blocked by atropine (Norcross, 1938). Consistent with these findings is the demonstration that the dilatation of pial vessels in response to a fall in blood pressure is abolished by atropine and other drugs with a similar action (Mchedlishvili & Nikolaishvili, 1970). It should be noted that these results do not by themselves confirm the presence of vasodilator nerve fibres or their possible role in the vascular response to hypotension since it is clear from other sources (Paton & Zar, 1968) that acetylcholine or atropine could have a direct action upon vascular smooth muscle. For the demonstration that the dilator nerves are truly cholinergic, it would be necessary in addition to the evidence quoted above to show that the effect of stimulation of these fibres and the action of exogenous acetylcholine is potentiated by eserine or

The action of acetylcholine upon cerebral vessels

other anti-cholinesterase substances and that stimulation of the dilator nerves causes the liberation of an acetylcholine-like substance. There is no evidence in the literature that such experiments have yet been carried out. The evidence however is strongly suggestive that there is a system of cholinergic fibres to cerebral blood vessels which are probably dilator; that there are cholinergic-like terminals with agranular vesicles near smooth muscle cells of pial and other small cerebral arteries and that these dilator fibres are principally involved in maintaining a constant blood flow in the face of arterial hypotension and in the vascular response to hypoxia and hypercapnia.

In vitro, acetylcholine has consistently been found to cause constriction of vascular smooth muscle from carotid artery strips (Keatinge, 1966) and in lengths of middle cerebral artery (Nielsen & Owman, 1971). From the latter study, the potency of acetylcholine appears to be similar to that of adrenaline and isoprenaline. The contractile response to acetylcholine is rapid, Fig. 12.2 and is abolished by atropine, Fig. 12.5.

In smooth muscle of guinea-pig ileum, acetylcholine causes membrane depolarization and an increase in spontaneous spike discharge whether or not the cholinergic innervation is intact (Bülbring & Burnstock, 1960). The increase in tension within smooth muscle fibres appears to be proportional to the frequency of spike activity (Bülbring, 1957) and more recent studies have shown that acetylcholine causes contraction of smooth muscle partly by releasing calcium from cellular stores and partly by increasing the membrane permeability for calcium (Durbin & Jenkinson, 1961; Schatzmann, 1961). The overall effects of acetylcholine observed by Nielsen & Owman (1971) are thus similar to those seen in guinea-pig ileum although this does not mean that the mechanisms of contraction are the same. Indeed it remains something of a mystery why acetylcholine does cause contraction of vascular smooth muscle *in vitro* and dilatation *in vivo*. Even in studies *in vitro*, different types of responses have been observed. Thus Keatinge (1965) has shown that acetylcholine applied to carotid artery strips gives rise to an irregular spike discharge which is abolished by hexamethonium, nicotine and alpha-receptor blocking agents. This was interpreted as showing that, under these conditions, acetylcholine could activate adrenergic fibres present and cause release of transmitter. In studies on the

12. The pharmacology of cerebral vascular smooth muscle

pulmonary artery of the rabbit, Su & Bevan (1965), acetylcholine, histamine and 5-HT were found to cause only a slow depolarization.

A further possibility, raised by Furchgott (1967) is that with the methods of mounting vascular smooth muscle used in most preparations *in vitro*, smooth muscle relaxation may be difficult

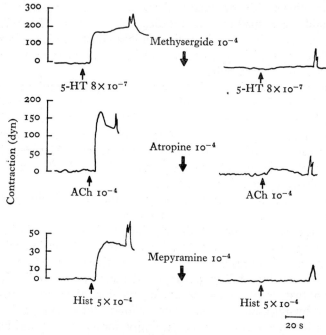

Fig. 12.5. Representative curves from three different arteries showing that the contractile response to 5-hydroxytryptamine (5-HT) is antagonized by methysergide, acetylcholine (ACh) by atropine, and histamine (Hist) by mepyramine. Powers of ten are molar concentrations. (From Nielsen & Owman (1971), *Brain Res.* **27**, 25–32, Fig. 4.)

to detect. This is unlikely to be the explanation since Keatinge (1966), for example, was able to demonstrate that amyl or sodium nitrite caused repolarization and relaxation of carotid artery strips which had previously contracted and repolarized in response to noradrenaline while Nielsen & Owman (1970) showed that in their type of preparation, the lengths of middle cerebral artery relaxed in response to a lowering of the temperature or pH of the perfusate or to the addition of histamine in the presence of mepyramine. The discrepancy between the response *in vivo* and *in vitro* of vascular smooth muscle to acetylcholine remains unexplained.

Histamine

There is general agreement in the literature that histamine causes dilatation of cerebral vessels as in other vascular beds (Sokoloff, 1959). The dilatation of vessels is more obvious than the increase in blood flow and this may be because of the associated hypotension: however, when perfusion pressure is held constant, histamine has been shown to cause an increase in blood flow at the same time as the pial vessels dilate (Finesinger & Putnam, 1933). In man, histamine has been found to cause an increase in cerebral blood flow provided that the systemic arterial pressure does not fall by more than 10–15 mmHg: with more severe hypotension, cerebral blood flow falls (Shenkin, 1951).

In vitro, histamine has been found to cause contraction and depolarization of smooth muscle of carotid artery strips (Keatinge, 1964) and contraction of smooth muscle has been confirmed by Nielsen & Owman (1971). These workers also showed that many of their cerebral artery preparations developed rapid tachyphylaxis to histamine so that satisfactory dose–response curves could be obtained by cumulative application, Fig. 12.1. The smooth muscle response to histamine was rapid as it was to noradrenaline and acetylcholine, Fig. 12.2, and was blocked by previous addition of mepyramine to the bath. In this situation, histamine was occasionally observed to cause dilatation of the vessel.

As with the cerebral vascular response to acetylcholine, it is unclear why histamine consistently causes smooth muscle constriction *in vitro* and dilatation *in vivo*. It should be noted that *in vitro* the potency of histamine is substantially less than that of adrenaline or isoprenaline and it is possible that histamine acts on smooth muscle, through a variety of receptors: mepyramine may well block only one series of these. The question as to whether the vascular response to histamine *in vivo* could be mediated indirectly, e.g. by changes in reflex or metabolic activity, has not been studied.

12. The pharmacology of cerebral vascular smooth muscle

Other vasoactive substances
Nitrites

Cerebral vessels have been found to participate in the widespread smooth muscle relaxation induced by amyl nitrite (Wolff, 1929) and glyceryl trinitrate (Schmidt & Hendrix, 1938). The effect is transient but persists despite the associated fall in arterial pressure. This relaxant effect of amyl nitrite has been confirmed *in vitro*

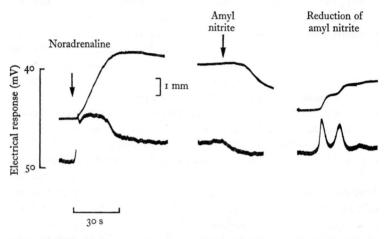

Fig. 12.6. Electrical and mechanical response of sheep carotid artery to noradrenaline 0.25 mg 100 ml^{-1} followed by amyl nitrite 10 ml 100 ml^{-1} and then by reduction of concentration of amyl nitrite with noradrenaline still present. Lower trace is electrical; upper trace mechanical. (From Keatinge (1966), *Circulation Res.* **18**, 641–9, Fig. 1.)

(Keatinge, 1966), and this is illustrated in Fig. 12.6 in which the relaxation is shown to be associated with repolarization. This action of amyl nitrite is similar to that seen in intestinal smooth muscle in response to adrenaline (Bülbring & Kuriyama, 1963) in which there is a rise in membrane potential and relaxation or in rat uterus in response to adrenaline (Edman & Schild, 1963). In the latter response, relaxation takes place after depolarization with high K^+ solutions but without electrical activity. It is also worth emphasizing that nitrites and other dilator agents have been found to cause repolarization and relaxation of vascular smooth muscle only after the muscle has been depolarized and contracted by some other agent, e.g. noradrenaline and in this

Other vasoactive substances

respect the response differs from that in intestinal smooth muscle where adrenaline can cause hyperpolarization and relaxation when the muscle is unstimulated (Bülbring, 1957).

Angiotensin and bradykinin

Angiotensin is known to be a powerful constrictor of peripheral blood vessels and although this action appears to be in part a direct one, there is evidence that it is also reflex. There is evidence from recent studies (Ferrario, Dickenson & McCubbin, 1970; Joy & Lowe, 1970) that angiotensin acts upon neurones in the medulla and that the efferent pathways are carried by parasympathetic and sympathetic fibres which regulate the degree of vagal inhibition and peripheral vasoconstriction, respectively (Scroop & Lowe, 1968; Scroop & Lowe, 1969). In none of these or other experiments, however, has the effect of angiotensin upon cerebral vessels been studied so that it is uncertain whether the effect of vertebral infusion of angiotensin was brought about by a direct action of this drug on medullary neurones or as a consequence of an intense and local vasoconstriction.

Certainly *in vitro*, angiotensin has been found to cause contraction of vascular smooth muscle in carotid artery strips (Keatinge, 1966) together with depolarization. This response was similar to that produced by bradykinin and both hormones caused such a rapid depolarization as to suggest propagation of action potentials through a limited number of smooth muscle cells. The possibility that angiotensin and bradykinin exerted their effects by releasing noradrenaline from sympathetic nerves, as does nicotine, was tested by observing the response after addition of hydergine which blocks the action of noradrenaline. The depolarization and contraction of smooth muscle was unaffected: nor were they affected by the addition of nicotine, hexamethonium bromide or atropine. It would seem, therefore, that, *in vitro*, the greater part if not all of the effect of angiotensin and bradykinin upon vascular smooth muscle is a direct one. It still has to be determined however whether this response is modified in cerebral vessels *in vivo*.

REFERENCES

Abramowicz, A. (1964). The pathogenesis of experimental periventricular cerebral necrosis and its possible relation to periventricular leucomalacia of birth trauma. *J. Neurol. Neurosurg. Psychiat.* **27**, 85–95.

Ackermann, T. (1858). Untersuchungen über den Einfluss der Erstickung auf die Menge des Blutes im Gehirn und in den Lungen. *Virchows Arch. path. Anat. Physiol.* **15**, 401–64.

Adamkiewicz, A. (1900). Zum Blutfässapparat der Ganglienzelle. *Anat. Anz.* **17**, 45–8.

Adams, R. D. (1958). Implications of the biology of the neuroglia and microglia cells for clinical neuropathology. In *Biology of neuroglia*, ed. Windle, W. F., pp. 245–63. Springfield, Ill.: Thomas.

Agnoli, A. (1968). Adaptation of CBF during induced chronic normoxic respiratory acidosis. *Scand J. clin. Lab. Invest.* Suppl. 102, VIII: *D*.

Agnoli, A., Prencipe, M., Priori, A. M., Bozzao, L. & Fieschi, C. (1969). Measurements of rCBF by intravenous injection of ^{133}Xe. A comparative study with the intra-arterial injection method. In *Cerebral blood flow*, ed. Brock, M., Fieschi, C., Ingvar, D. H., Lassen, N. A. & Schürmann, K., pp. 31–4, Berlin: Springer.

Ahlquist, R. P. (1948). A study of the adrenotropic receptors. *Am. J. Physiol.* **153**, 586–600.

(1965). Effects of the autonomic drugs on the circulatory system. In *Handbook of physiology*, section 2: Circulation, vol. 3, ed. Hamilton, W. F. & Dow, P., pp. 2457–75. Washington: American Physiological Society.

Alexander, F. G. & Cserna, S. (1913). Einfluss der Narkose auf den Gaswechsel des Gehirns. *Biochem. Z.* **53**, 100–15.

Alexander, F. G. & Révész, G. (1912). Ueber den Einfluss optischer Reize auf den Gaswechsel des Gehirns. *Biochem. Z.* **44**, 95–126.

Alexander, G. L., Cooper, R. A. & Crow, H. J. (1962). EEG, ECoG and oxygen availability (aO_2) in the cortex of a case of Sturge-Weber syndrome. *Electroen. Neurophysiol.* **14**, 284P.

Alexander, L. (1947). The vascular supply of the strio-pallidum. *Res. Publs Ass. Res. nerv. ment. Dis.* **21**, 77–132.

Alexander, S. C., Cohen, P. J., Wollman, H., Smith, T. C., Reivich, M. & Vander Molan, R. A. (1965). Cerebral carbohydrate metabolism during hypocarbia in man. Studies during nitrous oxide anaesthesia. *Anesthesiology* **26**, 624–32.

Alexander, S. C., Marshall, B. E. & Agnoli, A. (1968). Cerebral blood flow in the goat with sustained hypocarbia. *Scand. J. clin. Lab. Invest.* Suppl. 102, VIII: *C*.

References

Alexander, S. C., Wollman, H., Cohen, P. J., Chase, P. E. & Behar, M. (1964). Cerebrovascular response to P_{a,CO_2} during halothane anesthesia. *J. appl. Physiol.* **19**, 561–5.

Allen, G. & Morris, L. E. (1962). Central nervous system effects of hyperventilation during anaesthesia. *Br. J. Anaesth.* **34**, 296–305.

Alroy, G. G. & Flenley, D. C. (1967). The acidity of the cerebrospinal fluid in man with particular reference to chronic ventilatory failure. *Clin. Sci.* **33**, 335–43.

Altmann, F. (1947). Anomalies of the internal carotid artery and its branches; their embryological and comparative anatomical significance. *Laryngoscope, St Louis* **57**, 313–16.

Ambache, N. (1949). The nicotine action of substances supposed to be purely smooth muscle stimulating. *J. Physiol.* **110**, 164–72.

Andersen, H. (1966). Physiological adaptations in diving vertebrates. *Physiol. Rev.* **46**, 212–43.

Anrep, G. V. (1912). On local vascular reactions and their interpretation. *J. Physiol.* **45**, 318–27.

Anrep, G. V. & Saalfeld, E. von (1935). The blood flow through skeletal muscle in relation to its contraction. *J. Physiol.* **85**, 375–99.

Anrep, G. V. & Starling, E. H. (1925). Central and reflex regulation of the circulation. *Proc. R. Soc.* B **97**, 463–87.

Ariëns, E. J. & Simonis, A. M. (1964). A molecular basis for drug action. *J. Pharm. Pharmac.* **16**, 137–57.

Arnold, F. (1851). Handbuch der Anatomie des Menschen mit besonderer Rücksicht auf Physiologie und Praktische Medicin. Freiburg im Brezsgau: Emmerling & Herder.

Aronson, H. (1890). Ueber nerven und nervendigungen in der Pia mater. *Zentralblatt fur die med. Wissenschaften* **28**, 594–5.

Aserinsky, E. & Kleitman, N. (1953). Regularly occurring periods of eye motility and concomitant phenomena, during sleep. *Science, N.Y.* **118**, 273–4.

Ask-Upmark, E. (1935). The carotid sinus and cerebral circulation. An anatomical, experimental and clinical investigation. Including some observations on rete mirabile caroticum. *Acta Psychiat. Kbh.* Suppl. 6, 1–374.

— (1944). Über die makro-anatomische zu beobachtendens Schutzoorrichtungen des Gehirns gegan vaskulare insulte. *Acta path. microbiol. scand.* Suppl. 54, 530–8.

— (1953). On the entrance of the carotid artery into the cranial cavity in *Stenops gracilis* and *Otolicnus crassicaudatus*. *Acta anat.* **19**, 101–3.

Asmussen, E. (1943). CO_2-breathing and the output of the heart. *Acta physiol. scand.* **6**, 176–86.

Auckland, K. (1965). Hydrogen polarography in measurements of local blood flow: theoretical and empirical basis. *Acta neurol. scand.* Suppl. 14, 42–5.

Auckland, K., Bower, B. F. & Berliner, R. W. (1964). Measurement of local blood flow with hydrogen gas. *Circulation Res.* **14**, 164–87.

Axelrod, D. R. & Pitts, R. F. (1952). Relationship of plasma pH and anion pattern to mercurial diuresis. *J. clin. Invest.* **31**, 171–9.

Bailey, P. & Bremer, F. (1938). A sensory cortical representation of the vagus nerve. *J. Neurophysiol.* **1**, 405–12.

References

Bainton, C. R., Carcelan, B. A. & Severinghaus, J. W. (1965). Carotid chemoreceptor insensitivity in Andean natives. *J. Physiol.* **177**, 30–31 P.

Baird, H. W. & Garfunkel, J. M. (1953). A method for the measurement of cerebral blood flow in infants and children. *J. Pediat.* **42**, 570–5.

Bakay, L. (1956). *The blood–brain barrier with special regard to the use of radioactive isotopes.* Springfield, Illinois: Thomas.

Bakay, L., Hueter, T. F., Ballantine, H. T. & Sosa, D. (1956). Ultrasonically produced changes in the blood–brain barrier. *Archs Neurol. Psychiat., Chicago* **76**, 457–67.

Baldwin, B. A. (1960). The correlation between the vascular supply of the brain and cerebral function in ruminants. Ph.D. Thesis, University of London.

(1964). The anatomy of the arterial supply to the cranial regions of the sheep and ox. *Am. J. Anat.* **115**, 101–18.

Baldwin, B. A. & Bell, F. R. (1963 a). The anatomy of the cerebral circulation in the sheep and ox. The dynamic distribution of the blood supplied by the carotid and vertebral arteries to cranial regions. *J. Anat.* **97**, 203–15.

(1963 b). The effect of temporary reduction in cephalic blood flow on the EEG of sheep and calf. *Electroen. Neurophysiol.* **15**, 465–73.

Baldy-Moulinier, M. & Frerebeau, Ph. (1968). Blood flow of the cerebral cortex in intracranial hypertension. *Scand J. clin. Lab. Invest.* Suppl. 102, V: G.

Banker, B. Q. & Larroche, J. C. (1962). Periventricular leukomalacia of infancy. *Archs Neurol., Chicago* **7**, 386–410.

Barcroft, H. (1963). Circulation in skeletal muscle. In *Handbook of physiology*, section 2, vol. 2, pp. 1353–85. Washington: American Physiological Society.

Barcroft, H., Foley, T. H. & McSwiney, R. R. (1970). Experiments on the liberation of phosphate from active human muscle, and on the action of phosphate on human blood vessels. *J. Physiol.* **210**, 34 P.

Barcroft, J. (1914). The respiratory function of the blood. p. 73, Cambridge University Press.

Bartholinus, T. (1684). Anatomae quartum renovata. p. 675, Lyons: Huguetan.

Batson, O. V. (1940). Function of vertebral veins and their role in spread of metastases. *Ann. Surg.* **112**, 138–49.

(1942). Role of vertebral veins in metastatic processes. *Ann. intern. Med.* **16**, 38–45.

(1944). Anatomical problems concerned in the study of cerebral blood flow. *Fedn Proc.* **3**, 139–44.

Baust, W. (1967). Local blood flow in different regions of the brain-stem during natural sleep and arousal. *Electroen. Neurophysiol.* **22**, 365–72.

Bayliss, L. E. (1952). Rheology of blood and lymph. In *Deformation and flow in biological systems.* ed. Frey-Wyssling, A. pp. 354–418. Amsterdam: North Holland Publ. Co.

Bayliss, W. M. (1902). On the local reactions of the arterial wall to changes of internal pressure. *J. Physiol.* **28**, 220–31.

(1908). The excitation of vasodilator nerve fibres in depressor reflexes. *J. Physiol.* **37**, 264–77.

Bayliss, W. M., Hill, L. & Gulland, G. L. (1895). On intracranial pressure and the cerebral circulation. *J. Physiol.* **18**, 334–62.

References

Becker, F. C. & Olsen, O. (1914). Metabolism and mental work. *Skand. Arch. Physiol.* **31**, 81–197.

Bedford, T. H. B. (1935). The effects of increased intracranial venous pressure on the pressure of the cerebral spinal fluid. *Brain*, **58**, 427–47.

— (1936). The effect of prolonged occlusion of the external jugular veins on the cerebral spinal fluid and the torcular venous pressures in the dog. *Brain* **59**, 324–36.

Bekaert, J. & Demeester, G. (1954). Influence of potassium concentration of blood on potassium level of cerebrospinal fluid. *Expl Med. Surg.* **12**, 480–501.

Benedikt, M. (1874). Ueber die Innervation des plexus chorioideus inferior. *Virchows Arch. path. Anat. Physiol.* **59**, 395–400.

Benitez, H. H., Murray, M. R. & Woolley, D. W. (1955). Effects of serotonin and certain of its antagonists upon oligodendroglial cells. *Proc. 2nd Int. Congr. Neuropath. Lond.* **2**, 423–8.

Bennett, H. S., Bassett, D. L. & Beecher, H. K. (1944). Influence of anesthesia (ether, cyclopropane, sodium evipal) on the circulation under normal and shock conditions. *J. clin. Invest.* **23**, 181–208.

Bennett, H. S., Luft, J. H. & Hampton, J. C. (1959). Morphological classification of vertebrate blood capillaries. *Am. J. Physiol.* **196**, 381–90.

Berg, D. van den & Drift, J. H. van der (1963). Vertebro-basilaire angiografie via catheterisatie van de arteria brachialis. *Nederl. T. Geneesk.* **107**, 1743–8.

Bergel, D. H., Caro, C. G. & McDonald, D. A. (1960). The input impedance of the pulmonary vascular bed. *J. Physiol.* **154**, 18 P.

Bergel, D. H., McDonald, D. A. & Taylor, M. G. (1958). A method for measuring arterial impedance using a differential manometer. *J. Physiol.* **141**, 17 P.

Berger, H. (1924). Zur innervation des piamater und Gehirngefässe. *Arch. f. Psych.* **70**, 216–20.

— (1929). Über das Elektrenkephalogramm des Menschen. *Arch. Psychiat. NervKrankh.* **87**, 527–70.

Bering, E. A. (1955). Choroid plexus and arterial pulsation of cerebrospinal fluid. *Archs Neurol. Psychiat.*, Chicago **73**, 165–72.

Bernard, C. (1851). Influence du grand sympathetique sur la sensibilité et sur la calorification. *C. r. Séanc. Soc. Biol.* **3**, 163–5.

— (1853). Recherches expérimentales sur le grand sympathetique et specialement sur l'influence que la section de ce nerf exerce sur la chaleur animale. *C. r. Séanc. Soc. Biol.* **5**, 77–107.

— (1862). *Recherches expérimentales sur les nerfs vasculaires et calorifiques du grand sympathetique.* Paris: Mallet-Bachelier.

Berne, R. M. (1964). Metabolic regulation of blood flow. *Circulation Res.* **15**, Suppl. 1, 261–7.

Bessman, A. N., Alman, R. W. & Fazekas, J. F. (1952). Effect of acute hypotension on cerebral hemodynamics and metabolism of elderly patients *Archs intern. Med.* **89**, 893–8.

Betz, E. (1965). Zur Registrierung der lokalen Gehirndurchblutung mit Warmeleitsonden. *Pflügers Arch. ges. Physiol.* **284**, 278–84.

— (1968). Measurements of local blood flow by means of heat clearance. In *Blood flow through organs and tissues*, ed. Bain, W. H. & Harper, A. M., pp. 169–76. Edinburgh: Livingstone.

References

Betz, E., Braasch, D. & Hensel, H. (1961). The effect of 2,6-bis(di-(2 hydroxethyl)amino)-4,8-dipiperidinopyrimido(5,4-d)pyrimidine on the circulation of the myocardium, brain, kidney, liver, skin and musculature. *Arzneimittel-Forsch.* **11**, 333–6.
Betz, E., Gayer, J. & Weber, H. (1964). Fortlaugende Registrierung der lokalen Nierenrindendurchblutung mit Hilfe von Warmeleitsonden. *Z. Kreislaufforsch.* **53**, 524–9.
Betz, E. & Herrmann, E. (1966). Die fortlaufende Registrierung der Gehirndurchblutung beim Menschen mit flexiblen Warmeleitsonden. *Nervenarzt.* **37**, 173–5.
Betz, E. & Heuser, D. (1967). Cerebral cortical blood flow during changes of acid–base equilibrium of the brain. *J. appl. Physiol.* **23**, 726–33.
Betz, E., Ingvar, D. H., Lassen, N. A. & Schmail, F. W. (1966). Regional blood flow in the cerebral cortex, measured simultaneously by heat and inert gas clearance. *Acta physiol. scand.* **67**, 1–9.
Betz, E. & Kozak, R. (1967). Der Einfluss der Wasserstoffionenkonzentration der Gehirnrinde auf die Regulation der corticalen Durchblutung. *Pflügers Arch. ges. Physiol.* **293**, 56–67.
Biddulph, C., van Fossan, D. D., Criscuolo, D. & Clark, R. T. (1958). Lactic acid concentration of brain tissues of dogs exposed to hypoxemia and/or hypocapnia. *J. appl. Physiol.* **13**, 486–90.
Biedl, A. & Reiner, M. (1900). Studien über Hirncirculation und Hirnodem. *Pflügers Arch. ges. Physiol.* **69**, 158–94.
Bielschowsky, M. (1928). Nervengewebe. In *Nervensystem, Handbuch der mikroscopischen Anatomie des Menschen,* ed. Möllendorff, W. von, vol. 4, part 1. Berlin: Springer.
Bigelow, W. G., Lindsay, W. K., Harrison, R. C., Gordon, R. A. & Greenwood, W. F. (1950). Oxygen transport and utilization in dogs at low body temperatures. *Am. J. Physiol.* **160**, 125–37.
Bill, A. (1962). Studies of the heated thermocouple principle for determinations of blood flow in tissues. *Acta physiol. scand.* **55**, 111–26.
Binet, L., Cachera, R., Fauvert, R. & Strumza, M. V. (1937). Anoxémie et circulation cérébrale. *C. r. Séanc. Soc. Biol.* **126**, 166–9.
Binswanger, O. (1879). Anatomische Untersuchungen über die Ursprungstelle und den Anfangsteil der Carotis interna. *Arch. Psychiat.* **9**, 351–68.
Biscoe, T. J. & Millar, R. A. (1968). Effects of inhalation anaesthetics on carotid body chemoreceptor activity. *Br. J. Anaesth.* **40**, 2–12.
Biscoe, T. J., Purves, M. J. & Sampson, S. R. (1970). The frequency of nerve impulses in single carotid body chemoreceptor afferent fibres recorded *in vivo* with intact circulation. *J. Physiol.* **208**, 121–32.
Bizzi, E., Pompeiano, O. & Somogzi, I. (1964). Vestibular nuclei: activity of single neurons during natural sleep and wakefulness. *Science, N.Y.* **145**, 414–15.
Bochenek, A. (1899). Über die nervendigungen in den plexus chorioidei des Frosches. *Bulletin International (Academie des Sciences de Cracovie),* p. 346.
Bock, A. C. (1823). Darstellung der Venen des menslichen Körpers nach ihrer Structur, Vertheilung und Verlauf. Leipzig: Baumgärtner.
Bohr, D. F., Goulet, P. L. & Taquini, A. C. (1961). Direct tension recording

References

from smooth muscle resistance vessels from various organs. *Angiology* **12**, 478–85.

Bohr, V. C., Ralls, R. J. & Westermeyer, R. E. (1958). Changes in renal function during induced apnoea of diffusion respiration. *Am. J. Physiol.* **194**, 143–8.

Boissezon, P. de (1941). Le resau admirabile arteriel intracranien de l'agneau. *Bull. Soc. Hist. nat. Toulouse* **76**, 299–304.

Bollinger, A., Luthy, E. & Jenny, E. (1967). Volblutviskoitat bei verschiedenen Schergeschwindigkeiten und ihre Beeinflussung durch niedermolekulares Dextran. *Klin. Wschr.* **45**, 939–43.

Bolonyi, F. (1951). Étude sur la vascularisation du lobe frontal au point de vue phylogénétique. *Acta anat.* **12**, 110–16.

Bouckaert, J. J. & Heymans, C. (1935). On the reflex regulation of the cerebral blood flow and the cerebral vasomotor tone. *J. Physiol.* **84**, 367–80.

Bouckaert, J. J. & Jourdan, F. (1936). Recherches sur la physiologie et la pharmacodynamie des vaisseaux cérébraux; technique de perfusion de la circulation intracrânienne isolée chez le chien. *Archs int. Pharmacodyn. Thér.* **53**, 523–39.

Boyle, R. (1670). New pneumatical experiments about respiration. *Phil. Trans. R. Soc.* **5**, 2011–31.

Bozzao, L., Fieschi, C., Agnoli, A. & Nardini, M. (1968). Autoregulation of cerebral blood flow studied in the brain of cat. In *Blood flow through organs and tissues*, ed. Bain, W. H. & Harper, A. M., pp. 253–6. Edinburgh: Livingstone.

Brachet, J. L. (1837). *Recherches experimentales sur les fonctions du systeme nerveux ganglionnaire et son application à la pathologie.* Paris: Germer-Baillière.

Bradbury, M. W. B. & Kleeman, C. R. (1967). Stability of the potassium content of cerebrospinal fluid and brain. *Am. J. Physiol.* **213**, 519–28.

Bradbury, M. W. B. & Stulcova, B. (1970). Efflux mechanism contributing to the stability of the potassium concentration in cerebrospinal fluid. *J. Physiol.* **208**, 415–30.

Bradley, R. D. & Semple, S. J. G. (1962). A comparison of certain acid–base characteristics of arterial blood, jugular venous blood and cerebrospinal fluid in man and the effect on them of some acute and chronic acid–base disturbances. *J. Physiol.* **160**, 381–91.

Bradley, S. E. & Bing, R. J. (1942). Renal function in the harbor seal (*Phoca vitulina* L) during asphyxial ischemia and pyrogenic hyperemia. *J. cell. comp. Physiol.* **19**, 229–37.

Brea, J. B. (1956). Sulla persistenza della anastomosi carotido-basilare. *Sistema nerv.* **8**, 17–23.

Brebbia, D. R. & Altshuler, K. Z. (1965). Oxygen consumption rate and 'electroencephalographic' stage of sleep. *Science, N.Y.* **150**, 1621–3.

Bregeat, P., David, M., Fischgold, H. & Talairach, J. (1952). Opacification des vaisseaux orbitaires et de la choroid par l'angiographie carotidienne. *Revue neurol.* **87**, 549–51.

Breschet, G. (1832). *Recherches anatomiques, physiologiques et pathologiques sur le systeme vieneux.* Paris: Ballière.

References

Brickner, E. W., Dowds, E. G., Willitts, B. & Selkurt, E. E. (1956). Mesenteric blood flow as influenced by progressive hypercapnia. *Am. J. Physiol.* **184**, 275–81.

Bridges, T. J., Clark, K. & Yahr, M. D. (1958). Plethysmographic studies of the cerebral circulation: evidence for cranial nerve vasomotor activity. *J. clin. Invest.* **37**, 763–72.

Brightman, M. W. & Palay, S. L. (1963). The fine structure of ependyma in the brain of the rat. *J. cell Biol.* **19**, 415–39.

Brightman, M. W. & Reese, T. W. (1969). Junctions between intimately apposed cell membranes in the vertebrate brain. *J. cell Biol.* **40**, 648–77.

Britton, S. W., Corey, E. L. & Stewart, G. A. (1946). Effects of high acceleratory forces and their alleviation. *Am. J. Physiol.* **146**, 33–51.

Britton, S. W., Pertzoff, V. A., French, C. R. & Kline, R. F. (1947). Circulatory and cerebral changes and protective aids during exposure to acceleratory forces. *Am. J. Physiol.* **150**, 7–26.

Brodmann, K. (1909). *Vergleichende Lokalisationslehre der Grosshirninde.* Leipzig: J. A. Barth.

Brody, H. (1955). Organisation of the cerebral cortex. III. A study of ageing in the human cerebral cortex. *J. comp. Neurol.* **102**, 511–56.

Bronk, D. W. & Gesell, R. (1927). Regulation of respiration: effects of carbon dioxide, sodium bicarbonate and sodium carbonate on carotid and femoral flow of blood. *Am. J. Physiol.* **82**, 170–80.

Brookens, N. L., Ectors, L. & Gerard, R. W. (1936). Respiration of local brain regions. Technique and applications. *Am. J. Physiol.* **116**, 16–17.

Brooks, H. & Carroll, J. H. A. (1912). A clinical study of the effects of sleep and rest on blood pressure. *Archs intern. Med.* **10**, 97–102.

Brown-Sequard, C.-E. (1853). Note sur la découverte de quelques-uns des effets de la galvanisation du nerf grand sympathetique au cou. *Gaz. Med. de Paris* **9**, 22.

Bruens, J. H., Gastaut, H. & Giove, G. (1960). Electroencephalographic study of the signs of chronic vascular insufficiency of the Sylvian region in aged people. *Electroen. Neurophysiol.* **12**, 283–95.

Buck, R. C. (1958). The fine structure of endothelium of large arteries. *J. biophys. biochem. Cytol.* **4**, 187–90.

Buckell, M. (1964). Demonstration of substances capable of contracting smooth muscle in haematoma fluid from certain cases of ruptured cerebral aneurysm. *J. Neurol. Neurosurg. Psychiat.* **27**, 198–9.

Buhlmann, A., Scheitlin, W. & Rossier, P. H. (1963). The relations between blood and cerebrospinal fluid in disorders of acid–base equilibrium. *Schweiz. med. Wschr.* **93**, 427–32.

Bülbring, E. (1957). Changes in configuration of spontaneously discharged spike potentials from smooth muscle of the guinea-pig's taenia coli. The effect of electrotonic currents and of adrenaline, acetylcholine and histamine. *J. Physiol.* **135**, 412–25.

(1962). Electrical activity in intestinal smooth muscle. *Physiol. Rev.* **42** (Suppl. 5), 160–78.

Bülbring, E. & Burnstock, G. (1960). Membrane potential changes associated with tachyphylaxis and potentiation of the response to stimulating drugs in smooth muscle. *Br. J. Pharmac. Chemother.* **15**, 611–24.

References

Bülbring, E. & Kuriyama, H. (1963). Effects of changes in ionic environment on the action of acetylcholine and adrenaline on the smooth muscle cells of guinea-pig taenia coli. *J. Physiol.* **166**, 59–74.

Burn, J. H. & Epstein, H. G. (1959). Hypotension due to halothane. *Br. J. Anaesth.* **31**, 199–204.

Burn, J. H. & Rand, M. J. (1958). The action of sympatheticomimetic amines in animals treated with reserpine. *J. Physiol.* **144**, 314–36.

Burnam, J. F., Hickman, J. B. & McIntosh, H. D. (1954). Effect of hypocapnia on arterial blood pressure. *Circulation* **9**, 89–95.

Burnstock, G. (1958). The effects of acetylcholine on membrane potential, spike frequency, conduction velocity and excitability in the taenia coli of the guinea-pig. *J. Physiol.* **143**, 165–82.

Burrows, G. (1846). *On disorders of the cerebral circulation; and on the connection between affections of the brain and diseases of the heart.* London: Longman.

Burton, A. C. (1951). On the physical equilibrium of small blood vessels. *Am. J. Physiol.* **164**, 319–29.

(1952). In *Visceral circulation*, CIBA Foundation Symposium, ed. Wolstenholme, G. E. W. & O'Connor, M., p. 70. London: Churchill.

(1954). Relation of structure to function of the tissues of the wall of blood vessels. *Physiol. Rev.* **34**, 619–42.

Burton, A. C. & Stinson, R. H. (1960). The measurement of tension in vascular smooth muscle. *J. Physiol.* **153**, 290–305.

Busch, E. (1938). The innervation of the intracranial blood vessels. *Acta Psychiat. Kbh.* **13**, 131–8.

Butler, T. C. (1950). Theories of general anaesthesia. *Pharmac. Rev.* **2**, 121–60.

Caesar, R., Edwards, G. A. & Ruska, H. (1957). Architecture and nerve supply of mammalian smooth muscle. *J. biophys. biochem. Cytol.* **3**, 867–78.

Cahill, G. F. Jr. (1970). Starvation in man. *New Eng. J. Med.* **282**, 668–75.

Cain, S. M. (1965). Appearance of excess lactate in anaesthetized dogs during anemic and hypoxic hypoxia. *Am. J. Physiol.* **209**, 604–10.

Caldwell, P. C. (1956). Intracellular pH. *Int. Rev. Cytol.* **5**, 229–77.

(1958). Studies on the internal pH of large muscle and nerve fibres. *J. Physiol.* **142**, 22–62.

Callenfels, J. van der B. (1855). Über den Einfluss der vasomotorischen Nerven auf den Kreislauf und die Temperatur. *Ztschr. rationelle med.* **7**, 157–207.

Cameron, D. E. & Rosen, S. R. (1941). The reactivity of intracranial vessels in the aged. *Am. J. med. Sci.*, **201**, 871–6.

Cameron, I. R. & Kleeman, C. R. (1970). The effect of acute hyperkalaemia on the blood–c.s.f. potential difference. *J. Physiol.* **207**, 68P.

Cammermeyer, J. (1953). The agonal nature of cerebral red softening: reevaluation of morphologic findings. *Acta Psychiat. Neurol. Scand.* **28**, 9–25.

(1960). Reappraisal of the perivascular distribution of dendrocytes. *Am. J. Anat.* **106**, 197–219.

Campbell, A. C. (1937). The vascular architecture of the cat's brain. *Res. Publs Ass. Res. nerv. ment. Dis.* **17**, 719–25.

Cannon, W. B., Newton, H. F., Bright, E. M., Menkin, V. & Moore, R. M. (1929). Some aspects of the physiology of animals surviving complete exclusion of sympathetic nerve impulses. *Am. J. Physiol.* **89**, 84–107.

References

Canti, R., Bland, J. O. W. & Russell, D. S. (1937). Tissue culture of gliomata. *Res. Publs Ass. Res. nerv. ment. Dis.* **16**, 1–20.

Capon, A., Cleempoel, H., Lenaers, A. & Martin, Ph. (1968). Methodology of cerebral blood flow measurement. In *Cerebral circulation*: Progress in Brain Research, **30**, ed. Luyendijk, W., pp. 43–52. Amsterdam: Elsevier.

Carlsson, A. & Waldeck, B. (1963). β-hydroxylation of tyramine *in vivo*. *Acta Pharmac. (Kbh)* **20**, 371–4.

Carlyle, A. & Grayson, J. (1955). Blood pressure and the regulation of brain blood flow. *J. Physiol.* **127**, 15P.

Carrier, O., Walker, J. R. & Guyton, A. C. (1964). Role of oxygen in autoregulation of blood flow in isolated muscles. *Am. J. Physiol.* **206**, 951–4.

Carter, N. W., Rector, F. C., Campion, D. S. & Seldin, D. W. (1967). Measurement of intracellular pH with glass electrodes. *Fedn Proc.* **26**, 1322–6.

Casserius, J. (1645). Tabulae anatomicae. In *Opera quae extant*, Plate x. Amsterdam: Spigelius.

Cassin, S., Dawes, G. S. & Ross, B. B. (1964). Pulmonary blood flow and vascular resistance in immature foetal lambs. *J. Physiol.* **171**, 80–9.

Cassin, S., Gilbert, R. D. & Johnson, E. M. (1966). Capillary development during exposure to chronic hypoxia. *Brooks Air Force Base Techn. Report No. 66*, 16.

Casteels, R. (1970). The relation between the membrane potential and the ion distribution in smooth muscle. In *Smooth muscle*, ed. Bülbring, E., Brading, A. F., Jones, A. W. & Tomita, T., pp. 70–99. London: Arnold.

Castro, F. de (1928). Sur la structure et l'innervation de sinus carotidien de l'homme et des mammifères. Nouveaux faits sur l'innervation et la fonction de Glomus caroticum. Études anatomiques et physiologiques. *Trab. Lab. Invest. biol. Univ. Madr.* **25**, 331–80.

Cater, D. B., Garratini, S., Marina, F. & Silver, I. A. (1961). Changes in oxygen tension in brain and somatic tissues induced by vasodilator and vasoconstrictor drugs. *Proc. R. Soc. B.* **155**, 136–57.

Cater, D. B. & Silver, I. A. (1961). In *Reference electrodes*, ed. Ives, J. G. & Janz, J. G. New York: Academic Press.

Chambers, R. & Zweifach, B. W. (1944). Topography and function of mesenteric capillary circulation. *Am. J. Anat.* **75**, 173–205.

Chance, B., Cohen, P., Jopsis, F. & Schoener, B. (1962). Intracellular oxidation–reduction states *in vivo*. *Science, N.Y.* **137**, 499–507.

Chance, B. & Schoener, B. (1962). Correlation of oxidation–reduction changes of intracellular reduced pyridine nucleotide and changes in electroencephalogram of the rat in anoxia. *Nature, Lond.* **195**, 956–8.

Charachon, R. & Latarjet, M. (1952). Les injections de matières plastiques appliquées à l'étude de l'artère auditive interne. *C. r. Assoc. Anat.* **3**, 436–41.

Chesler, A. & Himwich, H. E. (1944). Comparative studies of the rates of oxidation and glycolysis in the cerebral cortex and brain stem of the rat. *Am. J. Physiol.* **141**, 513–17.

Chorobski, J. & Penfield, W. (1932). Cerebral vasodilator nerves and their pathway from the medulla oblongata. *Archs Neurol. Psychiat., Chicago* **28**, 1257–89.

Christensen, K., Polley, G. H. & Lewis, E. (1952). The nerves along the

References

vertebral artery and innervation of the blood vessels of the hind brain of the cat. *J. comp. Neurol.* **96**, 71–91.

Chute, A. L. & Smyth, D. H. (1939). Metabolism of the isolated perfused cat's brain. *Q. Jl exp. Physiol.* **29**, 379–94.

Clark, L. C. J. & Bargerson, L. M. (1959). Detection and direct recording of right to left shunts with a hydrogen electrode catheter. *Surgery, St Louis* **46**, 797–804.

Clark, S. L. (1928). Nerve endings in the choroid plexus of the fourth ventricle. *J. comp. Neurol.* **47**, 1–16.

(1929). Innervation of the blood vessels of the medulla and spinal cord. *J. comp. Neurol.* **48**, 247–65.

(1934). Innervation of the choroid plexuses and the blood vessels within the central nervous system. *J. comp. Neurol.* **60**, 21–31.

Cobb, S. & Finesinger, J. E. (1932). Cerebral circulation: XIX. The vagal pathway of the vasodilator fibres. *Archs Neurol. Psychiat.*, Chicago **28**, 1243–56.

Cobb, S. & Fremont-Smith, F. (1931). The cerebral circulation. XVI. Changes in the human retinal circulation and in the pressure of the cerebrospinal fluid during inhalation of a mixture of carbon dioxide and oxygen. *Archs Neurol. Psychiat.*, Chicago **26**, 731–6.

Cobb, S. & Talbott, J. H. (1927). Studies in cerebral circulation. II. A quantitative study of cerebral capillaries. *Trans. Ass. Am. Physns.* **42**, 255–62.

Cohen, M. W., Gerschenfeld, H. M. & Kuffler, S. W. (1968). Ionic environment of neurones and glial cells in the brain of an amphibian. *J. Physiol.* **197**, 363–80.

Cohen, P. J., Alexander, S. C., Smith, T. C., Reivich, M. & Wollman, H. (1967). Effects of hypoxia and normocarbia on cerebral blood flow and metabolism in conscious man. *J. appl. Physiol.* **23**, 183–9.

Cohen, P. J., Wollman, H., Alexander, S. C., Chase, P. E. & Behar, M. G. (1964). Cerebral carbohydrate metabolism in man during halothane anaesthesia. *Anesthesiology* **25**, 185–91.

Collip, J. B. & Backus, P. L. (1920). The alkali reserve of the blood plasma, spinal fluid and lymph. *Am. J. Physiol.* **51**, 551–67.

Conn, H. L. (1955). Measurement of organ blood flow without blood sampling. *J. clin. Invest.* **34**, 916.

Cooper, A. (1836). Some experiments and observations on tying the carotid and vertebral arteries and the pneumogastric, phrenic and sympathetic nerves. *Guy's Hosp. Rep.* **1**, 457–75.

Cooper, D. B., Crow, H. J., Walter, W. G. & Winter, A. L. (1960). Studies of the level of 'available oxygen' in human brain. *Electroen. Neurophysiol.* **12**, 760.

(1965). Variations of occipital blood flow, oxygen availability and the EEG during reading and flicker in man. *Electroen. Neurophysiol.* **19**, 315.

Cooper, R., Moskalenko, Y. E. & Walter, W. G. (1964). The pulsation of the human brain. *J. Physiol.* **172**, 54P.

Cormack, R. S., Cunningham, D. J. C. & Gee, J. B. L. (1958). The effect of carbon dioxide on the respiratory response to want of oxygen in man. *Q. Jl exp. Physiol.* **43**, 303–19.

Cotev, S., Cullen, D. & Severinghaus, J. W. (1968). Cerebral ECF acidosis induced by hypoxia at normal and low PCO_2. *Scand. J. clin. Lab. Invest.* Suppl. 102, III: E.

References

Courtice, F. C. (1940). The effect of raised intracranial pressure on the cerebral blood flow. *J. Neurol. Psychiat., Lond.* **3**, 293-305.
— (1941). The effect of oxygen lack on the cerebral circulation. *J. Physiol.* **100**, 198-211.
Cow, D. (1911). Some reactions of surviving arteries. *J. Physiol.* **42**, 125-43.
Craigie, E. H. (1920). On the relative vascularity of various parts of the central nervous system of the albino rat. *J. comp. Neurol.* **31**, 429-64.
— (1921). The vascularity of the cerebral cortex of the albino rat. *J. comp. Neurol.* **33**, 193-211.
— (1924). Changes in the vascularity in the brain stem and cerebellum of the albino rat between birth and maturity. *J. comp. Neurol.* **38**, 27-48.
— (1925). Postnatal changes in vascularity in the cerebral cortex of the male albino rat. *J. comp. Neurol.* **39**, 301-24.
— (1930). Vascular supply of archicortex of rat: albino rat (*Mus norvegicus albinus*). *J. comp. Neurol.* **51**, 1-11.
— (1955). Vascular patterns of the developing nervous system. In *Biochemistry of the developing nervous system*, ed. Waelsch, P. New York: Academic Press.
Cramer, P. (1873). Die reflectorische der piaarterien. Inaugural address, Dorpat.
Crompton, M. R. (1964). The pathogenesis of cerebral infarction following rupture of cerebral berry aneurysm. *Brain*, **87**, 491-510.
Cross, B. A. & Silver, I. A. (1962). Some factors affecting oxygen tension in the brain and other organs. *Proc. R. Soc.* B **156**, 483-99.
Cruveilhier, J. (1834). *Anatomie descriptive*. Paris: Bechet jeune.
Cserr, H. (1965). Potassium exchange between cerebrospinal fluid, plasma and brain. *Am. J. Physiol.* **209**, 1219-26.
Cserr, H. & Pappenheimer, J. R. (1964). Potassium exchanges between cerebrospinal fluid and blood. *Fedn. Proc.* **23**, 211.
Cunningham, D. J. C., Hey, E. N. & Lloyd, B. B. (1958). The effect of intravenous infusion of noradrenaline on the respiratory response to carbon dioxide in man. *Q. Jl. exp. Physiol.* **43**, 394-9.
Cushing, H. (1901). Concerning a definite regulating mechanism of the vasomotor centre which controls blood pressure during cerebral compression. *Johns Hopkins Hosp. Bull.* **12**, 290-2.
— (1902). Some experimental and clinical observations concerning states of increased intracranial tension. *Am. J. med. Sci.* **124**, 375-400.
Cusick, P. L., Benson, O. O., Jr. & Boothby, W. M. (1940). Effect of anoxia and of high concentrations of oxygen on the retinal vessels: preliminary report. *Proc. Mayo Clin.* **15**, 500-2.
Cyon, E. de & Ludwig, C. F. (1866). Die Reflexe eines der sensiblen Nerven des Herzens auf die motorischen der Blutgefässe. *Arb. a.d. Physiol. Anst. zu Leipzig*, **1**, 128-49.
Czerny, A. (1896). Zur Kenntnis des physiologischen Schlafes. *Jahrb. f. Kinderheilk.* **11**, 337-42.
Dahl, E., Flora, G. & Nelson, E. (1965). Electron microscopic observations on normal human intracranial arteries. *Neurology, Minneapolis* **15**, 132-40.
Dahl, E. & Nelson, E. (1964). Electron microscopic observations on human intracranial arteries. *Archs Neurol. Psychiat., Chicago* **10**, 158-64.

References

Daly, I. de B. & Hebb, C. (1966). *Pulmonary and bronchial vascular systems.* pp. 118–57. London: Arnold.

Daniel, P. M., Dawes, J. D. K. & Pritchard, M. M. L. (1953). Studies of the carotid rete and its associated arteries. *Phil. Trans. R. Soc.* **237**, 173–208.

Darrow, C. W. (1943). Physiological and clinical tests of autonomic function and autonomic balance. *Physiol. Rev.* **23**, 1–36.

Darrow, C. W. & Gellhorn, E. (1939). The effects of adrenaline on the reflex excitability of the autonomic nervous system. *Am. J. Physiol.* **127**, 243–51.

Darrow, C. W. & Graff, C. G. (1945). Relation of electroencephalogram to photometrically observed vasomotor changes in the brain. *J. Neurophysiol.* **8**, 449–61.

Darrow, C. W., Green, J. R., Davis, E. W. & Garol, H. W. (1944). Parasympathetic regulation of high potential in the electroencephalogram. *J. Neurophysiol.* **7**, 217–26.

Darrow, C. W. & Pathman, J. H. (1943). The role of blood pressure in electroencephalographic changes during hyperventilation. *Fedn Proc.* **2**, 9.

 (1944). Relation of heart rate to slow waves in the electroencephalogram during overventilation. *Am. J. Physiol.* **140**, 583–8.

Darrow, C. W. & Solomon, A. P. (1940). Mutism and resistance behaviour in psychotic patients. *Am. J. Psychiat.* **96**, 1441–54.

Daugherty, R. M., Scott, J. B., Dabney, J. M. & Haddy, F. J. (1967) Local effects of O_2 and CO_2 on limb, renal and coronary vascular resistances. *Am. J. Physiol.* **213**, 1102–10.

Daugherty, R. M., Scott, J. B. & Haddy F. J. (1967). Effects of generalized hypoxemia and hypercapnia on forelimb vascular resistance. *Am. J. Physiol.* **213**, 1111–14.

Davies, D. F. & Shock, N. W. (1950). Age changes in glomerular filtration rate, effective renal plasma flow and tubular excretory capacity in adult males. *J. clin. Invest* **29**, 496–507.

Davies, P. W. & Brink, F. Jr. (1942). Microelectrodes for measuring local oxygen tension in animal tissues. *Rev. scient. Instrum.* **13**, 524–32.

Davies, P. W. & Bronk, D. W. (1957). Oxygen tension in mammalian brain. *Fedn Proc.* **16**, 689–92.

Davies, P. W., Grenell, R. G. & Bronk, D. W. (1948). The time course of *in vivo* oxygen consumption of cerebral cortex following electrical stimulation. *Fedn Proc.* **7**, 25.

Davies, P. W. & Remond, A. (1946). Oxygen consumption of the cerebral cortex of the cat during metrazole convulsions. *Res. Publs Ass. Res. nerv. ment. Dis.* **26**, 205–17.

Davis, D. D. & Storey, E. H. (1943). The carotid circulation in the domestic cat. *Zoological Series Field Museum of Natural History* **28**, 527.

Davis, E. W., McCulloch, W. S. & Roseman, E. (1944). Rapid changes in the oxygen tension of cerebral cortex during induced convulsions. *Am. J. Psychiat.* **100**, 825–9.

Davis, R. C. (1938). The relation of muscle action potentials to difficulty and frustration. *J. exp. Psychol.* **23**, 141–58.

Davson, H. (1960). Intracranial and intraocular fluids. In *Handbook of physiology*, section 1, volume III. *Neurophysiology*. Washington D.C.: American Physiological Society.

References

Davson, H. (1963). The cerebral spinal fluid. *Ergbn. Physiol.* **52**, 20–73.
— (1967). *Physiology of the cerebrospinal fluid.* pp. 337–82, London: Churchill.
Dawes, G. S., Fox, H. E., Leduc, B. M., Liggins, G. C. & Richards, R. T. (1970). Respiratory movements and paradoxical sleep in the foetal lamb. *J. Physiol.* **210**, 47–8P.
Day, M. E. (1923). The influence of mental activities upon vascular processes. *J. comp. Psychol.* **3**, 333–78.
Deal, C. P. & Green, H. D. (1954). Effects of pH upon blood flow and peripheral resistance in muscular and cutaneous vascular beds in hindlimb of the pentobarbitalized dog. *Circulation Res.* **2**, 148–54.
Decker, K. (1955). Die Arteria ophthalmica im Karotisangiogramm. *Fortschr. Rontgenstr.* **82**, 667–73.
Defares, J. G., Osborn, J. J. & Hara, H. H. (1963). Theoretical synthesis of the cardiovascular system. Study I. The controlled system. *Acta physiol. pharmacol. néerl.* **12**, 189–265.
Defares, J. G. & Van Der Waal, H. J. (1969). A method for the determination of systemic arterial compliance in man. *Acta physiol. pharmacol. néerl.* **15**, 320–43.
Del Rio-Hortega, P. (1922). Son homologables le glia de escasas radiaciones ye le célula de Schwann. *Bol. Soc. Españ. biol.* **10**, 25–8.
— (1928). Tercera aportacíon al conocimiento morfologica e interpretación functional de la oligodendroglia. *Mem. Real. Soc. Españ. Hist. Nat.* **14**, 5–122.
— (1930). Concepts histogenique, morphologique et physiopathologique de la microglie. *Revue neurol.* **37**, 956–86.
— (1932). Microglia. In *Cytology and cellular pathology of the nervous system*, ed. Penfield, W. New York: Hoeber.
Dempsey, E. W. & Luse, S. (1958). Fine structure of the neuropil in relation to neuroglia cells. In *Biology of Neuroglia*, ed. Windle, W. F. Springfield, Ill.: Thomas.
Dempsey, E. W. & Wislocki, G. B. (1955). An electron microscope study of the blood-brain barrier in the rat, employing silver nitrate as a vital stain. *J. biophys. biochem. Cytol.* **1**, 245–56.
Detar, R. & Bohr, D. F. (1968). Oxygen and vascular smooth muscle contraction. *Am. J. Physiol.* **214**, 241–4.
Devine, C. E. & Simpson, F. O. (1967). The fine structure of sympathetic neuromuscular contacts in the rat. *Am. J. Anat.* **121**, 153–73.
— (1968). Localization of tritiated norepinephrine in vascular sympathetic axons of the rat intestine and mesentery by electron microscope radiography. *J. cell Biol.* **38**, 184–92.
Dick, D. A. T. (1959). The rate of diffusion of water in the protoplasm of living cells. *Expl Cell Res.* **17**, 5–12.
Diemer, K. (1963). Eine verbesserte Modellvorstellung zur Sauerstoffversorgung des Gehirns. *Naturwissenschaften*, **50**, 617–18.
— (1964). Über die Entwicklung der Gefassversorgung des Gehirns im Sauglingsalter. *Mschr. Kinderheilk.* **112**, 240–2.
— (1965a). Über die Sauerstoffdiffusion im Gehirn. I. Mitteilung: Raumliche Vorstellung und Berechnung der Sauerstoffdiffusion. *Pflügers Arch. ges. Physiol.* **285**, 99–108.

References

(1965b). II. Mitteilung. Die Sauerstoffdiffusion bei O_2-Mangelzustanden. *Pflügers Arch. ges. Physiol.* **285**, 109–118.

(1968). Capillarisation and oxygen supply to the brain. In *Oxygen transport in blood and tissue*, ed. Lübbers, D. W., Luft, U. C., Thews, G. & Witzleb, E., pp. 118–23. Stuttgart: Thieme.

Diji, A. (1959). Local vasodilator action of carbon dioxide on blood vessels of the hand. *J. appl. Physiol.* **14**, 414–16.

Diji, A. & Greenfield, A. D. M. (1957). The local effect of CO_2 on the blood vessels of the human skin. *J. Physiol.* **140**, 42 P.

Dilenge, D. & David, M. (1964). Place de l'angiographie dans l'étude de l'hémodynamique cerebrale. *Neurochir.* **10**, 567–86.

Djojosugito, A. M., Folkow, B., Öberg, B. & White, S. (1970). A comparison of blood viscosity measured *in vitro* and in a vascular bed. *Acta physiol. scand.* **78**, 70–84.

Dole, V. P., Emerson, K., Philips, R. A., Hamilton, P. & Van Slyke, D. D. (1946). The renal extraction of oxygen in experimental shock. *Am. J. Physiol.* **145**, 337–45.

Donahue, J. & Pappas, G. D. (1961). The fine structure of capillaries in the cerebral cortex of the rat at various stages of development. *Am. J. Anat.* **108**, 331–48.

Donaldson, H. H. & Hatai, S. (1931). On the weight of the parts of the brain and on the percentage of water in them according to brain weight and to age in albino and in wild Norway rats. *J. comp. Neurol.* **53**, 263–307.

Donders, F. C. (1849). De bewegingen der hersenen en de veranderingen der vaatvulling van de pia mater, ook bij gesloten onuitzet-beren schedel regtstreeks onderzocht. *Onderzoek. ged. inh. physiol. Lab. d. Utrecht. Hoogeoch* ii, 97–128.

(1850). De bewegingen der Hersenen en de veranderingen der vaatvulling van de 'Pia Mater', ook bij gesloten onuitzetberen schedel regtstreeks onderzocht. *Nederlandsche Lancet*, Series 2, **5**, 521–53.

(1859). *Physiologie des Menschen*. Vol. 1, ed. 2, p. 139. Leipzig: Hirtzel.

Donnan, F. G. (1911). Theorie der Membrangleichgewichte und Membranpotentiale bei Vor handensein von nicht dialysierenden. Elektrolyten. Ein Beitrag zur physikalisch-chemischen Physiologie. *Z. Elektrochem.* **17**, 572–81.

Dosekun, F. O., Grayson, J. & Mendel, D. (1960). The measurement of metabolic and vascular responses in liver and muscle with observations on their responses to insulin and glucose. *J. Physiol.* **150**, 581–606.

Dowgjallo, N. D. (1932). Beitrage zur Lehre von der Innervation des peripherischen Blutgefassystems. *Z. Anat. EntwGesch.* **97**, 9–54.

Draper, W. B. & Whitehead, R. W. (1944). Diffusion respiration in dog anaesthetized by pentothal sodium. *Anesthesiology* **5**, 262–73.

Dripps, R. D. & Comroe, J. H., Jr. (1947). The effect of the inhalation of high and low oxygen concentrations on respiration, pulse rate, ballistocardiogram and arterial oxygen saturation (oximeter) of normal individuals. *Am. J. Physiol.* **149**, 277–91.

Duff, G. L., McMillan, G. C. & Ritchie, A. C. (1957). The morphology and early atherscleroticlesions of the aorta demonstrated by the surface technique

References

in rabbits fed cholesterol together with a description of the anatomy of the intima of the rabbit's aorta and the 'spontaneous' lesions which occur in it. *Am. J. Path.* **33**, 845–73.

Dumke, P. R. & Schmidt, C. F. (1943). Quantitative measurements of cerebral blood flow in the macacque monkey. *Am. J. Physiol.* **138**, 421–31.

Dunning, H. S. & Wolff, H. G. (1936). The relative vascularity of the trigeminal ganglion and nerve, cerebral cortex and white matter. *Trans. Am. Neurol. Ass.* **62**, 150–4.

(1937). The relative vascularity of various parts of the central and peripheral nervous system of the cat and its relation to function. *J. comp. Neurol.* **67**, 433–50.

Durbin, R. P. & Jenkinson, D. H. (1961). The calcium dependence of tension development in depolarized smooth muscle. *J. Physiol.* **157**, 90–6.

Easton, J. D. & Palvolgyi, R. (1968). The dissociation of cerebral vasoconstrictor response to hypocapnia and hypertension. *Scand. J. clin. Lab. Invest.* Suppl. 102, v: J.

Echlin, F. A. (1942). Vasospasm and focal cerebral ischaemia. An experimental study. *Archs Neurol. Psychiat.*, Chicago **47**, 77–96.

(1965). Spasm of basilar and vertebral arteries caused by experimental subarachnoid haemorrhage. *J. Neurosurg.* **23**, 1–11.

Ecker, A. & Riemenschneider, P. A. (1951). Arteriographic demonstration of spasm of the intracranial arteries with special reference to saccular arterial aneurysms. *J. Neurosurg.* **8**, 660–7.

Edman, K. A. P. & Schild, H. O. (1963). Calcium and the stimulant and inhibitory effects of adrenaline in depolarized smooth muscle. *J. Physiol.* **169**, 404–11.

Eklöf, B., Ingvar, D. H., Kågström, E. & Olin, T. (1971). Persistence of cerebral blood flow autoregulation following chronic bilateral cervical sympathectomy in the monkey. *Acta physiol. scand.* **82**, 172–6.

Elfvin, L.-G. (1958). The ultrastructure of unmyelinated fibres in the splenic nerve of the cat. *J. Ultrastruct. Res.* **1**, 428–54.

Ellenberger, W. & Baum, H. (1943). *Handbuch der vergleichenden Anatomie der Haustiere*. Revised by O. Zietschmann, E. Ackerknecht and H. Grau, 18th edition. Berlin: Springer-Verlag.

Elliott, K. A. C. & Heller, J. H. (1957). In *Metabolism of the nervous system*. ed. Richter, D. London: Pergamon.

Elliott, K. A. C. & Jasper, H. H. (1949). Physiological salt solutions for brain surgery. Studies of local pH and vessel reactions to buffered and unbuffered isotonic solutions. *J. Neurosurg.* **6**, 140–52.

Eltherington, L. G., Stoff, J., Hughes, T. & Melmon, K. L. (1968). Constriction of human umbilical arteries. Interaction between oxgen and bradykinin. *Circulation Res.* **22**, 747–52.

Emanuel, D. A., Fleishman, M. & Haddy, F. J. (1957). The effect of pH change upon renal vascular resistance and urine flow. *Circulation Res.* **5**, 607–11.

Emerson, G. A. (1935). Effects of various anesthesias on autoxidation rate of surviving brain tissue. *Proc. Soc. Exp. Biol. N.Y.* **33**, 171–7.

Emerson, T. E. & Heath, C. (1969). Effects of local alterations of PCO_2 and

References

pH on cerebrovascular resistance in isolated dogs' heads. *Proc. Soc. exp. Biol. Med.* **130**, 318–22.

Espagno, J. & Lazorthes, Y. (1965). Measurement of regional cerebral blood flow in man by local injections of Xenon 133. *Acta neurol. scand.* Suppl. 14, 58–62.

Euler, C. von and Soderberg, U. (1956). The relation between gamma motor activity and the electroencephalogram. *Experientia* **12**, 278.

Evans, C. L. & Saaman, A. (1936). Simultaneous ligature of the carotid and vertebral arteries in the dog. *J. Physiol.* **87**, 33P.

Evans, M. G. (1942). Bilateral jugular ligation following bilateral suppurative mastoiditis. *Ann. Otol. Rhinol. Lar.* **51**, 615–25.

Evarts, E. V. (1962). Activity of neurons in visual cortex of the cat during sleep with low voltage fast EEG activity. *J. Neurophysiol.* **25**, 812–16.

(1964). Temporal patterns of discharge of pyramidal tract neurons during sleep and waking in the monkey. *J. Neurophysiol.* **27**, 152–71.

Fåhraeus, R. & Lindqvist, T. (1931). The viscosity of the blood in narrow capillary tubes. *Am. J. Physiol.* **96**, 562–8.

Fairchild, H. M., Ross, J. & Guyton, A. C. (1966). Failure of recovery from reactive hyperemia in the absence of oxygen. *Am. J. Physiol.* **210**, 490–2.

Falck, B., Hillarp, N-A., Thieme, G. & Torp, A. (1962). Fluorescence of catechol amines and related compounds condensed with formaldehyde. *J. Histochem. Cytochem.* **10**, 348–54.

Falck, B., Mchedlishvili, G. I. & Owman, Ch. (1965). Histochemical demonstration of adrenergic nerves in the cortex-pia of rabbit. *Acta pharmac. tox*, **23**, 133–42.

Falck, B., Nielsen, K. C. & Owman, Ch. (1968). Adrenergic innervation of the pial circulation. *Scand. J. clin. Lab. Invest.* Suppl. 102, VI: B.

Farquhar, M. G. & Hartmann, J. F. (1956). Electron microscopy of cerebral capillaries. *Anat. Rec.* **124**, 288–9.

(1957). Neuroglial structure and relationships as revealed by electron microscopy. *J. Neuropath. exp. Neurol.* **16**, 18–39.

Fazekas, J. F., Alman, R. W. & Bessman, A. N. (1952). Cerebral physiology of the aged. *Am. J. med. Sci.* **223**, 245–57.

Fazekas, J. F., Alman, R. W. & Parrish, A. E. (1951). Irreversible post-hypoglycaemic coma. *Am. J. Med. Sci.* **222**, 640–3.

Fazekas, J. F. & Bessman, A. N. (1953). Coma mechanisms. *Am. J. Med.* **15**, 804–12.

Fazekas, J. F. & Himwich, H. E. (1943). Anaerobic survival of adult animals. *Am. J. Physiol.* **139**, 366–70.

Fazekas, J. F., Kleh, J. & Parrish, A. E. (1955). The influence of shock on cerebral hemodynamics and metabolism. *Am. J. med. Sci.* **229**, 41–5.

Fazekas, J. F., Kleh, J. & Witkin, L. (1953). Cerebral hemodynamics and metabolism in subjects over 90 years of age. *J. Am. Geriatrics Soc.* **1**, 836–839.

Fedoruk, S. & Feindel, W. (1960). Measurement of brain circulation time by radio-active iodinated albumin. *Can. J. Surg.* **3**, 312–18.

Feigl, E. & Folkow, B. (1963). Cardiovascular responses in 'diving' and during brain stimulation in ducks. *Acta physiol. scand.* **57**, 99–110.

References

Feinberg, I., Lane, M. H. & Lassen, N. A. (1960). Senile dementia and cerebral oxygen uptake measured on the right and left sides. *Nature, Lond.* **188**, 962-4.

Feindel, W. (1965). *Thomas Willis. The anatomy of the brain and nerves.* Montreal: McGill University Press.

Fencl, V., Miller, T. B. & Pappenheimer, J. R. (1966). Studies on the respiratory response to disturbances of acid–base metabolism balance, with deductions concerning the ionic composition of cerebral interstitial fluid. *Am. J. Physiol.* **210**, 459-72.

Fencl, V., Vale, J. R. & Broch, J. R. (1968). Cerebral blood flow and pulmonary ventilation in metabolic acidosis and alkalosis. *Scand J. clin. Lab. Invest.* Suppl. 102, VIII: B.

Fenishel, I. R. & Horowitz, S. B. (1963). The transport of non-electrolytes as a diffusional process in the cytoplasm. *Acta physiol. scand.* **60**, Suppl. 221.

Ferrario, C. M., Dickenson, C. J. & McCubbin, J. W. (1970). Central vasomotor stimulation by angiotensin. *Clin. Sci.* **39**, 239-45.

Ferris, E. B., Jr. (1941). Objective measurement of relative intracranial blood flow in man with observations concerning the hydrodynamics of the craniovertebral system. *Archs Neurol. Psychiat., Chicago* **46**, 377-401.

Ferris, E. B., Jr., Engels, G. L., Stevens, C. D. & Logan, M. (1946). The validity of internal jugular venous blood in studies of cerebral metabolism and blood flow in man. *Am. J. Physiol.* **147**, 517-21.

Fick, A. (1855). Ueber Diffusion. *Annln. Phys.* **94**, 59-86.

Field, E. J., Grayson, J. & Rogers, A. F. (1951). Observations on the blood flow in the spinal cord of the rabbit. *J. Physiol.* **144**, 56-70.

Field, J., Fuhrman, F. A. & Martin, A. W. (1944) Effect of temperature on the oxygen consumption of brain tissue. *J. Neurophysiol.* **7**, 117-26.

Fieschi, C., Agnoli, A., Battistini, N., Bozzao, L. (1966). Relationships between cerebral transit time of non-diffusible indicators and cerebral blood flow. A comparative study with Krypton 85 and radio-albumin. *Experientia* **22**, 189-90.

Fieschi, C., Agnoli, A. & Galbo, E. (1963). Effects of CO_2 on cerebral hemodynamics in normal subjects and in cerebrovascular disease studied by carotid injection of radioalbumin. *Circulation Res.* **13**, 436-77.

Fieschi, C., Bozzao, L. & Agnoli, A. (1965). Regional clearance of hydrogen as a measure of cerebral blood flow. *Acta neurol. scand.* Suppl. **14**, 46-52.

Fieschi, C., Bozzao, L., Agnoli, A. & Kety, S. S. (1964). Misurazioni regionali del flusso sanguigno cerebrale mediante registrazioni in profondata della curve di 'clearance' di hydrogeno. *Boll. Soc. ital. Biol. sper.* **40**, 1505-9.

Fieschi, C., Isaacs, G. & Kety, S. S. (1968). On the question of heterogeneity of the local blood flow in grey matter of the brain. In *Blood flow through organs and tissues*, ed. Bain, W. H. & Harper, A. M., p. 226. Edinburgh: Livingstone.

Finesinger, J. E. & Cobb, S. (1933). Cerebral circulation. XXVII. Action of the pial arteries of the convulsants caffeine, absinth, camphor and picrotoxin. *Archs Neurol. Psychiat., Chicago* **30**, 980-1002.

(1935). The cerebral circulation. XXXIV. The action of narcotic drugs on the pial vessels. *J. Pharmac. exp. Ther.* **53**, 1-33.

References

Finesinger, J. & Putnam, T. J. (1933). Cerebral circulation. XXIII. Induced variations in volume flow through the brain perfused at constant pressure. *Archs Neurol. Psychiat., Chicago* **30**, 775–94.

Finn, H., Kao, F. F., Mei, S. S. & Harmel, M. H. (1968). Csf-blood potential in cats and its modification by sodium gammahydroxybutyrate. *Arch. int. Pharmacodyn.* **176**, 319–25.

Finnerty, F. A., Guillaudeu, R. L. & Fazekas, J. F. (1957). Cardiac and cerebral hemodynamics in drug induced postural collapse. *Circulation Res.* **5**, 34–9.

Finnerty, F. A., Witkin, L. & Fazekas, J. F. (1954). Cerebral hemodynamics during cerebral ischemia induced by acute hypotension. *J. clin. Invest.* **33**, 1227–32.

Fleischauer, K. (1961). Regional differences in the structure of the ependyma and subependymal layers of the cerebral ventricles of the cat. In *Regional neurochemistry*, ed. Kety, S. S. & Elkes, J., pp. 279–83. New York: Pergamon.

Fleishman, M., Scott, J. & Haddy, F. J. (1957). Effect of pH change upon systemic large and small vessel resistance. *Circulation Res.* **5**, 602–6.

Flexner, L. B. (1952). Physiologic development of the cortex of the brain and its relationship to its morphology, constitution and enzyme systems. *Res. Publs Ass. Res. nerv. ment. Dis.* **32**, 297–304.

Florey, H. (1925). Microscopical observation on the circulation of the blood in the cerebral cortex. *Brain*, **48**, 43–64.

Floyd, W. F. & Neil, E. (1952). Influence of the sympathetic innervation of the carotid bifurcation in chemoreceptor and baroreceptor activity in the cat. *Archs int. Pharmacodyn. Thér.* **91**, 230–40.

Fofanow, L. L. & Tschalussow, M. A. (1913). Ueber die Beziehungen des N. depressor zu den motorischen Zentren. *Pflügers Arch. ges. Physiol*, **151**, 543–82.

Fog, M. (1934). *Om pia arterernes Vasomotoriske Reactioner.* Doctoral thesis. University of Copenhagen.

— (1937). Cerebral circulation. The reaction of the pial arteries to a fall in blood pressure. *Archs Neurol. Psychiat., Chicago* **37**, 351–64.

— (1938). The relationship between the blood pressure and the tonic regulation of the pial arteries. *J. Neurol. Psychiat.* **1**, 187–97.

— (1939a). Cerebral circulation. I. Reaction of pial arteries to epinephrine by direct application and intravenous injection. *Archs Neurol. Psychiat., Chicago* **41**, 109–18.

— (1939b). Cerebral circulation: reaction of pial arteries to increase in blood pressure. *Archs Neurol. Psychiat., Chicago* **41**, 260–8.

— (1968). Autoregulation of cerebral blood flow and its abolition by local hypoxia and/or trauma. *Scand J. clin. Lab. Invest.* Suppl. 102, v: B.

Foix, C. & Hillemand, P. (1925). Les artères de l'axe encéphalique jusqu'au diencéphale inclusivement. *Revue neurol.* **32**, 705–39.

Folkow, B. (1949). Intravascular pressure as a factor regulating the tone of the small vessels. *Acta physiol. scand.* **17**, 289–310.

— (1955). Nervous control of the blood vessels. *Physiol. Rev.* **35**, 629–63.

— (1964). Description of the myogenic hypothesis. *Circulation Res.* **15**: Suppl: 279–87.

Folkow, B. & Uvnäs, B. (1948). The distribution and functional significance

References

of sympathetic vasodilators to the hindlimbs of the cat. *Acta physiol. scand.* **15**, 389–400.

(1949). The chemical transmission of nerve impulses to the hind limbs of the dog. *Acta physiol. scand.* **17**, 191–4.

Forbes, H. S. (1928). Cerebral circulation. I. Observation and measurement of pial vessels. *Archs Neurol. Psychiat., Chicago* **19**, 751–61.

Forbes, H. S. & Cobb, S. (1938). Vasomotor control of cerebral vessels. *Res. Publs Ass. Res. nerv. ment. Dis.* **18**, 201–17.

Forbes, H. S., Finley, H. K. & Nason, G. I. (1933). Cerebral circulation. XXIV. *A.* Action of epinephrine on pial vessels. *B.* Action of pituitary and pitressin on pial vessels. *C.* Vasomotor response in the pia and in the skin. *Archs Neurol. Psychiat., Chicago* **30**, 957–79.

Forbes, H. S. & Nason, G. I. (1935). Cerebral circulation; vascular responses to hypertonic solutions and withdrawal of cerebrospinal fluid. *Archs Neurol. Psychiat., Chicago* **34**, 533–47.

Forbes, H. S., Nason, G. I., Cobb, S. & Wortman, R. C. (1937). Cerebral circulation. XLV. Vasodilation in pia following stimulation of the geniculate ganglion. *Archs Neurol. Psychiat., Chicago* **37**, 776–81.

Forbes, H. S., Nason, G. I. & Wortman, R. C. (1937). Cerebral circulation XLIV. Vasodilation in the pia following stimulation of the vagus, aortic and carotid sinus nerves. *Archs Neurol. Psychiat., Chicago* **37**, 334–50.

Forbes, H. S., Schmidt, C. F. & Nason, G. I. (1939). Evidence of vasodilator innervation in the parietal cortex of the cat. *Am. J. Physiol.* **125**, 216–19.

Forbes, H. S. & Wolff, H. G. (1928). Cerebral circulation. III. The vasomotor control of cerebral vessels. *Archs Neurol. Psychiat., Chicago* **19**, 1057–86.

Forrester, A. C., McDowall, D. G., Harper, A. M. & Nisbet, I. (1964). The effect of hypothermia on cerebral blood flow at constant arterial carbon dioxide tension. *Proc. 3rd World Congress Anesthesiology*, **3**, 129–34.

Forster, R. P. & Nyboer, J. (1955). Effect of induced apnoea on cardiovascular renal functions in rabbit. *Am. J. Physiol.* **183**, 149–54.

Francois-Franck, C.-E. (1887). *Sur les Fonctions Motrices du Cerveau.* p. 199. Paris.

Franklin, K. J., McGee, L. E. & Ullman, E. A. (1951). Effects of severe asphyxia on kidney and urine flow. *J. Physiol.* **112**, 43–53.

Freeman, J. & Ingvar, D. H. (1968). Elimination by hypoxia of cerebral blood flow autoregulation and EEG relationship. *Expl Brain Res.* **5**, 61–71.

Fremont-Smith, F. & Forbes, H. S. (1927). Intra-ocular and intracranial pressure. *Archs Neurol. Psychiat., Chicago* **18**, 550–64.

Freygang, W. H. & Sokoloff, L. (1958). Quantitative measurements of regional circulation in central nervous system by use of radioactive inert gas. In *Advances in biological and medical physics*, vol. 6, p. 263. New York: Academic Press.

Freyhan, F. A., Woodford, R. B. & Kety, S. S. (1951). Cerebral blood flow and metabolism in psychoses of senility. *J. nerv. ment. Dis.* **113**, 449–56.

Friede, R. (1953). Über die tropische Funktion der Glia. *Virchows Arch. path. Anat.* **324**, 15–26.

Froman, C. & Crampton-Smith, A. (1967). Metabolic acidosis of the cerebrospinal fluid associated with subarachnoid haemorrhage. *Lancet*, i, 965–7.

References

Frugoni, P. (1952). Persistenza della anastomosi carotidio basilare. *Chirurgia* **7**, 327–35.

Fulton, J. F. (1928). Observations upon the vascularity of the human occipital lobe during visual activity. *Brain* **51**, 310–20.

Funaki, S. (1961). Spontaneous spike-discharges of vascular smooth muscle. *Nature, Lond.* **191**, 1102–3.

Furchgott, R. F. (1967). The pharmacological differentiation of adrenergic receptors. *Ann. N.Y. Acad. Sci.* **139**, 553–70.

Gaertner, G. & Wagner, J. (1887). Über den Hirnkreislauf. *Wien. med. Wschr.* **37**, 601–39.

Ganter, G. (1928). Über die Vorgänge im Kreislauf bei der Arbeit. *Arch. exp. Pathol. Pharmakol.* **138**, 276–300.

Garfunkel, J. M., Baird, H. W. & Ziegler, J. (1954). The relationship of oxygen consumption to cerebral functional activity. *J. Pediat.* **44**, 64–72.

Garry, R. C. (1928). The effect of oxygen lack upon surviving smooth muscle. *J. Physiol.* **66**, 235–48.

Gaskell, P. & Burton, A. C. (1953). Local postural vasomotor reflexes arising from the limb veins. *Circulation Res.* **1**, 27–39.

Gaskell, T. W. H. (1880). On the tonicity of the heart and blood vessels. *J. Physiol.* **3**, 48–75.

Gasser, H. S. (1958). Comparison of the structure, as revealed with the electron microscope, and the physiology of the unmedullated fibres in the skin nerves and in the olfactory nerves. *Expl Cell Res.* Suppl. 5, 3–17.

Gayda, T. (1914). Sul ricambio gassoso dell'encefalo. *Arch Fisiol.* **12**, 215–44.

Geddes, I. C. & Gray, T. C. (1959). Hyperventilation in the maintenance of anesthesia. *Lancet* ii, 4–6.

Geiger, A. (1958). Correlation of brain metabolism and function by the use of a brain perfusion method *in situ. Physiol. Rev.* **38**, 1–20.

Geiger, A., Gombos, G. & Otzuki, S. (1963). The effect of hypoxaemia on the metabolic pattern of the perfused brain of cats. In *Selective vulnerability of the brain in hypoxaemia*, ed. Schade, J. P. & McMenemey, W. H., pp. 295–304. Philadelphia: Davis.

Geiger, A. & Magnes, J. (1947). The isolation of the cerebral circulation and the perfusion of the brain in the living cat. *Am. J. Physiol.* **149**, 517–37.

Gerard, R. W. (1938). Brain metabolism and circulation. *Res. Publs Ass. Res. nerv. ment. Dis.* **18**, 316–45.

Gesell, R., Brassfield, C. R. & Hamilton, J. A. (1942). An acid neurohumoral mechanism of nerve cell activation. *Am. J. Physiol.* **136**, 604–8.

Gesell, R. & Bronk, D. W. (1926). Some effects of alveolar carbon dioxide tension on the carotid and femoral flow of blood. *Proc. Soc. exp. Biol. Med.* **24**, 255–6.

Gibbs, E. L. & Gibbs, F. A. (1934). The cross section areas of the vessels that form the torcular and the manner in which flow is distributed to the right and left lateral sinus. *Anat. Rec.* **59**, 419–26.

Gibbs, E. L., Gibbs, F. A., Lennox, W. G. & Nims, L. F. (1942). Regulation of cerebral carbon dioxide. *Archs Neurol. Psychiat., Chicago* **47**, 879–89.

—— (1943). The value of carbon dioxide in counteracting the effects of low oxygen. *J. Aviat. Med.* **14**, 250.

References

Gibbs, F. A. (1933a). Cerebral blood flow preceding and accompanying experimental convulsions. *Archs Neurol. Psychiat., Chicago* **30**, 1003-10.

(1933b). A thermoelectric blood flow recorder in the form of a needle. *Proc. Soc. Exp. Biol. Med.* **31**, 141-6.

Gibbs, F. A., Gibbs, E. L. & Lennox, W. G. (1935a). The cerebral blood flow in man as influenced by adrenaline, caffein, amyl nitrite and histamine. *Am. Heart J.* **10**, 916-24.

(1935b). Changes in human cerebral blood flow consequent on alterations in blood gases. *Am. J. Physiol.* **111**, 557-63.

(1935c). The cerebral blood flow during sleep in man. *Brain*, **58**, 4-48.

Gibbs, F. A., Lennox, W. G. & Gibbs, E. L. (1934). Cerebral blood flow preceding and accompanying epileptic seizures in man. *Archs Neurol. Psychiat., Chicago* **32**, 257-72.

Gibbs, F. A., Maxwell, H. & Gibbs, E. L. (1947). Volume flow of blood through the human brain. *Archs Neurol. Psychiat., Chicago* **57**, 137-44.

Gibbs, F. A., Williams, D. & Gibbs, E. L. (1940). Modification of the cortical frequency spectrum by changes in CO_2, blood sugar and O_2. *J. Neurophysiol.* **3**, 49-58.

Gibson, Q. H. (1959). The kinetics of reactions between haemoglobin and gases. *Prog. Biophys.* **9**, 1-53.

Gillespie, R. D. (1924). The relative influence of mental and muscular work on the pulse rate and blood pressure. *J. Physiol.* **58**, 425-532.

Glass, H. I. & Harper, A. M. (1963). Measurement of regional blood flow in cerebral cortex of man through intact skull. *Br. med. J.* i, 593.

Gleichmann, U., Ingvar, D. H., Lassen, N. A., Lübbers, D. S., Siesjö, B. K. & Thews, G. (1962). Regional cerebral cortical metabolic rate of oxygen and carbon dioxide, related to the EEG in the anaesthetised dog. *Acta physiol. scand.* **55**, 82-94.

Gleichmann, U., Ingvar, D. H., Lübbers, D. W., Seisjö, B. K. & Thews, G. (1962). Tissue PO_2 and PCO_2 of the cerebral cortex, related to blood gas tensions. *Acta physiol. scand.* **55**, 127-38.

Goldensohn, E. S., Whitehead, R. W., Parry, T. M., Spencer, J. H., Grover, R. F. & Draper, W. B. (1951). Effect of diffusion respiration and high concentrations of CO_2 on cerebral spinal fluid pressure of anaesthetised dogs. *Am. J. Physiol.* **165**, 334-40.

Gollwitzer-Meier, K. L. & Eckhardt, P. (1935). Weitere Untersuchungen über den Nerveneinfluss auf die Hirndurchblutung. *Arch. exp. Path. Pharmak.* **177**, 501-18.

Gollwitzer-Meier, K. & Schulte, H. (1932). Das Verhalten der Hirndurchblutung bei Reizung der Sinusnerven. *Arch. exp. Path. Pharmak.*, **165**, 685-95.

Goodrich, C. (1965). Effect of chronic acidosis and alkalosis on rat CSF-blood potential. *Physiologist.* **8**, 178.

Gotoh, F., Meyer, J. S. & Tomita, M. (1966). Hydrogen method for determining cerebral blood flow in man. *Archs Neurol. Chicago* **15**, 549-59.

Gotoh, F., Tazaki, Y. & Meyer, J. S. (1961). Transport of gases through brain and their extravascular vasomotor action. *Expl Neurol.* **4**, 48-58.

Gottstein, U. & Held, D. K. (1969). The effect of haemodilution caused by low molecular weight dextran on human cerebral blood flow and metabolism.

References

In *Cerebral blood flow*, ed. Brock, M., Fieschi, C., Ingvar, D. H., Lassen, N. A. & Schürmann, K., pp. 104–5. Berlin: Springer.

Graf, K., Golenhofen, K. & Hensel, H. (1957). Fortlaufende Registrierung der Leberdurchblutung mit der Wärmeleitsonde. *Pflügers Arch. ges. Physiol.* **264**, 44–60.

Graf, K. & Rosell, S. (1958). Untersuchungen zur fortlaufenden Durchblutungs registrierung mit Wärmleitsonden. *Acta physiol. scand.* **42**, 51–73.

Granholm, L., Lukjanova, L. & Siesjö, B. K. (1968). Evidence of cerebral hypoxia in pronounced hyperventilation. *Scand. J. clin. Lab. Invest.* Suppl. 102, VI: C.

Granholm, L. & Siesjö, B. K. (1968). Signs of cerebral hypoxia in hyperventilation. *Experientia* **24**, 337–8.

Grant, F. C., Spitz, E. B., Shenkin, H. A., Schmidt, C. F. & Kety, S. S. (1947). The cerebral blood flow and metabolism in idiopathic epilepsy. *Trans. Am. neurol. Ass.* **72**, 82–6.

Grayson, J. (1952). Internal calorimetry in the determination of thermal conductivity and blood flow. *J. Physiol.* **118**, 54–72.

Green, H. D. & Dension, A. N., Jr. (1956). Absence of vasomotor responses to epinephrine and arterenol in isolated intracranial circulation. *Circulation Res.* **4**, 565–73.

Green, H. D. & Krepchar, J. H. (1959). Control of peripheral resistance in major systemic vascular beds. *Physiol. Rev.* **39**, 617–86.

Green, H. D., Lewis, R. N. & Nickerson, N. D. (1943). Quantitation of changes in vasomotor tone. Change of vasomotor tone as cause of Traube Hering waves. *Proc. Soc. exp. Biol. Med.* **53**, 228–9.

Green, H. D., Lewis, R. N., Nickerson, N. D. & Heller, A. L. (1944). Blood flow, peripheral resistance and vascular tonus with observations on relationship between blood flow and cutaneous temperature. *Am. J. Physiol.* **141**, 518–36.

Greenfield, J. C. & Tindall, G. T. (1965). Effect of acute increase in intracranial pressure on blood flow in the internal carotid artery in man. *J. clin. Invest.* **44**, 1343–51.

Grillo, M. A. (1966). Electron microscopy of sympathetic tissues. *Pharmac. Rev.* **18**, 387–99.

Grino, A. & Billet, E. (1949). The diagnosis of orbital tumours by angiography. *Am. J. Ophthal.* **32**, 879–911.

Gros, Cl., Minvielle, J. & Vlahovitch, B. (1956). Anastomoses arterielles intracraniennes. Etude arteriographique et clinique. *Neurochir.* **2**, 281–302.

Grunewald, W. (1968). Theoretical analysis of the oxygen supply in tissue. In *Oxygen transport in blood and tissue*, ed. Lübbers, D. W., Luft, U. C., Thews, G. & Witzleb, E., pp. 100–14. Stuttgart: Thieme.

Gulland, G. L. (1898). The occurrence of nerves on intracranial blood vessels. *Br. med. J.* ii, 781–2.

Gurdjian, E. S. & Thomas, L. M. (1961). Human pial circulation. *Archs Neurol., Paris* **5**, 111–18.

Gurdjian, E. S., Stone, W. E., Webster, J. W. (1944). Cerebral metabolism in hypoxia. *Archs Neurol. Psychiat., Chicago* **51**, 472–7.

Gurdjian, E. S., Webster, J. E., Martin, F. A. & Thomas, L. M. (1958). Cinephotomicrography of the pial circulation. *Archs Neurol. Psychiat., Chicago* **80**, 418–35.

References

Gurdjian, E. S., Webster, J. E. & Stone, W. E. (1944). Cerebral metabolism in experimental head injury. *War. Med.* **6**, 173–9.

(1949). Cerebral constituents in relation to blood gases. *Am. J. Physiol.* **156**, 149–57.

Guyot, H. (1829). Essai sur les vaisseaux sanguins de cerveau. *J. de physiol. exper.* **9**, 29–43.

Guyton, A. C., Ross, J. M., Carrier, O. & Walker, J. (1964). Evidence for tissue oxygen demand as the major factor causing autoregulation. *Circulation Res.* **15**, 1–60–8.

Haddy, F. J. & Scott, J. B. (1968). Metabolically linked vasoactive chemicals in local regulation of blood flow. *Physiol. Rev.* **48**, 688–707.

Hafkenshiel, J. H., Crumpton, C. W. & Friedland, C. K. (1954). Cerebral oxygen consumption in essential hypertension: constancy with age, severity of disease, sex and variations of blood constituents, as observed in 101 patients. *J. clin. Invest.* **33**, 63–8.

Hafkenschiel, J. H. & Friedland, C. K. (1952). Physiology of the cerebral circulation in essential hypertension: the effects of inhalation of 5 % carbon dioxide oxygen mixtures on cerebral hemodynamics and oxygen metabolism. *J. Pharmac. exp. Ther.* **106**, 391–2.

Hagen, E. & Wittkowski, W. (1969). Licht- und elecktronenmikroskopische Untersuchungen zur Innervation der Piagefässe. *Z. Zellforsch. mikrosk. Anat.* **95**, 429–44.

Häggendal, E. (1965). Blood flow autoregulation of the cerebral grey matter with comments on its mechanism. *Acta neurol. scand.* Suppl. 14, 104–10.

(1968). Elimination of autoregulation during arterial and cerebral hypoxia. *Scand. J. clin. Lab. Invest.* Suppl. 102, v: D.

Häggendal, E., Löfgren, J., Nilsson, N. J. & Zwetnow, N. (1966). Die Gehirndurchblutung bei experimentellen Liquordrückanderungen. *Verhandl. Intern. Neurochirurgen-Kongress*, Bad. Dürkheim.

Häggendal, E. & Johansson, B. (1965). Effects of arterial carbon dioxide tension and oxygen saturation on cerebral blood flow autoregulation in dogs. *Acta physiol. scand.* **66**, Suppl. 258, 27–53.

Häggendal, E., Nilsson, N. J. & Norbäck, B. (1965). On the components of Kr^{85} clearance curves from the brain of the dog. *Acta physiol. scand.* Suppl. 258, 5–25.

(1966). Effect of blood corpuscle concentration on cerebral blood flow. *Acta chir. scand.* Suppl. 364, 3–12.

Häggendal, E. & Norbäck, B. (1966). Effect of viscosity on cerebral blood flow. *Acta chir. scand.* Suppl. 364, 13–22.

Hain, R. F., Westhaysen, P. V. & Swank, R. L. (1952). Haemorrhagic cerebral infarction by arterial occlusion; experimental study. *J. Neuropath.* **11**, 34–43.

Haining, J. L., Turner, M. D., Pantall, R. M. (1970). Local cerebral blood flow in young and old rats during hypoxia and hypercapnia. *Am. J. Physiol.* **218**, 1020–4.

Haller, A. von (1755). *Dissertation on the sensible and irritable parts of animals.* pp. 20 and 21. London: Nourse.

(1774). *Bibliotheca Anatomica.* Tiguri: apud Orell, Gessner, Fuessli.

References

Hamelberg, W., Sprouse, J. H., Mahaffey, J. E. & Richardson, J. A. (1960). Catechol amine levels during light and deep anesthesia. *Anesthesiology* **21**, 297–302.

Handley, C. A., Sweeney, H. M., Scherman, Q. & Severance, R. (1943). Metabolism of perfused dog's brain. *Am. J. Physiol.* **140**, 190–6.

Hansen, D. B., Sultzer, M. R., Freygang, W. H. & Sokoloff, L. (1957). Effects of low O_2 and high CO_2 concentrations in inspired air on local cerebral circulation. *Fedn Proc.* **16**, 54.

Harmel, M. H., Hafkenschiel, J. H., Austin, G. M., Crumpton, C. W. & Kety, S. S. (1949). The effect of bilateral stellate ganglion block on the cerebral circulation in normotensive and hypertensive patients. *J. clin. Invest.* **28**, 415–18.

Harper, A. M.
 (1965). Physiology of cerebral blood flow. *Br. J. Anaesth.* **37**, 225–35.
 (1966). Autoregulation of cerebral blood flow: influence of the arterial blood pressure on the blood flow through the cerebral cortex. *J. Neurol. Neurosurg. Psychiat.* **29**, 398–403.

Harper, A. M. & Bell, R. A. (1963). The effect of metabolic acidosis and alkalosis on the blood flow through the cerebral cortex. *J. Neurol. Neurosurg. Psychiat.* **26**, 341–4.

Harper, A. M. & Glass, H. I. (1965). Effect of alterations in the arterial carbon dioxide tension on the blood flow through the cerebral cortex at normal and low arterial blood pressures. *J. Neurol. Neurosurg. Psychiat.* **28**, 449–52.

Harper, A. M., Glass, H. I., Glover, M. (1961). Measurement of blood flow in the cerebral cortex of dogs by the clearance of Krypton 85. *Scott. med. J.* **6**, 12–17.

Harper, A. M., Ledingham, I. McA. & McDowall, D. G. (1965). The influence of hyperbaric oxygen on the blood flow and oxygen uptake of the cerebral cortex in hypovolaemic shock. In *Hyperbaric oxygenation*, ed. Ledingham, I. McA., pp. 342–346. Edinburgh: Livingstone.

Harper, A. M., Rowan, J. O. & Jennett, W. B. (1968). Oldendorf's non-diffusible indicator transit approach. *Scand. J. clin. Lab. Invest.* Suppl. 102, XI: B.

Harris, H. A. (1941). A note on the clinical anatomy of the veins, with special reference to the spinal veins. *Brain* **64**, 291–300.

Harvey, J. & Rasmussen, T. (1951). Occlusion of the middle cerebral artery. *Archs Neurol. Psychiat., Chicago* **66**, 20–9.

Hassin, G. B. (1929). The nerve supply of the cerebral blood vessels. A histological study. *Archs Neurol. Psychiat., Chicago* **22**, 375–91.

Hassler, O. (1962a). A systematic investigation of the physiological intima cushions associated with the arteries in 5 human brains. *Acta Soc. Med. upsal.* **67**, 35–41.
 (1962b). Physiological intima cushions in the large cerebral arteries of young individuals. 1. Morphological structure and possible significance for the circulation. *Acta path. microbiol. scand.* **55**, 19–27.

Haynes, R. H. (1961). The rheology of blood. *Trans. Soc. Rheology* **5**, 85–101.

Hedlund, S., Nylin, G. & Regnström, O. (1962). The behaviour of the cerebral circulation during muscular exercise. *Acta physiol. scand.* **54**, 316–24.

References

Hegedis, S. A. & Shackelford, R. T. (1965). A comparative anatomical study of the cranocervical systems in mammals. *Am. J. Anat.* **116**, 375–86.

Heidenreich, J., Erdmann, W., Metzger, H. & Thews, G. (1970). Local hydrogen-clearance and PO_2-measurements in micro-areas of the rat brain. *Experientia* **26**, 257–9.

Heisey, S. R., Held, D. & Pappenheimer, J. R. (1962). Bulk flow and diffusion in the cerebrospinal fluid system of the goat. *Am. J. Physiol.* **203**, 775–81.

Held, D., Fencl, D. & Pappenheimer, J. R. (1964). Electrical potential of the C.S.F. *J. Neurophysiol.* **27**, 942–59.

Held, K., Gottstein, U. & Niedermayer, W. (1969). CBF in non-pulsatile perfusion. In *Cerebral blood flow*, ed. Brock, M., Fieschi, C., Ingvar, D. H., Lassen, N. A. & Schürmann, K., pp. 94–5. Berlin: Springer.

Hellinger, F. R., Bloor, B. M. & McCutchen, J. J. (1962). Total cerebral blood flow and oxygen consumption using dye-dilution method: study of occlusive arterial disease and cerebral infarction. *J. Neurosurg.* **19**, 964–70.

Henderson, V. E. (1930). Present status of theories of narcosis. *Physiol. Rev.* **10**, 171–220.

Henle, J. (1841). *Allgemeine Anatomie*. Leipzig: Voss.

Henry, J. P., Gauer, O. H., Kety, S. S. & Kramer, K. (1951). Factors maintaining cerebral circulation during gravitational stress. *J. clin. Invest.* **30**, 292–300.

Hensel, H. & Bender, F. (1956). Bestimmung der Hautdurchblutung am Menschen mit einer electriscken Warmeleitmesser. *Pflügers Arch. ges. Physiol.* **263**, 603–14.

Hensel, H. & Ruef, J. (1954). Fortlaufende Registrierung der Muskeldurchblutung am Menschen mit einer Calorimetersonde. *Pflügers Arch. ges. Physiol.* **259**, 267–80.

Hering, H. E. (1923). Der Karotisdruckversuch. *Münch. med. Wschr.* **70**, 1287–90.

—— (1924). Der Sinus caroticus an der Ursprungsstelle der Carotis interna als Ausgangsort eines hemmenden Herzreflexes und eines depressorischen Gefässreflexes. *Münch. med. Wschr.* **71**, 701–4.

—— (1925). Ueber die Wand des Sinus caroticus als Reizempfänger und den Sinusnerv als zentripetale Bahn fur die Sinusreflexe. *Dt. med. Wschr.* **51**, 1140–1.

—— (1927). Über die Blutdrucksregulierung bei Aenderung der Körperstellung vermittelst der Blutdruckszügler und das Zustandekommen der Ohnmacht beim plötzlichen Uebergang vom Liegen zum Stehen. *Münch. med. Wschr.* **74**, 1611–13.

—— (1932). *Der Blutdruckszüglertonus in seiner Bedeutung für den Parasympatheticustonus und Sympatheticustonus*. Leipzig: Thieme.

Herrmann, E. (1968). Application of heat clearance method for measurements of cerebral blood flow in man. In *Blood flow through organs and tissues*, ed. Bain, W. H. & Harper, A. M. pp. 182–9. Edinburgh: Livingstone.

Hess, A. (1955). The fine structures and morphological organization of non-myelinated fibres. *Proc. R. Soc.* B **144**, 496–506.

Hess, H. H. & Pope, A. (1960). Intralaminar distribution of cytochrome oxidase activity in human frontal isocortex. *J. Neurochem.* **5**, 207–17.

Heyman, A., Patterson, J. L. & Duke, T. W. (1952). Cerebral circulation and

References

metabolism in sickle cells and other chronic anaemias, with observations on the effects of oxygen inhalation. *J. clin. Invest.* **31**, 824–8.

Heyman, A., Patterson, J. L., Duke, T. W. & Battey, L. L. (1953). The cerebral circulation and metabolism in arteriosclerotic and hypertensive cerebrovascular disease. *New Engl. J. Med.* **249**, 223–9.

Heyman, A., Patterson, J. L. & Jones, R. W. (1951). Cerebral circulation and metabolism in uremia. *Circulation* **3**, 558–63.

Heymans, C. & Bouckaert, J. J. (1929). Le sinus carotidien, zone réflexogène régulatrice due tonus des vaisseaux céphaliques. *C. r. Séanc. Soc. Biol.* **100**, 202–4.

—— (1931). Observations chez le chien en hypertension artérielle chronique et expérimentale. *C. r. Séanc. Soc. Biol.* **106**, 471–3.

—— (1932). Sinus carotidien et régulation réflexe de la circulation artérielle encéphalobulbaire. *C. r. Séanc. Soc. Biol.* **110**, 996–9.

Heymans, J. F. & Heymans, C. (1927). Sur les modifications directes et sur la régulation réflexe de l'activité du centre respiratoire de la tête isolée du chien. *Archs int. Pharmacodyn. Thér.* **33**, 273–372.

Hickham, J. B., Schieve, J. F. & Wilson, W. P. (1953). The relation between retinal and cerebral vascular reactivity in normal and arteriosclerotic subjects. *Circulation* **7**, 84–7.

Hill, A. V. (1928). The diffusion of oxygen and lactic acid through tissues. *Proc. R. Soc.* **104**, 39–96.

Hill, L. (1895). The influence of the force of gravity on the circulation of the blood. *J. Physiol.* **18**, 15–53.

—— (1896). *The physiology and pathology of the cerebral circulation. An experimental research.* p. 208. London: Churchill.

Hill, L. & McLeod, J. J. R. (1900). A further enquiry into the supposed existence of cerebral vaso-motor nerves. *J. Physiol.* **26**, 394–404.

Hill, L. & Moore, B. (1894). Effects of compression of the common carotid artery. *Br. med. J.* i, 962–3.

Hills, B. A. (1967). Diffusion versus blood perfusion in limiting the rate of uptake of inert non-polar gases by skeletal rabbit muscle. *Clin. Sci.* **33**, 67–87.

Himwich, H. E. (1951). *Brain metabolism and cerebral disorders*, p. 129. Baltimore: Williams & Wilkins.

Himwich, H. E., Baker, Z. & Fazekas, J. F. (1939). The respiratory metabolism of infant brain. *Am. J. Physiol.* **125**, 601–6.

Himwich, H. E., Bernstein, A. O., Herrlich, H., Chesler, A. & Fazekas, J. F. (1941). Mechanisms for the maintenance of life in the newborn during anoxia. *Am. J. Physiol.* **135**, 387–91.

Himwich, H. E. & Fazekas, J. F. (1937). The effect of hypoglycaemia on the metabolism of the brain. *Endocrinology* **21**, 800–7.

—— (1941). Studies of the metabolism of the brain of infant and adult dogs. *Am. J. Physiol.* **132**, 454–9.

Himwich, H. E. & Nahum, L. H. (1932). The respiratory quotient of the brain. *Am. J. Physiol.* **101**, 446–53.

Himwich, W. A. & Himwich, H. E. (1946). Pyruvic acid exchange of the brain. *J. Neurophysiol.* **9**, 133–6.

References

Himwich, W. A., Homburger, E., Maresca, R. & Himwich, H. E. (1947). Brain metabolism in man; unanesthetized and in pentothal narcosis. *Am. J. Psychiat.* **103**, 689–96.

Hinck, V. C. & Gordy, P. D. (1964). Persistent primitive trigeminal artery. *Radiology* **83**, 41–5.

Hinsey, J. C. (1935). The anatomical relations of the sympathetic nervous system to visceral sensation. *Res. Publs Ass. Res. nerv. ment. Dis.* **15**, 105–80.

Hinshaw, L. B., Ballin, H. M., Day, S. B. & Carlson, C. H. (1959). Tissue pressure and autoregulation in the dextran-perfused kidney. *Am. J. Physiol.* **197**, 853–5.

Hinshaw, L. B., Day, S. B. & Carlson, C. H. (1959). Tissue pressure as a causal factor in the autoregulation of blood flow in the isolated perfused kidney. *Am. J. Physiol.* **197**, 309–12.

Hinshaw, L. B., Flaig, R. D., Logemann, R. L. & Carlson, H. (1960). Intrarenal venous and tissue pressure and autoregulation of blood flow in the perfused kidney. *Am. J. Physiol.* **198**, 891–4.

Hoche, A. (1899). Vergleichund – Anatomisches über die Blutversongung der Ruckenmarksubstanz. *Z. Morph. Anthr.* **1**, 241–8.

Hodes, R. & Magoun, H. W. (1942). Autonomic responses to electrical stimulation of forebrain and midbrain with special reference to pupil. *J. comp. Neurol.* **76**, 169–90.

Høedt-Rasmussen, K., Sveinsdottir, E. & Lassen, N. A. (1966). Regional cerebral blood flow in man determined by intra-arterial injection of radioactive inert gas. *Circulation Res.* **18**, 237–47.

Hohorst, H. J., Betz, E. & Weidner, A. (1968). Relation between energy rich substrates, tissue redox changes and EEG during and after hypoxia. *Scand. J. clin. Lab. Invest.* Suppl. 102 III:B.

Hokfelt, T. (1968). *Electron microscopic studies on peripheral and central monoamine neurons.* Thesis for degree of M.D., Stockholm.

Holmdahl, M. H. (1956). Pulmonary uptake of oxygen, acid–base metabolism and circulation during prolonged apnoea. *Acta chir. scand.* **111**, Suppl. 212, 421–4.

Holmes, E. G. (1930). Oxidations in central and peripheral nervous tissue. *Biochem. J.* **24**, 914–25.

—— (1932). Metabolic activity of cells of trigeminal ganglion. *Biochem. J.* **26**, 2005–9.

Homberger, E., Himwich, W. A., Etstein, B., York, G., Maresca, R. & Himwich, H. E. (1946). Effect of pentothal anesthesia on canine cerebral cortex. *Am. J. Physiol.* **147**, 343–5.

Hornbein, T. F., Griffo, Z. J. & Roos, A. (1961). Quantitation of chemoreceptor activity: interrelation of hypoxia and hypercapnia. *J. Neurophysiol.* **24**, 561–8.

Horstmann, E. (1960). Abstand und Durchmesser der Kapillaren im Zentralnervensystem verscheideener Wirbeltierklassen. In *Structure and function of the cerebral cortex*, ed. Tower, D. B. & Schadé, J. P., p. 59. Amsterdam: Elsevier.

Hou, C. L. (1926). On the amount of blood supply and the oxygen consumption of the brain. *J. Orient. Med.* **5**, 20–31.

References

Hovelaque, A. (1927). *Anatomie des Nerfs craniens et rachidiens et du systeme grand sympathique chez l'homme.* Paris: Doin et Cie.

Huber, G. C. (1899). Observations on the innervation of intracranial vessels. *J. comp. Neurol.* **9**, 1–34.

Huckabee, W. E. (1958). Relationships of pyruvate and lactate during anaerobic metabolism. Effect of breathing low oxygen gases. *J. clin. Invest.* **37**, 264–71.

(1961). Relationship of pyruvate and lactate during anaerobic metabolism. v. Coronary adequacy. *Am. J. Physiol.* **200**, 1169–76.

Hughes, J. R., King, B. D., Cutler, J. A. & Markello, R. (1962). The EEG in hyperventilated, lightly anaesthetized patients. *Electroen. Neurophysiol.* **14**, 274–7.

Humphreys, S. P. (1939). Anatomic relations of cerebral vessels and perivascular nerves. *Archs Neurol. Psychiat.*, Chicago **41**, 1207–21.

Hunter, F. E., Jr. & Lowry, O. H. (1956). Effects of drugs on enzyme systems. *Pharmac. Rev.* **8**, 89–135.

Hunter, W. (1900). On the presence of nerve fibres in the cerebral vessels. *J. Physiol.* **26**, 465–9.

Hürthle, K. (1889). Beiträge zue Haemodynamik. *Pflügers Arch. ges. Physiol.* **44**, 561.

Hutten, H., Schwartz, W. & Schulz, V. (1969). Dependence of $^{85}Kr(\beta)$ - clearance rCBF determination on the input function. In *Cerebral blood flow*, ed. Brock, M., Fieschi, C., Ingvar, D. H., Lassen, N. A. & Schürmann, K., pp. 1–3. Berlin: Springer.

Huttenlocker, P. R. (1961). Evoked and spontaneous activity in single units of medial brain stem during natural sleep and waking. *J. Neurophysiol.* **24**, 451–68.

Hutter, O. F. & DeMello, W. C. (1966). The anion conductance of crustacean muscle. *J. Physiol.* **183**, 11P.

Hutter, O. F. & Warner, A. E. (1967). The pH sensitivity of the chloride conductance of frog skeletal muscle. *J. Physiol.* **189**, 403–25.

Hyden, H. & Lange, P. W. (1965). In *Sleep mechanisms*, ed. Akert, K., Bally, C. & Schadé, J. P., p. 92. Amsterdam: Elsevier. *Prog. Brain Res.* **18**, 92–5.

Hyman, E. S. (1961). Linear system for quantitating hydrogen at a platinum electrode. *Circulation Res.* **9**, 1093–7.

Ibister, W. H., Schofield, P. F. & Torrance, H. B. (1966). Cerebral blood flow estimated by ^{133}Xe clearance technique. *Archs Neurol.* **14**, 512–21.

Ingvar, D. H. (1958). Cortical state of excitability and cortical circulation. In *Reticular formation of the brain.* Henry Ford Hosp. Symp. Detroit, pp. 381–408. Boston: Littlebrown.

(1961). Measurements of regional gaseous metabolism and blood flow in the cerebral cortex. In *Regional neurochemistry*, ed. Kety, S. S. & Elkes, J., pp. 118–25. New York: Pergamon.

Ingvar, D. H., Cronquist, S., Ekberg, R., Riseberg, J. & Høedt-Rasmussen, K. (1965). Normal values of regional cerebral blood flow in man, including flow and weight estimates of grey and white matter. *Acta neurol. scand.* Suppl. 14, 72–8.

Ingvar, D. H. & Lassen, N. A. (1962). Regional blood flow of the cerebral cortex determined by Krypton 85. *Acta physiol. scand.* **54**, 325–38.

References

Ingvar, D. H., Lübbers, D. W. & Siesjö, B. K. (1960). Measurement of oxygen tension on the surface of the cerebral cortex of the cat during hyperoxia and hypoxia. *Acta physiol. scand.* **48**, 373–81.

(1962). Normal and epileptic EEG patterns related to cortical oxygen tension in the cat. *Acta physiol. scand.* **55**, 210–24.

Ingvar, D. H. & Riseberg, J. (1967). Increase of regional cerebral blood flow during mental effort in normals and in patients with focal brain disorders. *Expl Brain Res.* **3**, 195–211.

Ingvar, D. H., Siesjö, B. K. & Hertz, Ch. (1959). Measurement of tissue carbon dioxide pressure in the brain. *Experientia* **15**, 306–8.

Ingvar, D. H. & Soderberg, U. (1956). A new method for measuring cerebral blood flow in relation to the electroencephalogram. *Electroen. Neurophysiol.* **8**, 403–12.

(1958). Cortical blood flow related to EEG patterns evoked by stimulation of the brain stem. *Acta physiol. scand.* **42**, 130–43.

Irving, L. (1938). Changes in the blood flow through the brain and muscles during the arrest of breathing. *Am. J. Physiol.* **122**, 207–14.

Irving, L. & Welch, M. S. (1935). The effect of the composition of the inspired air upon the circulation through the brain. *Q. Jl exp. Physiol.* **25**, 121–9.

Ishikawa, S., Handa, J., Meyer, J. S. & Huber, P. (1965). Hemodynamics of the circle of Willis and the Leptomeningeal anastomoses: and electromagetic flowmeter study of intracranial arterial occlusion in the monkey. *J. Neurol. Neurosurg. Psychiat.* **28**, 124–36.

Iwayama, T. (1970). Ultrastructural changes in the nerves innervating the cerebral artery after sympathectomy. *Z. Zellforsch. mikrosk. Anat.* **109**, 465–80.

Iwayama, T., Furness, J. B. & Burnstock, G. (1970). Dual adrenergic and cholinergic innervation of the cerebral arteries of the rat: an ultrastructural study. *Circulation Res.* **26**, 635–46.

Jacobson, I., Harper, A. M. & McDowall, D. G. (1963a). Relationship between venous pressure and cortical blood flow. *Nature, Lond.* **200**, 173–5.

(1963b). The effects of oxygen under pressure on cerebral blood flow and cerebral venous oxygen tension. *Lancet* ii, 549.

James, I. M. (1968). Changes in cerebral blood flow and in systemic arterial blood pressure following spontaneous subarachnoid haemorrhage. *Clin. Sci.* **35**, 11–22.

James, I. M., Millar, R. A. & Purves, M. J. (1969). Observations on the extrinsic neural control of cerebral blood flow in the baboon. *Circulation Res.* **25**, 77–93.

James, I. M., Xanalatos, C. & Nashat, S. (1971). The effect of CO_2 tension and blood pressure on cerebral metabolism and blood flow after isoprenaline infusion. In *Brain and blood flow*, ed. Ross Russell, R. S., pp. 229–32. London: Pitman.

Jasper, H. & Erickson, T. C. (1941). Cerebral blood flow and pH in excessive cortical discharge induced by metrazol and electrical stimulation. *J. Neurophysiol.* **4**, 333–47.

Jelsma, L. F. & McQueen, J. D. (1968). Changes in impedance pulsations and carotid blood flow during intracranial hypertension. In *Blood flow through organs and tissues*, ed. Bain, W. H. & Harper, A. M., pp. 328–35. Edinburgh: Livingstone.

References

Jenkner, F. L. (1957). Ueber den Wert des Schädelrheogrammes für die Diagnose zerebraler Gefäßströrungen. *Wien. klin. Wschr.* **69**, 619–20.

(1962). Rheoencephalography, a method for the continuous registration of cerebrovascular changes. Springfield, Ill.: Thomas.

Jennett, W. B., Harper, A. M. & Gillespie, F. C. (1966). Measurement of regional cerebral blood flow during carotid ligation. *Lancet* ii, 1162–3.

Jensen, K. B., Høedt-Rasmussen, K., Sveinsdottir, E., Stewart, B. M. & Lassen, N. A. (1966). Cerebral blood flow evaluated by inhalation of ^{133}Xe and extracranial recording: a methodological study. *Clin. Sci.* **30**, 485–94.

Jensen, P. (1904). Ueber die Blutversorgung des Gehirns. *Pflügers Arch. ges. Physiol.* **103**, 171–95.

Jewell, P. A. (1952). The anastomoses between internal and external carotid circulations in the dog. *J. Anat.* **86**, 83–94.

Joels, N. & White, H. (1968). The contribution of the arterial chemoreceptors to the stimulation of respiration by adrenaline and nor-adrenaline in the cat. *J. Physiol.* **197**, 1–23.

Johnson, A. E. & Gollan, F. (1965). Cerebral blood flow monitoring by external detection of inhaled radioactive xenon. *J. Nucl. Med.* **6**, Suppl. 9, 679–86.

Johnson, P. C. (1964). Review of previous studies and current theories of autoregulation. *Circulation Res.* 14–15 (Suppl. 1), I 2–9.[1]

Johnson, R. H., Lee, G. de J., Oppenheimer, D. R. & Spalding, J. M. K. (1966). Autonomic failure with orthostatic hypotension due to intermediolateral column degeneration. *Q. Jl Med.* **35**, 276–92.

Johnson, Sheena M. & Miller, K. W. (1970). Antagonism of pressure and anaesthesia. *Nature, Lond.* **228**, 75–6.

Johansson, B. & Bohr, D. F. (1966). Rhythmic activity in smooth muscle from small subcutaneous arteries. *Am. J. Physiol.* **210**, 801–6.

Jowett, M. & Quastel, J. H. (1937). Effects of hydroxymalonate on metabolism of brain. *Biochem. J.* **31**, 275–81.

Joy, M. D. & Lowe, R. D. (1970). The site of cardiovascular action of angiotensin. II. In the brain. *Clin. Sci.* **39**, 327–36.

Kaasik, A. E., Nilsson, L. & Siesjö, B. K. (1968). Acid–base and lactate/pyruvate changes in brain and CSF in asphyxia and stagnant hypoxia. *Scand. J. clin. Lab. Invest.* Suppl. 102, III:C.

Kabat, H. (1940). The greater resistance of very young animals to arrest of the brain circulation. *Am. J. Physiol.* **130**, 588–99.

Kajikawa, H. (1968). Fluorescence histochemical studies on the distribution of adrenergic nerve fibres to intracranial blood vessels. *Arch. Jap. Chir.* **37**, 473–82.

(1969). Mode of sympathetic innervation of the cerebral vessels demonstrated by the fluorescent histochemical technique in rats and cats. *Arch. Jap. Chir.* **38**, 227–35.

Kak, V. K. & Taylor, A. R. (1967). Cerebral blood flow in subarachnoid haemorrhage. *Lancet* i, 875–7.

Kanzow, E., Held, U. & Richtering, J. (1960). Bezeihungen zwischen EEG-aktivierung und Durchblutung der Hirnrinde des Hundes. *Pflügers Arch. ges. Physiol.* **272**, R13.

References

Kanzow, E., Krause, D. & Kuhnel, H. (1962). The vasomotor system of the cerebral cortex in the phases of desynchronized EEG activity during natural sleep in cats. *Pflügers Arch. ges. Physiol.* **274**, 593–607.

Kaplan, H. A. & Ford, D. H. (1966). *The brain vascular system.* Amsterdam: Elsevier.

Kappers, C. U. A. (1926). The relative weight of the braincortex in human races and in some animals and the asymmetry of the hemispheres. *J. nerv. ment. Dis.* **64**, 113–24.

Karrer, H. E. (1965). The ulstrastructure of mouse lung. General architecture of capillary and alveolar walls. *J. biophys. biochem. Cytol.* **2**, 241–52.

Kawamura, H. & Sawyer, C. (1965). Elevation in brain temperature during paradoxical sleep. *Science, N.Y.* **150**, 912–13.

Keatinge, W. R. (1964). Mechanism of adrenergic stimulation of mammalian arteries and its failure at low temperatures. *J. Physiol.* **174**, 184–205.

(1965). Sympathomimetic action of acetylcholine on cell membrane potential of blood vessels. *J. Physiol.* **181**, 75P.

(1966). Electrical and mechanical responses of vascular smooth muscle to vasodilator agents and vasoactive polypeptides. *Circulation Res.* **18**, 641–9.

(1968). Ionic requirements for arterial action potential. *J. Physiol.* **194**, 169–82.

Keller, C. J. (1930). Untersuchungen über die Gehirndurchblutung. I Mitteilung: Gibt es eine Autonomie der Gehirngefasse. *Arch. exp. Path. Pharmak.* **154**, 357–80.

Kellie, G. (1824a). An account of the appearances observed in the dissection of two of three individuals presumed to have perished in the storm of the 3rd and whose bodies were discovered in the vicinity of Leith on the morning of the 4th, November 1821 with some reflections on the pathology of the brain. *Trans. Edin. Med. Chirurg. Soc.* **1**, 84–122.

(1824b). Reflections on the pathology of the brain. *Trans. Edin. Med. Chirurg. Soc.* **1**, 123–69.

Kennedy, C. (1956). The cerebral metabolic rate in children. *Progr. Neurobiol. N.Y.* **1**, 230–8.

Kennedy, C., Grave, G. D., Jehle, J. & Sokoloff, L. (1969). Blood flow in cerebral white matter during maturation. *Neurology, Minneap.* **19**, 307.

Kennedy, C. & Sokoloff, L. (1957). An adaptation of the nitrous oxide method to the study of the cerebral circulation in children: normal values for cerebral blood flow and cerebral metabolic rate in childhood. *J. clin. Invest.* **36**, 1130–7.

Kennedy, C., Sokoloff, L. & Anderson, W. (1954). Cerebral blood flow and metabolism in normal children. *Am. J. Dis. Child.* **88**, 813.

Kety, S. S. (1950). Circulation and metabolism of the human brain in health and disease. *Am. J. Med.* **8**, 205–17.

(1951). Theory and applications of exchange of inert gas at lungs and tissues. *Pharmac. Rev.* **3**, 1–41.

(1956). Human cerebral blood flow and oxygen consumption as related to aging. *J. chron. Dis.* **3**, 478–86.

(1957a). Determinants of tissue oxygen tension. *Fedn Proc.* **16**, 666–70.

(1957b). The general metabolism of the brain *in vivo*. In *Metabolism of the nervous system*, ed. Richter, D., pp. 221–36. New York: Pergamon.

References

(1960). The cerebral circulation. In *Handbook of physiology*, vol. 3, ed. Field, J., pp. 1751–60. Washington: American Physiological Society.

(1965). Observations on the validity of a two compartmental model of the cerebral circulation. *Acta neurol. scand.* Suppl. 14, 85–7.

Kety, S. S., King, B. D., Horvath, S. M., Jeffers, W. A. & Hafkenshiel, J. H. (1950). The effect of an acute reduction in blood pressure by means of a differential spinal block on the cerebral circulation of hypertensive patients. *J. clin. Invest.* 29, 402–7.

Kety, S. S., Landau, W. M., Freygang, W. H., Rowland, L. P. & Sokoloff, L. (1955). Estimation of regional circulation in the brain by the uptake of an inert gas. *Fedn Proc.* 14, 85.

Kety, S. S., Polis, B. D., Nadler, C. S. & Schmidt, C. F. (1948). The blood flow and oxygen consumption of the human brain in diabetic acidosis and coma. *J. clin. Invest.* 27, 500–10.

Kety, S. S. & Schmidt, C. F. (1945). The determination of cerebral blood flow in man by the use of nitrous oxide in low concentrations. *Am. J. Physiol.* 143, 53–66.

(1946). The effects of active and passive hyperventilation on cerebral blood flow, cerebral oxygen consumption, cardiac output, and blood pressure of normal young men. *J. clin. Invest.* 25, 107–19.

(1948a). The nitrous oxide method for the quantitative determination of cerebral blood flow in man: theory, procedure, and normal values. *J. clin. Invest.* 27 476–483.

(1948b). The effect of altered arterial tensions of carbon dioxide and oxygen on cerebral blood flow and cerebral oxygen consumption of normal young men. *J. clin. Invest.* 27, 484–92.

Kety, S. S., Shenkin, H. A. & Schmidt, C. F. (1948a). Blood flow, vascular resistance, and oxygen consumption of brain in essential hypertension. *J. clin. Invest.* 27 511–14.

(1948b). The effects of increased intracranial pressure on cerebral circulatory effects in man. *J. clin. Invest.* 27, 493–9.

Kety, S. S., Woodford, R. B., Harmel, M. H., Freyhan, F. A., Appel, K. A. & Schmidt, C. F. (1948). Effects of barbiturate semi-narcosis, insulin coma and electroshock. *Am. J. Psychiat.* 104, 765–70.

Key, E. A. H. & Retzius, G. (1876). *Studien in der Anatomie des Nervensystems und des Bindesgewebe.* Stockholm: Samson & Wallin.

King, B. D., Sokoloff, L. & Wechsler, R. L. (1952). The effects of L-epinephrine and L-nor-epinephrine upon cerebral circulation and metabolism in man. *J. clin. Invest.* 31, 273–9.

Kjällquist, Å. (1970). The CSF/blood potential in sustained acid–base changes in the rat. With calculations of electro-chemical potential difference for H^+ and HCO_3^-. *Acta physiol. scand.* 78, 85–93.

Kjällquist, Å., Nardini, M. & Siesjö, B. K. (1969). The regulation of extra- and intracellular acid–base parameters in the rat brain during hyper- and hypocapnia. *Acta physiol. scand.* 76, 485–94.

Kjällquist, Å. & Siesjö, B. K. (1967). The CSF/Plasma potential in sustained acidosis and alkalosis in the rat. *Acta physiol. scand.* 71, 255–6.

References

Kjällquist, Å. & Siesjö, B. K. (1968). Regulation of CSF pH – influence of the CSF/plasma potential. *Scand. J. clin. Lab. Invest.* Suppl. 102, 1:C.

— (1969). A hypothesis for the regulation of extracellular pH in the brain. *J. Physiol.* 202, 4–5P.

Kjällquist, Å., Siesjö, B. K. & Zwetnow, N. (1969). Effects of increased intracranial pressure on cerebral blood flow and on cerebral venous PO_2, PCO_2, pH, lactate and pyruvate in dogs. *Acta physiol. scand.* 75, 267–75.

Kleinerman, J., Sancetta, S. M. & Hackel, D. B. (1958). Effects of high spinal anaesthesia on cerebral circulation and metabolism in man. *J. clin. Invest.* 37, 284–93.

Kleitman, N. (1939). *Sleep and wakefulness as alternating phases in the cycle of existence.* Chicago: University of Chicago Press.

Kobayashi, S., Rhoton, A. L. & Waltz, A. G. (1969). Effects of stimulating cervical sympathetic nerves in cortical blood flow and vascular reactivity. *Neurology, Minneapolis* 19, 307–8.

Kobayashi, S., Waltz, A. G. & Rhoton, A. L. (1971). Effects of stimulation of cervical sympathetic nerves on cortical blood flow and vascular reactivity. *Neurology, Minneapolis* 21, 297–302.

Koch, E. (1931). *Die reflektorische Selbsteuerung des Kreislaufs.* Dresden: Steinkopf.

Köllicker, A. (1856). *Handbuck der Gewebelehre des Menschen.* Leipzig: Engelmann.

— (1896). *Handbuch der Gewebelehre des Menschen.* Vol. 2, p. 835, Leipzig: Engelmann.

Koopmans, S. (1938). Function of the blood vessels in the brain. Effect of sympathetic and vagal fibres of neck. *Archs néerl. Physiol.* 23, 256–70.

— (1939). The function of the blood vessels in the brain. II. The effects of some narcotics, hormones and drugs. *Archs néerl. Physiol.* 24, 250–66.

Korey, S. R. & Orchen, M. (1959). Relative respiration of neuronal and glial cells. *J. Neurochem.* 3, 277–85.

Kovalcik, V. (1963). The response of the isolated ductus arteriosus to oxygen and anoxia. *J. Physiol.* 169, 185–97.

Krauspe, M. F. (1874). Ueber die reflectorische Beeinflussung der Piaarterien. *Virchows Arch. path. Anat. Physiol.* 59, 472–89.

Krayenbühl, H. (1958). Diagnostic value of orbital angiography. *Br. J. Ophthal.* 42, 180–90.

Krogh, A. (1918). The rate of diffusion of gases, with some remarks on the coefficient of invasion. *J. Physiol.* 52, 391–408.

— (1919). The number and distribution of capillaries in muscles with calculation of the oxygen pressure head necessary for supplying the tissue. *J. Physiol.* 52, 409–15.

Kubie, L. S. & Hetler, D. M. (1928). The cerebral circulation. IV. The action of hypertonic solutions. Part II. A study of the circulation in the cortex by means of colour photography. *Archs Neurol. Psychiat., Chicago* 20, 749–55.

Kuffler, S. W. & Nicholls, J. C. (1966). The physiology of neuroglial cells. *Ergeb. Physiol.* 57, 1–90.

Kuntz, A. (1934). *The autonomic system.* 2nd ed. pp. 157–60. Philadelphia: Lea & Febiger.

References

Kuntz, A., Hoffman, H. H. & Napolitano, L. M. (1957). Cephalic sympathetic nerves; components and surgical implications. *Archs Surg.*, *Chicago* **75**, 108–115.

Kurusu, M. & Hamada, I. (1929). Der histologische Nachweis der Gefässnerven des Gehirns. *Zentralbl. f.d. ges. Neurol. u. Psychiat.* **54**, 421.

Kussmaul, A. & Tenner, A. (1857). Ueber den Einfluss der Blutströmung in den grossen Gefässen des Halses auf die Wärme des ohrs beim Kaninchen und ihr Verhältniss zu dem Wärmeveränderingen, Welche durch Lähmung und Reizung des Sympatheticus bedingt Werden. *Untersuch. z. Naturl. d. Mensch. u.d. Thiere.* i, 99–132.

Lacroix, E. (1962). Influence of isoproterenol on splanchnic circulatory dynamics. *J. Physiol. Paris* **54**, 360–1.

Lambertsen, C. J., Kough, R. H., Cooper, D. Y., Emmel, G. L., Loeschcke, H. H. & Schmidt, C. F. (1953). Oxygen toxicity: effects in man of oxygen inhalation at 1 and 3.5 atmospheres upon blood gas transport, cerebral circulation and cerebral metabolism. *J. appl. Physiol.* **5**, 471–86.

Lambertsen, C. J., Semple, S. J. G., Smyth, M. G. & Gelfand, R. (1961). H^+ and PCO_2 as chemical factors in respiratory and cerebral circulatory control. *J. appl. Physiol.* **16**, 473–84.

Landau, W. M., Freygang, W. H., Jr., Rowland, L. P., Sokoloff, L. & Kety, S. S. (1955). The local circulation of the living brain; values in the unanesthetized and anesthetized cat. *Trans. Am. neurol. Assoc.* **80**, 125–9.

Landgren, S., Neil, E. & Zotterman, Y. (1952). The response of the carotid baroreceptors to the local administration of drugs. *Acta Physiol. scand.* **25**, 24–37.

Landis, C. (1925). Changes in blood pressure during sleep as determined by the Erlanger method. *Am. J. Physiol.* **73**, 551–5.

Langendorff, O. & Seelig, A. (1886). Ueber die in Folge von Athinungs hindernissen eintretenden Störungen der Respiration. *Pflügers Arch. ges. Physiol.* **39**, 223–37.

Langer, S. Z., Draskoczy, P. R. & Trendelenburg, U. (1967). Time course of the development of supersensitivity to various amines in the nictitating membrane of the pithed cat after denervation and decentralization. *J. Pharmac. exp. Ther.* **157**, 255–73.

Langfitt, T. W., Weinstein, J. D. & Kassell, N. F. (1965). Cerebral vasomotor paralysis produced by intracranial hypertension. *Neurology, Minneapolis* **15**, 622–41.

Langley, J. N. (1894). On the arrangement of the sympathetic nervous system based chiefly on observations upon pilomotor nerves. *J. Physiol.* **15**, 177–244.
— (1896). Observations on the medullated fibres of the sympathetic system and chiefly on those of the grey rami communicantes. *J. Physiol.* **20**, 55–76.
— (1900). Remarks on the degeneration of the upper thoracic rami communicantes chiefly in relation to commissural fibres in the sympathetic system. *J. Physiol.* **25**, 468–78.

Lassen, N. A. (1959). Cerebral blood flow and oxygen consumption in man. *Physiol. Rev.* **39**, 183–238.
— (1965). Blood flow of the cerebral cortex calculated from 85 Krypton beta clearance recorded over the exposed surface; evidence of inhomogeneity of flow. *Acta neurol. scand.* Suppl. 14, 24–8.

References

Lassen, N. A. (1968a). Neurogenic control of CBF. *Scand. J. clin. Lab. Invest.* Suppl. 102, VI:F.

(1968b). Kinetic and compartmental analysis of inert gas clearance curves for measurement of blood flow. In *Blood flow through organs and tissues*, ed. Bain, W. H. & Harper, A. M., pp. 210–14. Edinburgh: Livingstone.

Lassen, N. A., Fineberg, I. & Lane, M. H. (1960). Bilateral studies of cerebral oxygen uptake in young and aged normal subjects and in patients with organic dementia. *J. clin. Invest.* **39**, 491–500.

Lassen, N. A. & Høedt-Rasmussen, K. (1966). Human cerebral blood flow measured by two inert gas techniques: comparison of the Kety–Schmidt method and the intra-arterial injection method. *Circulation Res.* **19**, 681–8.

Lassen, N. A., Høedt-Rasmussen, K., Sorensen, S. C., Skinhøj, E., Cronquist, S., Bodforss, B. & Ingvar, D. H. (1963). Regional cerebral blood flow in man determined by a radioactive inert gas (Krypton 85). *Neurology, Minneapolis* **13**, 719–27.

Lassen, N. A. & Ingvar, D. H. (1961). Blood flow of the cerebral cortex determined by radioactive krypton[85]. *Experientia* **17**, 42–5.

Lassen, N. A. & Klee, A. (1965). Cerebral blood flow determined by saturation and desaturation with Krypton 85: an evaluation of the validity of the inert gas method of Kety and Schmidt. *Circulation Res.* **16**, 26–32.

Lassen, N. A. & Lane, M. H. (1961). Validity of internal jugular blood for study of cerebral blood flow and metabolism. *J. appl. Physiol.* **16**, 313–20.

Lassen, N. A. & Munck, O. (1955). The cerebral blood flow in man determined by the use of radioactive Krypton. *Acta physiol. scand.* **33**, 30–49.

Lassen, N. A., Munck, O. & Tottey, E. R. (1957). Mental function and cerebral oxygen consumption in organic dementia. *Archs Neurol. Psychiat., Chicago* **77**, 126–33.

Laubie, M. & Drouillat, M. (1967). Action de l'isoprotérénol sur l'hémodynamique et le métabolisme cérébral du chien. *Archs int. Pharmacodyn. Ther.* **170**, 93–8.

Lavrentieva, N. B., Mchedlishvili, G. I. & Plechkova, E. K. (1968). Distribution and activity of cholinesterase in the nervous structures of the pial arteries (a histochemical study). *Bull. biol. Med. exp. U.R.S.S.* **64**, 110–13.

Lazorthes, G. (1949). Les douleurs vasculaires cephaliques. *Presse méd.* **57**, 633–5.

(1961). *Vascularisation et circulation cérébrales*. Paris: Masson.

Lazorthes, G., Gaubert, J. & Poulhes, J. (1956). La distribution centrale et corticale de l'artère cérébrale antérieure. Etude anatomique et incidences neurochirurgicales. *Neurochir.* **2**, 237–53.

Lazorthes, G. & Reis, R. (1942). L'innervation des artères du cou et de la portion extracranienne de la tete. *Toulouse Medicale* **2**, 1–21.

Leão, A. A. & Morison, R. S. (1945). Proportion of spreading cortical depression. *J. Neurophysiol.* **8**, 33–45.

Lee, G. de J. & Dubois, A. B. (1955). Pulmonary capillary blood flow in man. *J. clin. Invest.* **34**, 1380–90.

Legait, E. (1945). Corpuscles nervaux dans la parois des artères du resau admirabile carotidien. *C. r. Séanc. Soc. Biol.* **139**, 340–1.

Lehmann, G. & Meesmann, A. (1924). Über das Bestehen eines Donnangleich-

References

gewichtes zwischen Blut und Kammerwasser bzw. Liquor Cerebrospinalis. *Pflügers Arch. ges. Physiol.* **205**, 210–32.

Leiden, E. (1866). Beiträge und Untersuchungen zur Physiologie und Pathologie des Gehirns. *Virchows Arch. path. Anat. Physiol.* **37**, 519–59.

Lende, R. A. (1960). Local spasm in cerebral arteries. *J. Neurosurg.* **17**, 90–113.

Lennox, W. G. (1931 a). The cerebral circulation. XIV. The respiratory quotient of the brain and extremities in man. *Archs Neurol. Psychiat.*, Chicago **26**, 719–24.

(1931 b). The cerebral circulation. XV. The effect of mental work. *Archs Neurol. Psychiat.*, Chicago **26**, 725–30.

Lennox, W. G. & Gibbs, E. L. (1932). The blood flow in the brain and the leg of man and the changes induced by alteration of blood gases. *J. clin. Invest.* **11**, 1155–77.

Lennox, W. G., Gibbs, F. A. & Gibbs, E. L. (1935). Relationship of unconsciousness to cerebral blood flow and to anoxemia. *Archs Neurol. Psychiat.*, Chicago **34**, 1001–13.

(1938). The relationship in man of cerebral activity to blood flow and to blood constituents. *Res. Publs Ass. Res. nerv. ment. Dis.* **18**, 274–97.

Leriche, R. & Fontaine, R. (1936). Résultats généraux de 1,256 sympathectomies. *Mém. Acad. de Chir.* **62**, 877–92.

Leusen, I. R. (1950a). pH sanguin, pH du liquide cephalo-rachidien et fonction respiratoire. *Archs int. Physiol.* **58**, 115–16.

(1950b). Influence du pH du liquide cephalo-rachidien sur la respiration. *Experientia* **6**, 272–4.

(1954a). Chemosensitivity of the respiratory centre. Influence of CO_2 in the cerebral ventricles on respiration. *Am. J. Physiol.* **176**, 39–44.

(1954b). Chemosensitivity of the respiratory centre. Influence of changes in H^+ and total buffer concentrations in the cerebral ventricles on respiration. *Am. J. Physiol.* **176**, 45–51.

Leusen, I. (1965). Aspects of the acid–base balance between blood and cerebrospinal fluid. In *Cerebrospinal fluid and the regulation of ventilation*, ed, Brooks, C. McC., Kao, F. F. & Lloyd, B. B. Oxford: Blackwell.

Leusen, I. & Demeester, G. (1966). Lactate and pyruvate in the brain of rats during hyperventilation. *Archs int. Physiol. Biochim.* **74**, 25–34.

Leusen, I., Lacroix, E. & Demeester, G. (1967). Lactate and pyruvate in the brain of rats during changes in acid–base balance. *Archs int. Physiol. Biochim.* **75**, 310–24.

Lever, J. D., Graham, J. D. P., Irvine, G. & Chick, W. J. (1965). The vesiculated axons in relation to arterial smooth muscle in the pancreas. A fine structural and quantitative study. *J. Anat.* **99**, 299–313.

Lever, J. D., Spriggs, T. L. & Graham, J. D. (1968). A formol-fluorescence, fine-structural and autoradiographic study of the adrenergic innervation of the vascular tree in the intact and sympathectomized pancreas of the cat. *J. Anat.* **103**, 15–34.

Levine, M. & Wolff, H. G. (1932). Cerebral circulation. Afferent impulses from blood vessels of the pia. *Archs Neurol. Psychiat.*, Chicago **28**, 140–50.

Levy, M. N., Dogeest, H. & Zieske, H. (1966). Effects of respiratory centre on the heart. *Circulation Res.* **18**, 67–78.

References

Lewin, L. (1920). *Kohlenoxydvergiftung*, p. 324, Berlin: Springer.

Lewis, B. M., Sokoloff, L. & Kety, S. S. (1955). Use of radioactive Krypton to measure rapid changes in cerebral blood flow. *Am. J. Physiol.* **183**, 638–9.

Lewis, B. M., Sokoloff, L., Wechsler, R. L., Wentz, W. B. & Kety, S. S. (1960). Method for continuous measurement of cerebral blood flow in man by means of radioactive krypton (Kr^{79}). *J. clin. Invest.* **39**, 707–16.

Libet, B., Fazekas, J. F. & Himwich, H. E. (1941). The electrical response of the kitten and adult cat brain to cerebral anaemia and analeptics. *Am. J. Physiol.* **132**, 232–8.

Liebermann, L. von (1926). Energiedarf und mechanisches Asquivalent der geistigen Arbeit. *Biochem. Z.* **173**, 181–9.

Lierse, W. (1961). Die Kapillarabstande in Verschiedenem Hirnregionen der katze. *Z. Zellforsch. mikrosk. Anat.* **54**, 199–206.

— (1963). Die Kapillardichte im Phinencephalon verschiedener Wirbeltiere und des Menschen. In *Progress in brain research*, vol. 3, ed. Bargmann, W. & Schadé, J. P. Amsterdam: Elsevier.

Lifshitz, K. (1963). Rheoencephalography. 1. Review of the technique. *J. nerv. ment. Dis.* **136**, 388–98.

Liljestrand, G. & Magnus, R. (1922). Die Wirkung des Kohlensäurebades beim Gesunden nebst Bemerkungen über den Einfluss des Hochgebirges. *Pflügers Arch. ges. Physiol.* **193**, 527–54.

Linzell, J. L. (1953). Internal calorimetry in the measurement of blood flow with heated thermocouples. *J. Physiol.* **121**, 390–402.

Linzell, J. & Waites, G. (1957). Effects of occluding the carotid and vertebral arteries in sheep and goats. *J. Physiol.* **138**, 20P.

Lloyd, B. B., Jukes, M. G. M. & Cunningham, D. J. C. (1958). The relation between alveolar pressure and the respiratory response to carbon dioxide in man. *Q. Jl exp. Physiol.* **43**, 214–27.

Loeschcke, H. H. (1956). Über Bestandspotentiale im Gebiete der Medulla Oblongata. *Pflügers Arch. ges. Physiol* **262**, 517–31.

Loeschke, H. H. & Koepchen, H. P. (1958a). Über das Verhalten der Atmung und des arteriellen Drucks bei Einbringen von Veratridin, Lobelin und Cyanid in der Liquor cerebrospinalis. *Pflügers Arch. ges. Physiol.* **266**, 586–610.

— (1958b). Beeinflussung von Atmung und Vasomotorik durch Einbringen von Novocain in die Liquorräume. *Pflügers Arch. ges. Physiol.* **266**, 611–27.

— (1958c). Versuch zur Lokalisation des Angriffsortes der Atmungs- und Kreislaufwirkung von Novocain im Liquor cerebrospinalis. *Pflügers Arch. ges. Physiol.* **266**, 628–41.

Loeschke, H. H., Koepchen, H. P. & Gertz, K. H. (1958). Über den Einfluss von Wasserstoffionenkonzentration und CO_2-Druck im Liquor cerebrospinalis auf die Atmung. *Pflügers Arch. ges. Physiol.* **266**, 569–85.

Loeschke, H. H., Mitchell, R. A., Katsaros, B., Perkins, J. F., Jr. & Konig, A. (1963). Interaction of intracranial chemosensitivity with peripheral afferents to the respiratory centres. *Ann. N.Y. Acad. Sci.* **109**, 651–60.

Lolley, R. N. & Samson, R. F. (1962). Cerebral high-energy compound; changes in anoxia. *Am. J. Physiol.* **202**, 77–9.

Loman, J. & Myerson, A. (1935). Studies in the dynamics of the human cranio-vertebral cavity. *Am. J. Psychiat.* **92**, 791–815.

References

Lorente de. Nó, R. (1927). Ein beitrag zur Kenntnis der Gefassverteilung in der Hirninde. *J. Physiol. Neurol. (Leipzig)* **35**, 19–27.

Lowry, O. H., Passonneau, J. V., Hasselberger, F. X. & Schultz, D. W. (1964). Effect of ischemia on known substrates and cofactors of the glycolytic pathway in brain. *J. biol. Chem.* **239**, 18–30.

Lübbers, D. W. (1968a). The oxygen pressure field of the brain and its significance for the normal and critical oxygen supply of the brain. In *Oxygen transport in blood and tissue*, ed. Lübbers, D. W., Luft, U. C., Thews, G. & Witzleb, E., pp. 124–39. Stuttgart: Thieme.

(1968b). Regional cerebral blood flow and microcirculation. In *Blood flow through organs and tissues*, ed. Bain, W. H. & Harper, A. M., pp. 162–6. Edinburgh: Livingstone.

Lübbers, D. W. & Baumgärtl, H. (1967). Herstellungstechnik von palladinierten Pt-Stichelektroden (1–5 μ Außendurchmesser) zur polarographischen Messung des Wasserstoffdnickes für die Bestimmung der Mikrozurkulation. *Pflügers Arch. ges. Physiol.* **294**, R39.

Lübbers, D. W., Ingvar, D. H., Betz, E., Fabel, H., Kessler, M. & Schmahl, F. W. (1964). Sauerstoffverbrauch der Grosshirnrinde in Schlaf – und Wachzustand beim Hund. *Pflügers Arch. ges. Physiol.* **281**, R58.

Lübbers, D. W., Kessler, M., Knaust, Kr., McDowall, D. G. & Wodick, R. (1966). Wasserstoff und Sauerstoff zur Messung der lokalen Gewebedurchblutung in situ mit der Platin-electrode. *Pflügers Arch. ges. Physiol.* **289**, R99.

Lubin, A. J. & Price, J. C. (1942). Effect of alkalosis and acidosis on cortical electrical activity and blood flow. *J. Neurophysiol.* **5**, 261–8.

Lubsen, N. (1941). Experimental studies on the cerebral circulation of the unanesthetized rabbit. III. The action of ergotamine tartrate and of some vasodilator drugs. *Archs néerl. Physiol.* **25**, 361–5.

Lucas, W., Kirschbaum, T. & Assali, N. S. (1966). Cephalic circulation and oxygen consumption before and after birth. *Am. J. Physiol.* **210**, 287–92.

Ludwigs, N. & Schneider, M. (1954). Über den Einfluss des Halssympathicus auf die Gehirndurchblutung. *Pflügers Arch. ges. Physiol.* **259**, 43–55.

Lumsden, C. E. (1955). The cytology and cell physiology of the neuroglia and of the connective tissue in the brain with reference to the blood brain barrier. *Proc. 2nd Int. Congr. Neuropath. Lond.* **2**, 373–6.

(1959). Nourishing brain cell. An inaugural lecture, 11th November 1957. Leeds: Leeds University Press.

Lumsden, C. E. & Pomerat, C. M. (1951). Normal oligodendrocytes in tissue culture. A preliminary report on the pulsatile glial cells in tissue culture from the corpus callosum of the normal adult rat brain. *Expl Cell Res.* **2**, 103–14.

Luse, S. A. & Harris, B. (1960). Electron microscopy of the brain in experimental edema. *J. Neurosurg.* **17**, 439–46.

Lushka, H. von (1850). *Die Nerven in der harten Hirnhaut*, p. 27. Tübingen: Laupp & Siebeck.

McArdle, L. & Roddie, I. C. (1958). Vascular responses to carbon dioxide during anaesthesia in man. *Br. J. Anaesth.* **30**, 358–66.

McArdle, L., Roddie, I. C., Shepherd, J. C. & Whelan, R. F. (1957). The effect of inhalation of 30 per cent carbon dioxide on the peripheral circulation of the human subject. *Br. J. Pharmac.* **12**, 293–6.

References

McCall, M. L. (1953). Cerebral circulation and metabolism in toxaemia of pregnancy. Observations on the effect of veratrum viride and apresoline (l-hydrazinophthalazine). *Am. J. Obstet. Gynec.* **66**, 1015–28.

McCall, M. L. & Taylor, H. W. (1952). Effects of barbiturate sedation on brain in toxæmia of pregnancy. *J. Am. med. Ass.* **149**, 51–4.

McDonald, D. A. (1960). *Blood flow in arteries.* London: Arnold.

McDowall, D. G. (1965). The effects of general anaesthetics on cerebral blood flow and cerebral metabolism. *Br. J. Anaesth.* **37**, 236–45.

—— (1966). Interrelationships between blood oxygen tension and cerebral blood flow. In *Oxygen measurements in blood and tissues*, ed. Payne, J. P. & Hill, D. W., pp. 205–14. London: Churchill.

—— (1967). The effects of clinical concentrations of halothane on the blood flow and oxygen uptake of the cerebral cortex. *Br. J. Anaesth.* **39**, 186–96.

McDowall, D. G. & Harper, A. M. (1968). The relationship between blood flow and extracellular pH of the cerebral cortex. In *Blood flow through organs and tissues*, ed. Bain, W. H. & Harper, A. M., pp. 261–78. Edinburgh: Livingstone.

McDowall, D. G., Harper, A. M. & Jacobson, I. (1963). Cerebral blood flow during halothane anaesthesia. *Br. J. Anaesth.* **35**, 394–402.

—— (1964). Cerebral blood flow during trichlorethylene anaesthesia: a comparison with halothane. *Br. J. Anaesth.* **36**, 11–18.

McElroy, W. T., Gerdes, A. J. & Brown, E. B. (1958). Effects of CO_2, bicarbonate and pH on performance of isolated perfused guinea pig hearts. *Am. J. Physiol.* **195**, 412–16.

McGinty, D. A. (1929). The regulation of respiration. xxv. Variations in the lactic acid metabolism in the intact brain. *Am. J. Physiol.* **88**, 312–25.

McHenry, L. C., Jr. (1964). Quantitative cerebral blood flow determination: application of Krypton-85 desaturation technique in man. *Neurology, Minneapolis* **14**, 785–93.

McHenry, L. C., Jr. (1966). Cerebral blood flow studies in middle cerebral and internal carotid artery occlusion. *Neurology, Minneapolis* **16**, 1145–50.

McHenry, L. C., Jr., Slocum, H. C., Bivens, H. E., Mayes, H. A. & Hayes, G. J. (1965). Hyperventilation in awake and anaesthetized man. *Archs Neurol. Psychiat., Chicago* **12**, 270–7.

McIlwain, H. (1953). Glucose level, metabolism, and response to electrical impulses in cerebral tissues from man and laboratory animals. *Biochem. J.* **55**, 618–24.

—— (1966). *Biochemistry and the central nervous system.* 3rd Ed. London: Churchill.

Mackey, W. A. & Scott, L. D. W. (1938). Treatment of apoplexy by infiltration of the stellate ganglion with novocain. *Br. med. J.* **2**, 1–4.

McNaughton, F. (1938). The innervation of the intracranial blood vessels and dural sinuses. *Res. Publs Ass. Res. nerv. ment. Dis.* **18**, 178–200.

Magendie, F. (1836). Experiences sur le liquide cephalo-rachidien. *J. Hebd. progr. sci. inst. med.* **1**, 83–5.

Mallett, B. L. & Veall, N. (1963). Investigation of cerebral blood flow in hypertension, using radioactive-xenon inhalation and extracranial recording. *Lancet*, i, 1081–2.

—— (1965). Measurement of cerebral clearance rates in man using xenon-133 inhalation and extracranial recording. *Clin. Sci.* **29**, 179–91.

References

Mangold, R., Sokoloff, L., Therman, P. O., Conner, E. H., Kleinerman, J. I. & Kety, S. S. (1955). The effects of sleep and lack of sleep on the cerebral circulation and metabolism of normal young men. *J. clin. Invest.* **34**, 1092–100.

Manley, E. M., Nash, C. B. & Woodbury, R. A. (1964). Cardiovascular responses to severe hypercapnia of short duration. *Am. J. Physiol.* **207**, 634–40.

Marshall, B. E., Gleaton, H. E., Hedden, M., Dripps, R. D. & Alexander, S. C. (1966). Cerebral blood flow of the goat during sustained hypocarbia. *Fedn Proc.* **25**, 462.

Martin, E. G. & Mendelhall, W. L. (1915). The response of the vasodilator mechanism to weak, intermediate and strong sensory stimulation. *Am. J. Physiol.* **38**, 98–107.

Maspes, P., Fasano, V. & Broggi, G. (1955). Aspect angiographique de la circulation collaterale dans les cas de thrombose de la carotide interne. *Neurochir.* **1**, 273–8.

Masserman, J. H. (1934). Cerebral spinal hydrodynamics. IV. Clinical experimental studies. *Archs Neurol. Psychiat., Chicago* **32**, 523–53.

Maynard, E. A., Schultz, R. L. & Pease, D. C. (1957). Electron microscopy of the vascular bed of the rat cerebral cortex. *Am. J. Anat.* **100**, 409–33.

Mchedlishvili, G. I. & Baramidze, D. G. (1967). The functional behaviour of cortical blood vessels under conditions of the convulsive activity. *Bull. Exp. Biol. Med.* **11**, 68–71.

Mchedlishvili, G. I., Baramidze, D. G. & Nikolaishvili, L. S. (1967). Functional behaviour of pial and cortical arteries in conditions of increased metabolic demand from the cerebral cortex. *Nature* **213**, 506–7.

Mchedlishvili, G. I. & Nikolaishvili, L. S. (1970). Evidence of a cholinergic nervous mechanism mediating the autoregulatory dilatation of the cerebral blood vessels. *Pflügers Arch. ges. Physiol.* **315**, 27–37.

Mchedlishvili, G. I., Ormotsadze, L. G., Nikolaishvili, L. S. & Baramidze, D. G. (1967). Reaction of different parts of the cerebral vascular system in asphyxia. *Expl Neurol.* **18**, 239–52.

Meier, P. & Zierler, K. L. (1954). On the theory of the indicator-dilution method for measurement of blood flow and volume. *J. appl. Physiol.* **6**, 731–44.

Mellander, S. & Johannson, B. (1968). Control of resistance, exchange and capacitance functions in the peripheral circulation. *Pharmac. Rev.* **20**, 117–96.

Mellander, S., Oberg, B. & Odelram, H. (1964). Vascular adjustments to increased transmural pressure in cat and man with special reference to shifts in capillary fluid transfer. *Acta physiol. scand.* **61**, 34–48.

Merker, H. & Opitz, E. (1949). Die Gefasse der Pia Mater hohenangepasster Kaninchen. *Pflügers Arch. ges. Physiol.* **251**, 117–22.

Merlet, C., Hoerter, J., Devilleneuve, C. & Tchobroutsky, C. (1970). Mise en évidence de mouvements respiratoires chez le foetus d'agneau *in utero* au cours du dernier mois de la gestation. *C. r. hebd. Séanc. Acad. Sci., Paris* **270**, 2462–4.

Merritt, H. H. & Fremont-Smith, F. (1937). *The cerebral spinal fluid.* Philadelphia: Saunders.

Merwarth, C. R. & Sieker, H. O. (1961). Acid–base changes in blood and cerebrospinal fluid during altered ventilation. *J. appl. Physiol.* **16**, 1016–18.

References

Metcalfe, J. C. & Burgen, A. S. V. (1968). Relaxation of anaesthetics in the presence of cytomembranes. *Nature, Lond.* **220**, 587–8.

Metcalfe, J. C., Seeman, P. & Burgen, A. S. V. (1968). The proton relaxation of benzyl alcohol in erythrocyte membranes. *Mol. Pharmac.* **4**, 87–95.

Meyer, A. & Pribam, E. (1875). In *Klinik der Nervenkrankheiten*, ed. Rosenthal, A., p. 97. Stuttgart: Ferdinand Enke.

Meyer, J. S. & Denny-Brown, D. (1955). Studies of cerebral circulation in brain injury; validity of combined local cerebral electropolography, thermometry and steady potentials as indicator of local circulatory and functional changes. *Electroen. Neurophysiol.* **7**, 511–28.

Meyer, J. S., Feng, H. C. & Denny-Brown, D. (1954). Polarographic study of cerebral collateral circulation. *Archs Neurol. Psychiat.*, Chicago **74**, 296–312.

Meyer, J. S. & Gastaut, H. (1961). *Cerebral anoxia and the electroencephalogram.* Springfield, Ill.: Thomas.

Meyer, J. S. & Gotoh, F. (1960). Metabolic and electroencephalographic effects of hyperventilation. *Archs Neurol.*, Paris **3**, 539–52.

(1961). Interaction of cerebral hemodynamics and metabolism. *Neurology, Minneapolis* **11**, 46–65.

(1964). Continuous recording of cerebral metabolism, internal jugular flow and EEG in man. *Trans. Am. neurol. Ass.* **89**, 151.

Meyer, J. S., Gotoh, F., Ebinhara, S. & Tomita, M. (1965). Effects of anoxia on cerebral metabolism and electrolytes in man. *Neurology, Minneapolis* **15**, 892–901.

Meyer, J. S., Gotoh, F. & Tazaki, Y. (1961). Inhibitory action of carbon dioxide and acetazoleamide in seizure activity. *Electroen. Neurophysiol.* **13**, 762–75.

Meyer, J. S., Ishikawa, S. & Lee, T. K. (1964). Electromagnetic measurement of internal jugular venous flow in the monkey. *J. Neurosurg.* **21**, 524–39.

Meyer, L. (1875). Über aneurysmatische Veränderungen der Carotis Interna Geisteskranker. *Arch Psychiat.* **6**, 84–109.

Meyer, M., Tschetter, T., Klassen, A. & Resch, J. (1969). Regional blood flow estimated by using particle distribution and isotope clearance methods. In *Cerebral blood flow*, ed. Brock, M., Fieschi, C., Ingvar, D. H., Lassen, N. A. & Schürmann, K., pp. 50–2. Berlin: Springer.

Michailow, S. (1911). Der Bau der Zentralen sympathischen Ganglien. *Int. Mschr. Anat. Physiol.* **28**, 26–115.

Mines, A. H., Morrill, C. G. & Sørensen, S. C. (1971). The effect of isocarbic metabolic acidosis in blood on $[H^+]$ and $[HCO_3^-]$ in csf with deductions about the regulation of active transport of H^+/HCO_3^- between blood and csf. *Acta physiol. scand.* **81**, 234–45.

Mines, A. H. & Sørensen, S. C. (1971). Changes in the electrochemical potential difference for HCO_3^- between blood and csf and in csf lactate concentration during isocarbic hypoxia. *Acta physiol. scand.* **81**, 225–33.

Mitchell, R. A. (1966). Cerebrospinal fluid and the regulation of respiration. In *Advances in respiratory physiology*, ed. Caro, C. G., pp. 1–47. London: Arnold.

Mitchell, R. A., Carman, C. T., Severinghaus, J. W., Richardson, B. W., Singer, M. M. & Schnider, S. (1965). Stability of cerebrospinal fluid pH in chronic acid–base disturbances in blood. *J. appl. Physiol.* **20**, 443–52.

References

Mitchell, R. A., Herbert, D. A. & Carman, C. T. (1965). Acid–base constants and temperature coefficients for cerebrospinal fluid. *J. appl. Physiol.* **20**, 27–30.

Mitchell, R. A., Loeschcke, H. H., Massion, W. H. & Severinghaus, J. W. (1963). Respiratory responses mediated through superficial chemosensitive areas on the medulla. *J. appl. Physiol.* **18**, 523–33.

Mitchell, R. A., Loeschcke, H. H., Severinghaus, J. W., Richardson, B. W. & Massion, W. H (1963). Regions of respiratory chemo-sensitivity on the surface of the medulla. *Ann. N.Y. Acad. Sci.* **109**, 661–81.

Mitchell, R. A. & Singer, M. M. (1965). Respiration and cerebral spinal fluid pH in metabolic acidosis and alkalosis. *J. appl. Physiol.* **20**, 905–11.

Mohamed, M. S. & Bean, J. W. (1951). Local and general alterations of blood CO_2 and influence of intestinal mobility in regulation of intestinal blood flow. *Am. J. Physiol.* **167**, 413–25.

Moisejeff, E. (1927). Zur kenntnis des Carotissinusreflexes. *Z. ges exp. Med.* **53**, 696–704.

Molnar, L. (1968). Role of brain stem in CBF regulation. *Scand J. clin. Lab. Invest.* Suppl. 102. VI:D.

Molnar, L. & Szanto, J. (1964). The effect of electrical stimulation of the bulbar vasomotor centre on the cerebral blood flow. *Q. Jl exp. Physiol.* **49**, 184–93.

Moniz, E. (1940). Die cerebrale Arteriographie und Phlebographie. *Handbuch der Neurologie* II. Berlin: Springer.

Monro, A. (1783). *Observations on the structure and functions of the nervous system.* pp. 2–7, Edinburgh: W. Creech.

Moore, J. W., Kinsman, J. M., Hamilton, W. F. & Spurling, G. (1929). Studies on the circulation. II. Cardiac output determination; comparison of the injection method with the direct Fick procedure. *Am. J. Physiol.* **89**, 331–9.

Moore, D. H. & Ruska, H. (1959). The fine structure of capillaries and small arteries. *J. biophys. biochem. Cytol.* **3**, 457–62.

Moskalenko, Y. E., Cooper, R., Crow, H. J. & Walter, G. (1964). Variations in blood volume and oxygen availability in the human brain. *Nature, Lond.* **202**, 159–61.

Mosso, A. (1881). *Ueber den Kreislauf des Blutes im menschlichen Gehirn.* Leipzig: Viet.

Mottschall, H.-J. & Loeschcke, H. H. (1963). Messungen des transmeningealen Potentials der Katze bei Aenderungen des CO_2 – Drucks und der H^+ – Ionenkonzentration im Blut. *Pflügers Arch. ges. Physiol.* **277**, 662–70.

Moyer, J. H., Miller, S. I. & Snyder, H. (1954). Effect of increased jugular pressure on cerebral hemodynamics. *J. appl. Physiol.* **7**, 245–7.

Moyer, J. H. & Morris, G. C. (1954). Cerebral hemodynamics during controlled hypotension induced by the continuous infusion of ganglion blocking agents (Hexamethonium, Pendiomide & Arfonad). *J. clin. Invest.* **33**, 1081-88.

Müller, C. (1921). Die Messung des Blutdrucks am Schlafenden als Klinische Methode II. *Acta med. scand.* **55**, 443–85.

Müller, O. & Siebeck, R. (1907). Ueber die vasomotoren des Gehirns. *Z. exp. Path. Ther.* **4**, 57-87.

References

Munck, O. & Lassen, N. A. (1957). Bi-lateral cerebral blood flow and oxygen consumption in man by use of Krypton 85. *Circulation Res.* **5**, 163–8.

Murtach, F., Stauffer, H. & Harley, A. (1955). A case of persistent carotid-basilar anastomosis. *J. Neurosurg.* **12**, 46–9.

Myerson, A., Halloran, R. D. & Hirsch, H. L. (1927). Technique for obtaining blood from the internal jugular vein and internal carotid artery. *Archs Neurol. Psychiat., Chicago* **17**, 807–8.

Nahas, G. G., Ligon, J. C. & Mehlman, B. (1960). Effects of pH changes on O_2 uptake and plasma catecholamine levels in dog. *Am. J. Physiol.* **198**, 60–6.

Nakano, J., McGiff, J. C., Zeckert, H., Guggenheim, F. G. & Wegria, R. (1961). Effect of isopropylarterenol hydrochloride (Isuprel[R]) on heart rate, mean arterial blood pressure, cardiac output, coronary blood flow, cardiac oxygen consumption, work and efficiency. *Archs int. Pharmacodyn. Ther.* **133**, 400–11.

Negovski, V. A. (1945). Agonal states and clinical death; problems in revival of organisms. VI. Resistance of central system to local anaemia. *Am. Rev. Soviet Med.* **3**, 147–67.

Neilsen, M. & Smith, H. (1951). Studies on the regulation of respiration in acute hypoxia. *Acta physiol. scand.* **24**, 293–313.

Neissing, K. (1952). Zellformen und Zellreaktionen der Mikroglia des Mäusegehirns. *Gegenbaurs Morph. Jahrbuch.* **92**, 102–22.

Nelson, E., Blinzinger, K. & Hager, H. (1961). Electron microscopic observations on subarachnoid and perivascular spaces of the Syrian Hamster brain. *Neurology, Minneapolis* **11**, 285–95.

Nelson, E. & Rennels, M. (1968). Electron microscopic studies on intracranial vascular nerves in the cat. *Scand. J. clin. Lab. Invest.* Suppl. 102, VI:A.

—— (1970). Neuromuscular contacts in intracranial arteries of the cat. *Science, N.Y.* **167** 301–2.

Nicoll, P. A. & Webb, R. L. (1955). Vascular patterns and active vasomotion as determiners of flow through minute vessels. *Angiology* **6**, 291–310.

Nielsen, K. C. & Owman, Ch. (1967). Adrenergic innervation of pial arteries related to the circle of Willis in the cat. *Brain Res.* **6**, 773–6.

—— (1971). Contractile response and amine receptor mechanisms in isolated middle cerebral artery of the cat. *Brain Res.* **27**, 33–42.

Nielsen, K. C., Owman, Ch. & Sporrong, B. (1971). Ultrastructure of the autonomic innervation in the main pial arteries of rats and cats. *Brain Res.* **27**, 25–32.

Nilsson, N. J. (1965). Observations on the clearance rate of β radiation from Krypton[85] dissolved in saline and injected in microliter amounts into the grey and white matter of the brain. *Acta neurol. scand.* **14**, 53–7.

Nims, L. F., Gibbs, E. L. & Lennox, W. G. (1942). Arterial and cerebral venous blood. Changes produced by altering arterial carbon dioxide. *J. biol. Chem.* **145**, 189–95.

Noell, W. & Schneider, M. (1941). Die Gehirndurchblutung bei Sauerstoffmangel und bei Übergang von Sauerstoffmangel auf reinen O_2 und auf O_2/CO_2 Gemische. *Luftfahrtmedizin* **5**, 234–50.

—— (1942a). Über die Durchblutung und die Sauerstoffversorgung des Gehirns

References

im akuten Sauerstoffmangel. I. Die Gehirndurchblutung. *Pflügers Arch. ges. Physiol.* **246**, 182–200.

(1942b). Über die Durchblutung und die Sauerstoffversorgung des Gehirns im akuten Sauerstoffmangel. III. Die arteriovenöse Sauerstoff- und Kohlensäuredifferenz. *Pflügers Arch. ges. Physiol.* **246**, 207–49.

(1944). Über die Durchblutung und Sauerstoffversorgung des Gehirns. IV. Die Rolle der Kohlensaure. *Pflügers Arch. ges. Physiol.* **247**, 514–27.

Noordergraaf, A. (1964). Analog of the arterial bed. In *Pulsatile blood flow*, ed. Attinger, E. O., pp. 373–87. New York: McGraw-Hill.

Norcross, N. C. (1938). Intracerebral blood flow: an experimental study. *Archs Neurol. Psychiat.*, Chicago **40**, 291–9.

Nothnagel, H. (1867). Die vasomotorischen Nerven der Gehirngefässe. *Virchows Arch. path. Anat. Physiol.* **40**, 203–13.

Novack, P., Shenkin, H. A., Bortin, L., Goluboff, B. & Soffe, A. M. (1953). The effects of carbon dioxide inhalation upon the cerebral blood flow and cerebral oxygen consumption in vascular disease. *J. clin. Invest.* **32**, 696–702.

Novikoff, A. B. (1959). Lysosomes and physiology and pathology of cells. *Biomet. Bull.* **117**, 385.

Nyboer, J. (1959). *Electrical impedance plethysmography*. Springfield, Ill.: Thomas.

Nygard, J. W. (1937). Cerebral circulation prevailing during sleep and hypnosis. *Psychol. Bull.* **34**, 727.

Nylin, G. & Blomer, H. (1955). Studies on distribution of cerebral blood flow with thorium B-labelled erythrocytes. *Circulation Res.* **3**, 79–85.

Nylin, G., Hedlund, S. & Regnström, O. (1961). Studies of the cerebral circulation with labelled erythrocytes in healthy man. *Circulation Res.* **9**, 664–74.

Nylin, G., Silfverskiold, B. P., Lofstedt, S., Regnström, O. & Hedlund, S. (1960). Studies on cerebral blood flow in man, using radioactive-labelled erythrocytes. *Brain* **83**, 293–335.

Obrist, W. D., Sokoloff, L., Lassen, N. A., Lane, M. H., Butler, R. N. & Feinberg, I. (1963). Relation of EEG to cerebral blood flow and metabolism in old age. *Electroen. Neurophysiol.* **15**, 610.

Obrist, W. D., Thompson, H. K., King, C. H. & Wang, H. S. (1967). Determination of regional cerebral blood flow by inhalation of ^{133}Xenon. *Circulation Res.* **20**, 124–35.

O'Connell, J. E. A. (1943). The vascular factor in intracranial pressure and the maintenance of the cerebrospinal fluid circulation. *Brain* **66**, 204–28.

Oertel, P. (1922). Über die persistenz embryonaler Verbindungen zwischen der A. carotis interna und der A. vertebralis. *Anat. Anz.* **55**, 281–95.

Ohgushi, N. (1968). Adrenergic fibres to the brain and spinal cord vessels in the dog. *Arch. Jap. Chir.* **37**, 294–303.

Oldendorf, W. H. (1962). Measurement of the mean transit time of cerebral circulation by external detection of an intravenously injected radioisotope. *J. nucl. Med.* **3**, 382–98.

Oldendorf, W. H. & Kitano, M. (1965). Isotope study of brain blood turnover in vascular disease. *Archs Neurol.*, Chicago **12**, 30–8.

Olsen, N. S. & Klein, J. R. (1947a). Effect of cyanide on the concentration of lactate and phosphates in the brain. *J. biol. Chem.* **167**, 739–46.

References

Olsen, N. S. & Klein, J. R. (1947b). Effect of convulsive activity upon the concentration of brain glucose, glycogen, lactate, and phosphates. *J. biol. Chem.* **167**, 747–55.

Olson, R. E. (1963). 'Excess lactate' and anaerobiosis (editorial). *Ann. Intern. Med.* **59**, 960–3.

Opitz, E. (1948). Über die Sauerstoffversorgung des Zentralnervensystems. *Naturwissenschaften* **35**, 80–8.

— (1951). Increased vascularization of the tissue due to acclimatization to high altitude and its significance for oxygen transport. *Expl Med. Surg.* **9**, 389–403.

Opitz, E. & Palme, F. (1944). Darstellung der Hohenanpassung im Gebirge durch Sauerstoffmangel. II. Und III Mitteilung. *Pflügers Arch. ges. Physiol.* **248**, 298–329; 330–75.

Opitz, E. & Schneider, M. (1950). Über die Sauerstoffversorgung des Gehirns und den Mechanisms von Mangelwirkungen. *Ergbn. Physiol.* **46**, 126–260.

Opitz, E. & Thorn, W. (1949). Überlebenszeit und Erholungszeit des Warmblütergehirns unter dem Einfluss der Höhenanpassung. *Pflügers Arch. ges. Physiol.* **251**, 369–87.

Orden, L. S. van, Bloom, F. E., Barnett, R. J. & Giarmon, N. J. (1966). Histochemical and functional relationships of catecholamines in adrenergic nerve endings. I. Participation of granular vesicles. *J. Pharmac. exp. Ther.* **154**, 185–99.

Padget, D. H. (1957). The development of the cranial venous system from the view point of comparative anatomy. *Contr. Embryol.* **611**, 36, 79–140.

Page, W. F., German, W. J. & Nims, L. F. (1951). The nitrous oxide method for measurement of cerebral blood flow and cerebral gaseous metabolism in dogs. *Yale J. Biol. Med.* **23**, 462–73.

Paillard, M., Sraer, F., Leviel, F. & Claret, M. (1971). Direct measurement of intracellular pH in crab and rat muscle. *J. Physiol.* **216**, 50P.

Palay, S. L. (1956). Synapses in the central nervous system. *J. biophys. biochem. Cytol.* Suppl. 2, 193–202.

Pappenheimer, J. R. (1941a). Vasoconstrictor nerves and oxygen consumption in the isolated perfused hindlimb muscles of the dog. *J. Physiol.* **99**, 182–200.

— (1941b). Blood flow, arterial oxygen saturation and oxygen consumption in the isolated perfused hindlimb of the dog. *J. Physiol.* **99**, 283–303.

Pappenheimer, J. R. (1965). The ionic composition of cerebral extracellular fluid and its relation to control of breathing. *Harvey Lectures, Series* 61. Academic Press N.Y.

Pappenheimer, J. R., Fencl, V., Heisey, S. R. & Held, D. (1965). Role of cerebral fluids in control of respiration as studied in unanaesthetized goats. *Am. J. Physiol.* **208**, 436–50.

Pappenheimer, J. R., Heisey, S. R., Jordan, E. F. & Downer, J. de C. (1962). Perfusion of the cerebral ventricular system in unanaesthetized goats. *Am. J. Physiol* **203**, 763–74.

Parker, F. (1958). An electron microscope study of coronary arteries. *Am. J. Anat.* **103**, 247–73.

Parker, H. R. & Purves, M. J. (1967). Some effects of maternal hyperoxia and hypoxia on the blood gas tensions and vascular pressures in foetal sheep. *Q. Jl exp. Physiol.* **52**, 205–21.

References

Parratt, J. R. (1965). Blockade of sympathetic β-receptors in the myocardial circulation. *Br. J. Pharmacol.* **24**, 601–11.
Parrish, A. E., Kleh, J. & Fazekas, J. F. (1957). Renal and cerebral hemodynamics with hypotension. *Am. J. med. Sci.* **233**, 35–9.
Passow, H. (1961). Zusammenwirken von Membranstruktur und Zellstoffwechsel bei der Regulierung der Ionenpermeabilität roter Blutkörperchen. In *Biochemie des Aktiven Transports* (Colloq. Ges. Chem. Mosbach). Berlin: Springer.
Paton, W. D. M. & Zar, A. M. (1968). The origin of acetyl-choline released from guinea-pig intestine and longitudinal muscle strips. *J. Physiol.* **194**, 13–34.
Patterson, J. L., Heymann, A., Battey, L. L. & Ferguson, R. W. (1955) Threshold of response of cerebral vessels of man to increase in blood carbon dioxide. *J. clin. Invest.* **34**, 1857–64.
Patterson, J. L., Heyman, A. & Duke, T. W. (1952). Cerebral circulation and metabolism in chronic pulmonary emphysema. With observations on the effects of inhalation of oxygen. *Am. J. Med.* **12**, 382–7.
Pauli, H. G., Vorburger, C. & Reubi, F. (1962). Chronic derangements of cerebrospinal fluid acid–base components in man. *J. appl. Physiol.* **17**, 993–8.
Pease, D. C. (1955). Electron microscopy of the vascular bed of the kidney cortex. *Anat. Rec.* **121**, 701–21.
Pease, D. C. & Molinari, S. (1960). Electron microscopy of muscular arteries; pial vessels of the cat and monkey. *J. Ultrastruct. Res.* **3**, 447–68.
Pease, D. C. & Paule, W. J. (1960). Electron microscopy of elastic arteries; the thoracic aorta of the rat. *J. Ultrastruct. Res.* **3**, 469–83.
Pease, D. C. & Schultz, R. L. (1958). Electron microscopy of the rat cranial meninges. *Am. J. Anat.* **102**, 301–24.
Penfield, W. (1928). Neuroglia and microglia. The interstitial tissue of the central nervous system. In *Special cytology*, ed. Cowdry, E. V. New York: Hoeber.
—— (1932). Intracerebral vascular nerves. *Archs Neurol. Psychiat.*, Chicago **27**, 30–44.
Penrod, K. E. (1949). Oxygen consumption and cooling rates in immersion hypothermia in the dog. *Am. J. Physiol.* **157**, 436–44.
Pentschew, A. (1958). Intoxikationen. In *Nervensystem*, ed, Scholz, W Handbuch der speziellen patholoischen Anatomie und Histologie, ed Lubarsch, O., Henke, F. und Rössle, R., vol. 13, part 2, section B Berlin: Springer.
Perry, J. (1950). *Chemical engineers' handbook*. New York: McGraw-Hill.
Pfeiffer, R. A. (1928). *Die Angioarchitektonik der Grosshirnrinde*. Berlin: Springer
Pflüger, E. (1872). Ueber die Diffusion des Sauerstoffs, den Ort und die Gesetze der Oxidationsprocesse im thierischen Organismus. *Pflügers Arch. ges. Physiol.* **6**, 43–64.
—— (1875). Beitrage zur Lehre von der Respiration. 1. Ueber die physiologische Verbrennung in der lebendigen Organismen. *Pflügers Arch. ges. Physiol.* **10**, 251–367.
Pierce, E. C., Lambertsen, C. J., Deutch, S., Chase, P. E., Linde, H. W., Dripps, R. D. & Price, H. L. (1962). Cerebral circulation and metabolism

References

during thiopental anesthesia and hyperventilation in man. *J. clin. Invest.* **41**, 1664–71.

Plattner, F., Galehr, O. & Kodera, Y. (1928). Über das Schicksal des Acetylcholin im Blute. IV. Die Abhangigkert der Acetylcholinzerstörung von der Wasserstoffionenkonzentration. *Pflügers Arch. ges. Physiol.* **219**, 678–87.

Plum, F. & Posner, J. B. (1967). Blood and cerebrospinal fluid lactate during hyperventilation. *Am. J. Physiol.* **212**, 864–70.

Politoff, A. & Macri, F. (1966). Pharmacoligic differences between isolated, perfusated arteries of the choroid plexus and of the brain parenchyma. *Int. J. Neuropharmac.* **5**, 155–62.

Poljak, S. (1927). An experimental study of the association of callosal and projection fibres of the cerebral cortex of the cat. *J. comp. Neurol.* **44**, 197–258.

Pomerat, C. M. (1952). Dynamic neurogliology. *Tex. Rep. Biol. Med.* **10**, 885–913.

—— (1960). Dynamic activities of cellular elements of the nervous system. Alfred Korzybski Mem. Symposium. *New York Gen. Semantics Bull.* Nos. 24–25.

Pontén, U. (1966). Acid–base changes in rat brain tissue during acute respiratory acidosis and baseosis. *Acta physiol. scand.* **68**, 152–63.

Pontén, U. & Siesjö, B. K. (1966). Gradients of CO_2 tension in the brain. *Acta physiol. scand.* **67**, 129–40.

—— (1967). Acid–base relations in arterial blood and cerebrospinal fluid of the unanaesthetized rat. *Acta physiol. scand.* **71**, 89–95.

Prados, M., Strowger, B. & Feindel, W. (1945a). Studies on cerebral edema. I. Reaction of the brain to air exposure: pathologic changes. *Archs Neurol. Psychiat., Chicago* **54**, 163–74.

—— (1945b). Studies on cerebral edema. II. Reaction of the brain to exposure to air: physiologic changes. *Archs Neurol. Psychiat., Chicago* **54**, 290–300.

Preswick, G., Reivich, M. & Hill, I. D. (1965). The EEG effects of combined hyperventilation and hypoxia in normal subjects. *Electroen. Neurophysiol.* **18**, 56–64.

Price, H. L. & Helrich, M. (1955). Effect of cyclopropane, diethyl ether, nitrous oxide, thiopental and hydrogen ion concentration on myocardial function of dog heart-lung preparation. *J. Pharmac. exp. Ther.* **115**, 206–16.

Price, H. L., Linde, H. W., Jones, R. E., Black, G. W. & Price, M. L. (1959). Sympatho-adrenal responses to general anaesthesia in man and their relation to hemodynamics. *Anesthesiology* **20**, 563–75.

Price, M. L. & Price, H. L. (1962). Effects of general anesthetics on contractile responses of rabbit aortic strips. *Anesthesiology* **23**, 16–20.

Prothero, J. & Burton, A. C. (1962). The physics of capillary flow. II. The capillary resistance to flow. *Biophys. J.* **2**, 199–211.

Purves, M. J. (1970a). The effect of hypoxia, hypercapnia and hypotension upon carotid body blood flow and oxygen consumption in the cat. *J. Physiol.* **209**, 395–416.

—— (1970b). The role of the cervical sympathetic nerve in the regulation of oxygen consumption of the carotid body of the cat. *J. Physiol.* **209**, 416–31.

Purves, M. J. & James, I. M. (1969). Observations on the control of cerebral blood flow in the sheep fetus and newborn lamb. *Circulation Res.* **25**, 651–67.

References

Putnam, T. J. & Ask-Upmark, E. (1934). Cerebral circulation. XXIX. Microscopic observations on the living ependyma of the cat. *Archs Neurol. Psychiat., Chicago* **31**, 72–80.

Quastel, J. H. & Wheatley, A. H. M. (1932). Narcosis and oxidations of brain. *Proc. R. Soc. B* **112**, 60–79.

Radigan, L. R. & Robinson, S. (1949). Effects of environmental heat stress and exercise on renal blood flow and filtration rate. *J. appl. Physiol.* **2**, 185–91.

Rall, D. P., Oppelt, W. W. & Patlak, C. S. (1962). Extracellular space of brain as determined by diffusion of inulin from the ventricular system. *Life Sci. Oxford* **2**, 43–8.

Ramon Y Cajal, S. (1909). *Histologie du système nervaux de l'homme et des vertébrés.* Vol. 1. Paris: A. Maloine.

— (1913). Contribución al conocimiento de la neuroglia del cerebro humano. *Trab. Lab. Invest. biol. Univ. Madr.* **11**, 255–315.

Ranson, S. W. & Billingsley, P. R. (1916). Vasomotor reactions from stimulation of the floor of the fourth ventricle. Studies in vasomotor reflex arcs. *Am. J. Physiol.* **41**, 85–90.

— (1918). The superior cervical ganglion and the cervical portion of the sympathetic trunk. *J. comp. Neurol.* **29**, 313–58.

Rapela, C. E. & Green, H. D. (1964). Autoregulation of canine cerebral blood flow. *Circulation Res.* **14**, 1–205–11.

— (1968). Autoregulation of cerebral blood flow during hypercarbia and during hypercarbia and controlled (H^+). *Scand. J. clin. Lab. Invest.* Suppl. 102 v: C.

Rapela, C. E., Green, H. D. & Denison, A. B. (1967). Baroreceptor reflexes and autoregulation of cerebral blood flow in the dog. *Circulation Res.* **21**, 559–68.

Rapport, M. M., Green, A. A. & Page, I. H. (1948). Crystalline serotonin. *Science* **108**, 329–30.

Rauber, A. (1872). *Über den sympathischen Grenzstrang des menshcher Kopfes.* pp. 98–103. München: Lentner.

Ravina, A. F. (1811). Specimen de motu cerebri. *Memorie Accad. Sci. Torino* **20**, 61–93, Memoires presentées.

Ray, B. S. & Wolff, H. G. (1940). Experimental studies on headache. Pain sensitive structures of the head and their significance in headache. *Archs Surg., Chicago* **41**, 813–56.

Reid, A. (1616). *A description of the body of man.* London: W. Iaggard.

Reid, G. (1952). Circulatory effects of 5-hydroxytryptamine. *J. Physiol.* **118**, 435–53.

Reigal, F. & Jolly, F. (1871). Ueber die Veranderungen der Piagefasse in Folge von Reisung sensibler Nerven. *Virchows Arch. path. Anat. Physiol.* **52**, 218–30.

Rein, H. (1929). Über Besonderheiten der Blutzirkulation in der Arteria carotis. *Z. Biol.* **89**, 307–18.

— (1931). Vasomotorische regulationen. *Ergebn. Physiol., Biol. Chem. Expl Pharmakol.* **32**, 28–72.

— (1941). *Lehrbuch d. Physiol,* p. 418. Berlin: Springer.

Reiner, J. M. (1947). The effect of age on the carbohydrate metabolism of tissue homogenates. *J. Geront.* **2**, 315–20.

References

Reivich, M. (1964). Arterial PCO_2 and cerebral hemodynamics. *Am. J. Physiol.* **206**, 25–35.

(1969). Observations on experimental models of cerebral clearance curves. In *Research on the cerebral circulation*, ed. Meyer, J. S. Springfield: Thomas.

Reivich, M., Dickson, J., Clark, J., Hedden, M. & Lambertsen, C. J. (1968). Role of hypoxia in cerebral circulatory and metabolic changes during hypocarbia in man: studies in hyperbaric milieu. *Scand. J. clin. Lab. Invest.* Suppl. 102, IV:B.

Reivich, M., Isaacs, G., Evarts, E. & Kety, S. S. (1968). The effect of slow wave sleep and REM sleep upon regional cerebral blood flow in cats. *J. Neurochem.* **15**, 301–6.

Reivich, M., Jehle, J., Sokoloff, L. & Kety, S. S. (1969). Measurement of regional cerebral blood flow with antipyrine ^{14}C in awake cats. *J. appl. Physiol.* **27**, 296–300.

Reivich, M., Slater, R. & Sano, N. (1969). Further studies on experimental models of cerebral clearance curves. In *Cerebral blood flow*, ed. Brock, M., Fieschi, C., Ingvar, D. H., Lassen, N. A. & Schürmann, K. pp. 4–7. Berlin: Springer.

Rhodin, J. A. (1967). The ultrastructure of mammalian arterioles and precapillary sphincters. *J. Ultrastruct. Res.* **18**, 181–223.

Rich, M., Scheinberg, P. & Belle, M. S. (1953). Relationship between cerebrospinal fluid pressure changes and cerebral blood flow. *Circulation Res.* **1**, 389–95.

Richardson, D. W., Wasserman, A. J. & Patterson, J. L. (1961). General and regional circulating responses to change in blood pH and carbon dioxide tension. *J. clin. Invest.* **40**, 31–43.

Richardson, K. C. (1960). Studies on the structure of autonomic nerves in the small intestine, correlating the silver impregnated image in light microscopy with the permanganate fixed ultrastructure in electron microscopy. *J. Anat.* **94**, 457–72.

Ridley, H. (1700). An experiment to discover the cause of the motion of the dura mater. *Phil. Trans. R. Soc.* **5**, 199–202.

Risberg, J. & Ingvar, D. H. (1968). Regional changes in cerebral blood volume during mental activity. *Expl Brain Res.* **5**, 72–8.

Risteen, W. A. & Volpitto, P. P. (1946). Role of stellate ganglion block in certain neurologic disorders. *South. Med. J.* **39**, 431–5.

Ritchie, J. M. (1967). The oxygen consumption of mammalian non-myelinated nerve fibres at rest and during activity. *J. Physiol.* **188**, 309–29.

Robertson, F. (1899). On a new method of obtaining a black reaction in certain tissue elements of the central nervous system. *Scott. med. surg. J.* **4**, 23–30.

Robin, E. D., Whaley, R. D., Crump, C. H., Bickelmann, A. G. & Travis, D. M. (1958). Acid–base relations between spinal fluid and arterial blood with special reference to control of ventilation. *J. appl. Physiol.* **13**, 385–92.

Rohnstein, R. (1900). Zur Frage nach dem Vorhandensein von Nerven an dem Blutgefassen der grossen Nervencentren. *Arch. fur. mikr. Anat.* **55**, 576–84.

References

Roos, A. (1965). Intracellular pH and intracellular buffering power of the cat brain. *Am. J. Physiol.* **209**, 1233–46.

Rosenblum, W. I. (1969a). Contractile effects of barium on the rat aorta. *Fedn Proc.* **28**, 829.

(1969b). Cerebral arteriolar spasm inhibited by β-adrenergic blocking agents. *Archs Neurol.*, Paris **21**, 296–302.

Rosendorf, C. & Cranston, W. I. (1969). Application of the ^{133}xenon clearance method to the measurement of local blood flow in the conscious animal. In *Cerebral blood flow*, ed. Brock, M., Fieschi, C., Ingvar, D. H., Lassen, N. A. & Schürmann, K. Berlin: Springer.

Rosenstein, A. (1935). *Handbuch der Neurologie*, vol. 1, p. 237. Berlin: Springer.

Rosomoff, H. L. (1956). Some effects of hypothermia on the normal and abnormal physiology of the nervous system. *Proc. R. Soc. Med.* **49**, 358–64.

Rosomoff, H. L. & Gilbert, R. (1955). Brain volume and cerebrospinal fluid pressure during hypothermia. *Am. J. Physiol.* **183**, 19–22.

Rosomoff, H. L. & Holaday, D. A. (1954). Cerebral blood flow and cerebral oxygen consumption during hypothermia. *Am. J. Physiol.* **179**, 85–8.

Ross, J. M., Fairchild, H. M., Weldy, J. F. & Guyton, A. C. (1962). Autoregulation of blood flow by oxygen lack. *Am. J. Physiol.* **202**, 21–4.

Roth, C. D. & Richardson, K. C. (1969). Electron microscopical studies in axonal degeneration in the rat iris following ganglionectomy. *Am. J. Anat.* **124**, 341–59.

Rougemont, J. de, Ames, A. III, Nesbett, F. B. & Hofmann, H. F. (1960). Fluid formed by choroid plexus; A technique for its collection and a comparison of its electrolyte composition with serum and cisternal fluids. *J. Neurophysiol.* **23**, 485–95.

Roughton, F. J. W. (1952). Diffusion and chemical reaction velocity in cylindrical and spherical systems of physiological interest. *Proc. R. Soc.* B **140**, 203–30.

Rowbotham, G. F. & Little, E. (1965). Circulations of the cerebral hemispheres. *Br. J. Surg.* **1**, 8–20.

Roy, C. S. & Brown, J. G. (1879). The blood pressure and its variations in the arterioles, capillaries and smaller veins. *J. Physiol.* **2**, 323–59.

Roy, C. S. & Sherrington, C. S. (1890). On the regulation of the blood supply of the brain. *J. Physiol.* **11**, 85–108.

Rudolph, A. M. & Heymann, M. A. (1967). The circulation of the fetus in utero. Methods for studying distribution of blood flow, cardiac output and organ blood flow. *Circulation Res.* **21**, 163–84.

Rushmer, R. F., Beckman, E. L. & Lee, D. (1947). Protection of cerebral circulation by the cerebrospinal fluid under the influence of radial acceleration. *Am. J. Physiol.* **151**, 355–65.

Ruska, H. & Ruska, C. (1961). Licht- und Elektronmicroskopie des peripheren neuro-vegitativen Systems im Himblick auf die Funktion. *Dt. med. Wschr.* **86**, 1697–701.

Ryder, H. W., Espey, F. F., Kimbell, F. D., Penka, E. J., Rosenauer, A., Podolsky, B. & Evans, J. P. (1952). Influence of changes in blood flow on the cerebro-spinal fluid pressure. *Archs Neurol. Psychiat., Chicago* **68**, 165–9.

References

Sachs, H., Share, L., Osinchak, J. & Carpi, A. (1967). Capacity of the neurohypophysis to release vasopressin. *Endocrinology* **81**, 755-70.

Sacks, W. (1958). Cerebral metabolism of isotopic lipid and protein derivatives in normal human subjects. *J. appl. Physiol.* **12**, 311-18.

(1965). Cerebral metabolism of doubly labelled glucose in humans *in vivo*. *J. appl. Physiol.* **20**, 117-30.

Sagawa, K. & Guyton, A. C. (1961). Pressure flow relationships in isolated canine cerebral circulation. *Am. J. Physiol.* **200**, 711-14.

Samarasinghe, D. D. (1965). The innervation of the cerebral arteries in the rat: an electron microscope study. *J. Anat.* **99**, 815-28.

Sandison, J. C. (1931). Observations on circulating blood cells, adventitial (rouget) and muscle cells, endothelium, and macrophages in transparent chamber of rabbit's ear. *Anat. Rec.* **50**, 355-79.

Sapirstein, L. A. (1958). Regional blood flow by fractional distribution of indicators. *Am. J. Physiol.* **193**, 161-8.

(1962). Measurement of the cephalic and cerebral blood flow fractions of the cardiac output in man. *J. clin. Invest.* **41**, 1429-35.

Sato, S. (1966). An electron microscopic study on the innervation of the intracranial artery of the rat. *Am. J. Anat.* **118**, 873-89.

Schafer, H. (1877). Über die aneurysmatische Erweiterung der Carotis interna an ihrem Ursprung. *Z. Psychiat.* **34**, 438-51.

Schatzmann, H. J. (1961). Calciumaufnahme und-abgabe am Darmmuskel des Meerschweinchens. *Pflügers Arch. ges. Physiol.* **274**, 295-310.

Scheibel, M. E. & Scheibel, A. B. (1955). The inferior olive. *J. comp. Neurol.* **102**, 77-131.

Scheinberg, P. (1950). Cerebral blood flow in vascular disease of the brain, with observations on the effect of stellate ganglion block. *Am. J. Med.* **8**, 139-47.

(1951). Cerebral blood flow and metabolism in pernicious anemia. *Blood* **6**, 213-27.

Scheinberg, P., Blackburn, I., Rich, M. & Saslaw, M. (1953). Effects of aging on cerebral circulation and metabolism. *Archs Neurol. Psychiat.*, Chicago **70**, 77-85.

Scheinberg, P., Blackburn, I., Saslaw, M., Rich, M. & Baum, G. (1953). Cerebral circulation and metabolism in pulmonary emphysema and fibrosis with observations on the effects of mild exercise. *J. clin. Invest.* **32**, 720-8.

Scheinberg, P., Bourne, B. & Reinmuth, O. M. (1965). Human cerebral lactate and pyruvate extraction. *Archs Neurol., Paris* **12**, 246-50.

Scheinberg, P. & Jayne, H. W. (1952). Factors influencing cerebral blood flow and metabolism. A review. *Circulation* **5**, 225-36.

Scheinberg, P. & Stead, E. A. (1949). The cerebral blood flow in male subjects as measured by the nitrous oxide technique: normal values for blood flow, oxygen utilization, glucose utilization and peripheral resistance, with observations on the effect of tilting and anxiety. *J. clin. Invest.* **28**, 1163-71.

Scheinberg, P., Stead, E. A., Jr., Brannon, E. S. & Warren, J. V. (1950). Correlative observations on cerebral metabolism and cardiac output in myxedema. *J. clin. Invest.* **29**, 1139-46.

References

Schieve, J. F. & Wilson, W. P. (1953a). The changes in cerebral vascular resistance of man in experimental alkalosis and acidosis. *J. clin. Invest.* **32**, 33-8.

(1953b). The influence of age, anaesthesia and cerebral arteriosclerosis on cerebral vascular activity to carbon dioxide. *Am. J. Med.* **15**, 171-4.

Schlote, W. (1959). Zur Gliaarchitektonik der menslichen Grosshirnrinde im Nissl-Bild. *Arch. Psychiat. Nervenkr.* **199**, 573-95.

Schmidt, C. F. (1928a). The influence of cerebral blood flow on respiration. I. The respiratory responses to changes in cerebral blood flow. *Am. J. Physiol.* **84**, 202-22.

(1928b). The influence of cerebral blood flow on respiration. II. The gaseous metabolism of the brain. *Am. J. Physiol.* **84**, 223-41.

(1928c). The influence of cerebral blood flow on respiration. III. The interplay of factors concerned in the regulation of respiration. *Am. J. Physiol.* **84**, 242-59.

(1934). The intrinsic regulation of the circulation in the hypothalamus of the cat. *Am. J. Physiol.* **110**, 137-52.

(1936). The intrinsic regulation of the circulation in the parietal cortex of the cat. *Am. J. Physiol.* **114**, 572-85.

(1950). *The cerebral circulation in health and disease.* Springfield, Illinois: C. C. Thomas.

(1960). Central nervous system circulation, fluid and barriers. In *Handbook of physiology*, vol. 3, ed. Field, J., pp. 1745-50. Washington: American Physiological Society.

Schmidt, C. F. & Hendrix, J. P. (1938). Action of chemical substances on cerebral blood-vessels. *Res. Publs Ass. Res. nerv. ment. Dis.* **18**, 229-76.

Schmidt, C. F., Kety, S. S. & Pennes, H. H. (1945). The gaseous metabolism of the brain of the monkey. *Am. J. Physiol.* **143**, 33-52.

Schmidt, C. F. & Pierson, J. C. (1934). The intrinsic regulation of blood vessels of the medulla oblongata. *Am. J. Physiol.* **108**, 241-63.

Schmidt, K. (1910). Die arteriellen Kopfgefasse des Rindes. *Int. Mschr. Anat. Physiol.* **27**, 187-264.

Schneider, M. (1953). Durchblutung und Sauerstoffversorgung des Gehirns. *Verh. Deutsch. ges. Kreislaufforsch.* 19 Darmstadt: Steinkopf.

(1963). Critical blood pressure in the cerebral circulation. In *Selective vulnerability of the brain in hypoxaemia*, ed. Schadé, J. P. & McMenemey, W. H. Oxford: Blackwell.

Schneider, M. & Schneider, D. (1934). Untersuchungen über die Regulierung der Gehirndurch-blutung; Einwirkung verschiedener Pharmaca auf die Gehirndurchblutung. *Arch. exp. Path. Pharmak.* **175**, 640-64.

Scholz, W. (1933). Einiges über progressive und regressive Metamorphosen der astrocytären Glia. *Z. ges. Neurol. Psychiat.* **147**, 489-504.

Schroeder, J. S., Robison, S. C., Miller, H. A. & Harrison, D. C. (1970). Effects of respiratory acidosis on circulatory response to isoproterenol. *Am. J. Physiol.* **218**, 448-52.

Schuller, M. (1874). Ueber die Einwirkung einiger Arzneimittel auf die Gehirngefässe. *Berl. klin. Wschr.* **25**, 294-6.

References

Schulten, M. W. Von (1884). Experimentelle Untersuchungen über die Circulationsverhaltnisse des Anges und über den Zusammenhang Zwischen den Circulationsverhaltnisse des Anges und des Gehirns. *Archs Ophthal.*, *N.Y.* **30**, 59–61.

Schultz, A. (1866). Zur Lehre von der Blutbewegung im Innern des Schädels. *St. Petersb. med. Ztschr.* **11**, 122–8.

Schultz, R. L., Maynard, E. A. & Pease, D. C. (1957). Electron microscopy of neurons and neuroglia of cerebral cortex and corpus callosum. *Am. J. Anat.* **100**, 369–408.

Schumacher, G. A., Ray, B. S. & Wolff, H. G. (1940). Experimental studies on headache. Further analysis of histamine headache and its pain pathways. *Archs Neurol. Psychiat., Chicago* **44**, 701–17.

Schwab, M. (1962a). Das Säure-Basen-Gleichgewicht im arteriellen Blut und Liquor cerebrospinalis bei chronischer Niereninsuffizienz. *Klin. Wschr.* **40**, 765–72.

(1962b). Das Säure-Basen-Gleichgewicht im arteriellen Blut und Liquor cerebrospinalis bei Herzinsuffizienz und Cor Pulmonale und seine Beeinflussing durch Carboanhydrase-Hemmung. *Klin. Wschr.* **40**, 1233–45.

Schwartz, W. B. & Relman, A. S. (1963). A critique of the parameters used in the evaluation of acid–base disorders. *New Engl. J. Med.* **268**, 1382–8.

Scroop, G. C. & Lowe, R. D. (1968). Central pressor effect of angiotensin mediated by the parasympathetic nervous system. *Nature, Lond.* **220**, 1331–2.

(1969). Efferent pathways of the cardiovascular response to vertebral artery infusions of angiotensin in the dog. *Clin. Sci.* **37**, 605–19.

Sechzer, P. H., Egbert, L. D., Linde, H. W., Cooper, D. Y., Dripps, R. D. & Price, H. L. (1960). Effect of carbon dioxide inhalation on arterial pressure, electrocardiogram and plasma concentration of catecholamines and 17-OH corticosteroids in man. *J. appl. Physiol.* **15**, 454–8.

Sensenbach, W., Madison, L. & Ochs, L. (1953). A comparison of the effects of l-nor-epinephrine, synthetic l-epinephrine and U.S.P. epinephrine upon cerebral blood flow and metabolism in man. *J. clin. Invest.* **32**, 226–32.

Severinghaus, J. W. & Carcelan, B. (1964). Cerebrospinal fluid in man native to high altitude. *J. appl. Physiol.* **19**, 319–21.

Severinghaus, J. W. (1965). Electrochemical gradients for hydrogen and bicarbonate ions across the blood–CSF barrier in response to acid–base balance changes. In *Cerebrospinal fluid and the regulation of ventilation*, ed, Brooks, C. McC., Kao, F. F. & Lloyd, B. B., pp. 247–58. Oxford: Blackwell.

Severinghaus, J. W. (1968). Outline of H^+-blood flow relationships in brain. *Scand. J. clin. Lab. Invest.* Suppl. 102, VIII: K.

Severinghaus, J. W., Chiodi, H., Eger, E. I., Brandstater, B. & Hornbein, T. F. (1966). Cerebral blood flow in man at high altitude. Role of cerebrospinal fluid pH in normalization of flow in chronic hypocapnia. *Circulation Res.* **19**, 274–82.

Severinghaus, J. W. & Lassen, N. (1967). Step hypocapnia to separate arterial from tissue PCO_2 in the regulation of cerebral blood flow. *Circulation Res.* **20**, 272–8.

References

Severinghaus, J. W., Mitchell, R. A., Richardson, B. W. & Singer, M. M. (1964). Respiratory control at high altitude suggesting active transport regulation of CSF pH. *J. appl. Physiol.* **18**, 1155–66.

Shalit, M. N., Reinmuth, O. M., Shimojyo, S. & Scheinberg, P. (1967a). Carbon dioxide and cerebral circulatory control. II. The intravascular effect. *Archs Neurol.*, *Chicago* **17**, 337–41.

(1967b). Carbon dioxide and cerebral circulatory control. III. The effects of brain stem lesions. *Archs Neurol. Chicago* **17**, 342–53.

Shalit, M. N., Shimojyo, S., & Reinmuth, O. M. (1967). Carbon dioxide and cerebral circulatory control. I. The extravascular effect. *Archs Neurol.*, *Chicago* **17**, 298–303.

Shanes, A. M. (1958). Electrochemical aspects of physiological and pharmacological action in excitable cells. *Pharmac. Rev.* **10**, 59–273.

Shapiro, W., Wasserman, A. J & Patterson, J. L. (1965). Human cerebrovascular response time to elevation of arterial carbon dioxide tension. *Archs Neurol.*, *Chicago* **13**, 130–8.

(1966). Human cerebrovascular response to combined hypoxia and hypercapnia. *Circulation Res.* **19**, 903–10.

Shapot, U. S. (1957). In *Metabolism of the nervous system*, ed. Richter, D. London: Pergamon.

Shelburne, S. A., Blain, D. & O'Hare, J. D. (1932). Spinal fluid in hypertension. *J. clin. Invest.* **11**, 489–96.

Shenkin, H. A. (1951). The effects of various drugs upon cerebral circulation and metabolism in man. *J. appl. Physiol.* **3**, 465–71.

(1953). The cerebral circulation in post-operative intracranial hypotension. *J. Neurosurg.* **10**, 48–51.

Shenkin, H. A., Cabieses, F. & Van den Noordt, G. (1951). Effect of bilateral stellate ganglionectomy upon the cerebral circulation in man. *J. clin. Invest.* **30**, 90–3.

Shenkin, H. A., Cabieses, F., Van den Noordt, G., Sayers, P. & Copperman, R. (1951). The hemodynamic effect of unilateral carotid ligation on cerebral circulation of man. *J. Neurosurg.* **8**, 38–45.

Shenkin, H. A., Harmel, M. H. & Kety, S. S. (1948). Dynamic anatomy of the cerebral circulation. *Archs Neurol. Psychiat.*, *Chicago* **60**, 240–52.

Shenkin, H. A., Novack, P., Goluboff, B., Soffe, A. M. & Bortin, L. (1953). The effects of ageing, arteriosclerosis and hypertension upon the cerebral circulation. *J. clin. Invest.* **32**, 459–65.

Shenkin, H. A., Spitz, E. B., Grant, F. C. & Kety, S. S. (1948). The acute effects on the cerebral circulation of the reduction of increased intracranial pressure by means of intravenous glucose or ventricular drainage. *J. Neurosurg.* **5**, 466–70.

Shepard, J. F. (1914). *The circulation and sleep*. New York: Macmillan.

Shepherd, J. T. (1963). *Physiology of the circulation in human limbs in health and disease*. pp. 73–87, W. B. Saunders: Philadelphia.

Shinohara, Y., Meyer, J. S., Kitamura, A., Toyoda, M. & Ryu, T. (1969). Measurement of hemispheric blood flow by intracarotid injection of hydrogen gas. *Circulation Res.* **25**, 735–45.

References

Siciliano, (1900). Les effets de la compressions des carotides sur la pression, sur le coeur et sur la respiration. *Archs ital. Biol.* **33**, 338.

Sieker, H. O. & Hickam, J. B. (1953). Normal and impaired retinal vascular reactivity. *Circulation*, **7**, 79–83.

Siesjö, B. K. (1961). A method for the continuous measurement of the carbon dioxide tension on the cerebral cortex. *Acta physiol. scand.* **51**, 297–313.

(1962a). The solubility of carbon dioxide in cerebral cortical tissue of cats. With a note on the solubility of carbon dioxide in water, 0.16M NaCL and cerebrospinal fluid. *Acta physiol. scand.* **55**, 325–41.

(1962b). The bicarbonate/carbonic acid buffer system of the cerebral cortex of cats as studied in tissue homogenates. II. The pK_i of carbonic acid at 37.5°C and the relation between carbon dioxide tension and pH. *Acta neurol. scand.* **38**, 121–41.

Siesjö, B. K., Brzezinski, J., Kjällquist, Å. & Pontén, U. (1967). Carbon dioxide and acid–base equilibria in brain tissue. In *Brain edema*, ed. Klatzo, J. & Setelberger, F., New York: Springer.

Siesjö, B. K., Granholm, L. & Kjällquist, Å. (1968). Regulation of lactate and pyruvate levels in the CSF. *Scand J. clin. Lab. Invest.* Suppl. 102, I: F.

Siesjö, B. K., Kaasik, A. E., Nilsson, L. & Pontén, U. (1968). Biochemical basis of tissue acidosis. *Scand. J. clin. Lab. Invest.* Suppl. 102, III: A.

Siesjö, B. K. & Kjällquist, Å. (1969). A new theory for the regulation of the extracellular pH in the brain. *Scand. J. clin. Lab. Invest.* **24**, 1–7.

Siesjö, B. K., Kjallquist, A., Pontén, U., Zwetnow, N. (1968). Extracellular pH in the brain and cerebral blood flow. In *Progress in brain research*, ed. Luyendik, W., vol. 30, pp. 93–8. Amsterdam: Elsevier.

Siesjö, B. K. & Pontén, U. (1966). Factors affecting the cerebrospinal fluid (CSF) bicarbonate concentration. *Experientia* **22**, 611–12.

Silver, I. A. (1965). Some observations on the cerebral cortex with an ultramicro, membrane covered, oxygen electrode. *Med. Electron. Biol. Engng.* **3**, 377–87.

Silverman, A. J., Busse, E. W. & Barnes, R. H. (1955). Studies in the process of ageing: electroencephalographic findings in 400 elderly subjects. *Electroen. Neurophysiol.* **7**, 67–74.

Sjörstrand, T. (1935). On the principles for the distribution of the blood in the peripheral vascular system. *Skand. Arch. Physiol.* **71**, Suppl. 1–150.

(1948). Brain volume, diameter of the blood vessels in the pia mater, and intracranial pressure in acute carbon monoxide poisoning. *Acta physiol. scand.* **15**, 351–61.

Skinhøj, E. (1968). CBF adaptation in man to chronic hypo- and hypercapnia and its relation to CSF pH. *Scand J. clin Lab. Invest.* Suppl. 102, VIII: A.

Skinner, N. S. & Costin, J. C. (1969). Role of O_2 and K^+ in abolition of sympathetic vasoconstriction in dog skeletal muscle. *Am. J. Physiol.* **217**, 438–44.

Smith, C. A. & Rasmussin, G. L. (1965). Degeneration in the efferent nerve endings in the cochlea after axonal section. *J. cell Biol.* **26**, 63–77.

Smith, D. J. & Coxe, J. W. (1951). Reactions of isolated pulmonary blood vessels to anoxia, epinephrine, acetylcholine and histamine. *Am. J. Physiol.* **167**, 732–7.

Smith, D. J. & Vane, J. R. (1966). Effects of oxygen on vascular and other smooth muscle. *J. Physiol.* **186**, 284–94.

References

Snyder, F., Hobson, J. A., Morrison, D. F. & Goldfranck, F. (1964). Changes in respiration, heart rate and systolic blood pressure in human sleep. *J. appl. Physiol.* **19**, 417–22.

Sohler, T. P., Lothrop, G. N. & Forbes, H. S. (1941). The pial circulation of normal, non-anaesthetized animals. II. The effects of drugs, alcohol and CO_2. *J. Pharmac. exp. Ther.* **71**, 331–5.

Sokoloff, L. (1956). Relation of cerebral circulation and cerebral metabolism to mental activity. *Progr. Neurobiol.*, *N.Y.* **1**, 216–29.

Sokoloff, L. (1957). Local blood flow in neural tissue. In *New research techniques of neuroanatomy*, pp. 51–61. Springfield: C. C. Thomas.

(1959). The action of drugs on the cerebral circulation. *Pharmac. Rev.* **11**, 1–85.

(1960). The effect of carbon dioxide on the cerebral circulation. *Anesthesiology* **21**, 664–73.

(1961). Local cerebral circulation at rest and during altered cerebral activity induced by anaesthesia or visual stimulation. In *Regional neurochemistry*, pp. 107–17, ed. Kety, S. S. & Elkes, J. New York: Pergamon.

Sokoloff, L., Mangold, R., Wechsler, R. L., Kennedy, C. & Kety, S. S. (1955). The effect of mental arithmetic on cerebral circulation and metabolism. *J. clin. Invest.* **34**, 1101–8.

Sokoloff, L., Wechsler, R. L., Mangold, R., Balls, K. & Kety, S. S. (1953). Cerebral blood flow and oxygen consumption in hyperthyroidism before and after treatment. *J. clin. Invest.* **32**, 202–8.

Solomon, A. P., Darrow, C. W. & Blaurock, M. (1939). Blood pressure and palmar sweat (galvanic) responses of psycholic patients before and after insulin and metrazol therapy. *Psychosom. Med.* **1**, 118–37.

Sonnenschein, R. R., Stein, S. N. & Perot, P. L. (1953). Oxygen tension of the brain during hyperoxic convulsions. *Am. J. Physiol.* **173**, 161–3.

Sørensen, S. C. (1970). Ventilatory acclimatization to hypoxia after total denervation of peripheral chemoreceptors. *J. appl. Physiol.* **28**, 836–9.

Sørensen, S. C. & Mines, A. H. (1970). Ventilatory responses to acute and chronic hypoxia in goats after sinus nerve section. *J. appl. Physiol.* **28**, 832–5.

Sørensen, S. C. & Severinghaus, J. W. (1970). The effect of cerebral acidosis on the csf-blood potential difference. *Am. J. Physiol.* **219**, 68–71.

Sorokina, Z. A. (1965). Measurement of the activity of hydrogen ions outside and within the nerve cells of ganglia of molluscs. *J. Evol. Biochem. Physiol.* **1**, 341–50.

Speden, R. N. (1967). Adrenergic transmission in small arteries. *Nature, Lond.* **216**, 289–90.

Spielmeyer, W. (1922). *Histopathologie des Nervensystems*. Berlin: Springer.

Spina, A. (1898). Experimentelle untersuchungen über die bildung des liquor cerebrospinalis. *Pflügers Arch. ges. Physiol.* **76**, 204–218.

Spina, A. (1900). Ueber den einfluss des hohen blutdrükes auf die Neubildung der Cerebrospinal flüssigkeit. *Pflügers Arch. ges. Physiol.* **80**, 370–407

Spoendlin, H. H. & Gacek, R. R. (1963). Electronmicroscopic study of the efferent and afferent innervation of the organ of corti in the cat. *Ann. Otol. Rhinol. Lar.* **72**, 660–86.

References

Stavraky, G. W. (1936). Response of cerebral blood vessels to electrical stimulation of the thalamus and hypothalamic regions. *Archs Neurol. Psychiat., Chicago* **35**, 1002–28.

Steedman, W. M. (1966). Micro-electrode studies on mammalian vascular muscle. *J. Physiol.* **186**, 382–400.

Stephen, C. R., Woodhall, B., Golden, J. B., Martin, R. & Nowill, W. K. (1954). Influence of anesthetic drugs and techniques on intracranial tension. *Anesthesiology* **15**, 365–77.

Sterzi, G. (1904). Die Blutgefasse des Ruckenmarks. Untersuchungen über ihre vergleichende Anatomie und Entwickelungsgeschichte. Anatomische Heft I Abteilung, part 74, vol. 24, S. 1–364.

Stevenson, L., Christensen, B. E. & Wortis, S. B. (1929). Some experiments in intracranial pressure during sleep and under certain other conditions. *Am. J. med. Sci.* **178**, 663–77.

Stewart, G. N. (1921). The pulmonary circulation time, the quantity of blood in the lungs and the output of the heart. *Am. J. Physiol.* **58**, 20–43.

Stohr, P. (1922a). Über die innervation der pia mater und des plexus chorioideus des menschen. *Z. Anat. EntwGesch.* **63**, 562–607.

— (1922b). Beobachtungen über die innervation der pia mater des Ruckenmarkes und der telae chorioideae beim Menschen. *Z. Anat. EntwGesch.* **64**, 555–64.

— (1928). *Mikroscopische Anatomie des vegetativen Nervensystems.* p. 183. Springer: Berlin.

Stone, J. E., Wells, J., Draper, W. B. & Whitehead, R. W. (1958). Changes in renal blood flow in dogs during inhalation of 30% carbon dioxide. *Am. J. Physiol.* **194**, 115–19.

Stone, W. E., Marshall, C. & Nims, L. F. (1941). Chemical changes in the brain produced by injury and by anoxia. *Am. J. Physiol.* **132**, 770–5.

Stone, W. E., Webster, J. E. & Gurdjian, E. S. (1945). Chemical changes in the cerebral cortex associated with convulsive activity. *J. Neurophysiol.* **8**, 233–40.

Stosseck, K. (1970). Bestimmung der Mikrodurchblutung im Gehirn durch lokale Gabe von gasförmigen Wasserstoff. *Pflügers Arch ges. Physiol.* **316**, R20.

Stosseck, K. & Acker, H. (1969). Grosse und Bedeutung des pialen Gasaustausches. *Pflügers Arch ges. Physiol.* **312**, R149.

Strickholm, A., Wallin, B. G. & Shrager, P. (1969). The pH dependency of relative ion permeabilities in the crayfish giant axon. *Biophys. J.* **9**, 873–83.

Su, C. & Bevan, J. A. (1965). The electrical response of pulmonary artery muscle to acetylcholine, histamine and serotonin. *Life Sci. Oxford* **4**, 1025–9.

— (1966). Electrical and mechanical responses of pulmonary artery muscle to neural and chemical stimulation. *Biblphie anat.* **8**, 30–4.

Sugar, O. & Gerard, R. W. (1938). Anoxia and brain potentials. *J. Neurophysiol.* **1**, 558–70.

Sullivan, W. J. & Dorman, P. J. (1955). Renal response to chronic respiration acidosis. *J. clin. Invest.* **34**, 268–76.

Szikla, G. & Zolnai, B. (1956). Der Nachweis der Gehirnangioarchitektur mittels mit Kunstharz verfertigten Korrosion-präparaten. *Anat. Anz.* **103**, 386–93.

References

Takenaka, F. (1963). Response of vascular strip preparations to noradrenaline and tyramine. *Jap. J. Pharmac.* **13**, 274–81.

Tandler, J. (1899). Zur vergleichenden Anatomie der Kopfarterien bei den Mammalia. *Denkschr. Akad. Wiss. Wien.* **67**, 677–784.

Tannenberg, J. (1926). Bau und Funktion der Blutkapillaren. *Frank. Z. Path.* **34**, 1–19.

Tarchanoff, J. (1894). Quelque observations sur le sommeil normal. *Archs ital. Biol.* **21**, 318.

Taxi, J. (1958). Recherches en vue de l'identification au microscope electronique des cellules interstitielles de Cajal. In *Fourth international conference on electron microscopy*, pp. 440–3. Berlin: Springer.

Taylor, A. R. & Bell, T. K. (1966). Showing a cerebral circulation after concussional head injury. *Lancet* ii, 178–80.

ten Cate, J. & Horsten, G. P. M. (1952). Sur l'influence de la ligature temporaire de l'aorte sur l'activité electrique de l'écorce cérébrale. *Archs int. Physiol.* **60**, 441–8.

Tenney, S. M. (1956a). Mechanism of hypertension during diffusion respiration. *Anesthesiology* **17**, 768–76.

— (1956b). Sympatho-adrenal stimulation by carbon dioxide and the inhibitory effects of carbonic acid upon epinephrine release. *Am. J. Physiol.* **187**, 341–6.

Tenney, S. M. & Ou, L. C. (1970). Physiological evidence for increased tissue capillarity in rats acclimatized to high altitude. *Respir. Physiol.* **8**, 137–50.

Thaemert, J. C. (1963). The ultrastructure and disposition of vesiculated nerve processes in smooth muscle. *J. cell Biol.* **16**, 361–77.

Thews, G. (1953). Über die Mathematische Behandlung physiologischer Diffusionsprozesse in zylinderformigen Objekten. *Acta biotheor. (Leiden)* **10**, 105–38.

— (1960a). Die Sauerstoffdiffusion im Gehirn. *Pflügers Arch. ges. Physiol.* **271**, 197–226.

— (1960b). Ein Verfahren zur Bestimmung des O_2-Diffusions-koeffizienten der O_2 Leitfähigkeit und des O_2 Löslichkeit-koeffizienten im Gehirngewebe. *Pflügers Arch. ges. Physiol.* **271**, 227–44.

— (1963). Implications to physiology and pathology of oxygen diffusion at the capillary level. In *Selective vulnerability of the brain*, ed. Schadé, J. P. & McMenemey, W. H., pp. 27–40. Oxford: Blackwell.

Thomas, R. C. (1971). pNa microelectrodes with sensitive glass inside the tip. In *Ion selective microelectrodes*, ed. Hebert, N. C. & Khuri, R. H. New York: Marcel Deccer.

Thompson, A. M., Cavert, H. M. & Lifson, N. (1958). Kinetics of distribution of D_2O and antipyrine in isolated perfused rat liver. *Am. J. Physiol.* **192**, 531–7.

Thorn, W., Scholl, H., Pfleiderer, G. & Mueldener, B. (1958). Metabolic processes in the brain at normal and reduced temperatures and under anoxic and ischaemic conditions. *J. Neurochem.* **2**, 150–65.

Tindall, G. T., Odom, G. L., Dillon, M. L., Cupp, H. B., Jr., Mahaley, M. S., Jr. & Greenfield, J. C., Jr. (1963). Direction of blood flow in the internal and external carotid arteries following occlusion of the ipsilateral common carotid artery. *J. Neurosurg.* **20**, 985–93.

References

Tobin, R. B. (1964). In vivo influences of hydrogen ions on lactate and pyruvate of blood. *Am. J. Physiol.* **207**, 601–5.

Tolani, A. J. & Talwar, G. P. (1963). Differential metabolism of various brain regions. Biochemical heterogeneity of mitochondria. *Biochem. J.* **88**, 357–62.

Tönnis, W. & Schiefer, W. (1959). Zirkulationsstörungen des Gehirns im Serienangiogramm. Heidleberg: Springer.

Toth, A. (1965). Kapillarvermehrung in der Hirninde der Ratte unter chronischem Sauerstoffmangel. *Naturwissenschaften* **52**, 135–6.

Traum, E. (1925). Beiträge zur Innervation der Dura Mater cerebri. *Z. ges. Anat.* **77**, 488–92.

Tschirigi, R. D. (1958). The blood–brain barrier. In *Biology of neuroglia*, ed. Windle, W. F. Springfield, Ill.: Thomas.

(1960). Chemical environment of the central nervous system. In *Handbook of physiology, Neurophysiology*, section 1, vol. 3, pp. 1865–90. Washington D.C.: American Physiological Society.

Tschirgi, R. D. & Taylor, J. L. (1958). Slowly changing bioelectric potentials associated with the blood–brain barrier. *Am. J. Physiol.* **195**, 7–22.

Turner, J., Lambertsen, C. J., Owen, S. G., Wendel, H. & Chiodi, H. (1957). Effects of .08 and .8 atmospheres of inspired PO_2 upon cerebral hemodynamics at a 'constant' alveolar PCO_2 of 43 mmHg. *Fedn. Proc.* **16**, 130.

Tyler, D. B. & van Harreveld, A. (1942). The respiration of the developing brain. *Am. J. Physiol.* **136**, 600–3.

Ueda, H., Hatano, S., Molde, T. & Gondoaira, T. (1965). Discussion on compartmental analysis of the human brain blood flow. *Acta Neurol. Scand.* **41**, Suppl. 14, 88–9.

Ussing, H. H. (1953). Transport through biological membranes. *A. Rev. Physiol.* **15**, 1–20.

(1960). The frog skin potential. *J. gen. Physiol.* **43**, 135–47.

Van Breeman, V. L. & Clemente, C. D. (1955). Silver deposition in the central nervous system and the haematoencephalic barrier studied with the electron microscope. *J. biophys. biochem. Cytol.* **1**, 161–6.

Van den Bergh, R. (1961). La vascularisation artérielle intracérébrale. *Acta neurol. psychiat. belg.* **61**, 1013–23.

(1964). Bijzonderheden over de anatomie van de hersenbloedvaten. *Belg. Tijschr. Geneesk.* **17**, 891–8.

Van den Bergh, R. & Vander Eecken, H. (1968). Anatomy and embryology of cerebral circulation. *Prog. Brain Res.* **30**, 1–26, ed. Luyendijk, W. Amsterdam: Elsevier.

Van Harreveld, A. (1947). The EEG after prolonged brain asphyxiation. *J. Neurophysiol.* **10**, 361–9.

Van Harreveld, A. & Stamm, J. S. (1952). Vascular concomitants of spreading cortical depression. *J. Neurophysiol.* **15**, 487–96.

Van Heijst, A. N. P., Visser, B. F. & Maas, A. H. J. (1961). A micro-method for the determination of pH and P_{CO_2} in human cerebro-spinal fluid. *Clin. Chim. Acta* **6**, 589–90.

Van Vaerenbergh, P., Demeester, G. & Leusen, I. (1965). Lactate in cerebrospinal fluid during hyperventilation. *Archs int. Physiol. Biochim.* **73**, 738–47.

References

Vaughan Williams, E. M. (1955). Individual effects of CO_2, bicarbonate and pH on the electrical and mechanical activity of the isolated rabbit auricles. *J. Physiol.* **129**, 90–110.

Veall, N. (1969). Comment on the paper by Agnoli et al. In *Cerebral blood flow*, ed. Brock, M., Fieschi, C., Ingvar, D. H., Lassen, N. A. & Schürmann, K., p. 35. Berlin: Springer.

Veall, N. & Mallett, B. L. (1965a). The partition of trace amounts of Xenon between human blood and brain tissues at 37°C. *Physics. Med. Biol.* **10**, 375–80.

(1965b). The two-compartment model using Xe^{133} inhalation and external counting. *Acta neurol scand.* **41**, Suppl. 14, 83–4.

(1966). Regional cerebral blood flow determination by ^{133}Xe inhalation and extracranial recording. *Clin. Sci.* **30**, 353–69.

Verity, M. A., Bevan, J. A. & Ostrom, R. J. (1966). Plurovesical nerve endings in the pulmonary artery. *Nature, Lond.* **211**, 537.

Vesalius, Andreas (1543). *De humani corporis fabrica*. Basel: Operinus.

Veslingius, J. (1651). *Syntagma anatomica*, p. 195. Padua: Frambotti.

Vogel, H. (1947). Die Geshwindigkeit des Blutes in der Lungen Kapillaren. *Helv. physiol. pharmac. Acta*, **5**, 105–21.

Volpitto, P. P. & Rustin, W. A. (1943). The use of stellate ganglion block in cerebral vascular occlusions. *Anesthesiology* **4**, 403–8.

Vujic, V. (1933). Schlaf und Liquordruck. Beitrag zur Physiologie und Pathologie des Schlafes. *Jahrb. f. Psychiat. u. Neurol.* **49**, 113–62.

Vulpian, A. (1875). *Leçons sur l'appareil vasomoteur (physiology et pathology) faites à la Faculté de médicine de Paris*, p. 108. Paris: H.-C. Carville.

Wagner, W. W., Jr. & Filley, G. F. (1963). Microcinephotography of the in vivo lung. *Motion Picture Sessions Fedn Proc.* Programme p. 6.

Wahl, M., Deetjen, P., Thurau, K., Ingvar, D. H. & Lassen, N. A. (1970). Micropuncture evaluation of the importance of perivascular pH for the arteriolar diameter on the brain surface. *Pflügers Arch. ges. Physiol.* **316**, 152–63.

Walker, J. & Brown, A. M. (1970). Unified account of the variable effects of carbon dioxide on nerve cells. *Science, N.Y.* **167**, 1502–4.

Waller, A. (1853). Neuvieme memoire sur le systeme nerveux. *C. r. Acad. Sci., Paris* **36**, 378–82.

Waltz, A. G. & Ray, C. D. (1965). Impedance cephalography ('rheoencephalography'). *Trans. Am. neurol. Ass.* **90**, 305–7.

Waltz, A. G., Sundt, T. M., Jr. & Owen, C. A., Jr. (1966). Effect of middle cerebral artery occlusion on cortical blood flow in animals. *Neurology, Minneapolis* **16**, 1185–90.

Waltz, A. G., Yamaguchi, R. & Regli, F. (1971). Regulatory responses of cerebral vasculature after sympathetic denervation. *Am. J. Physiol.* **221**, 298–302.

Wasserman, A. J. & Patterson, J. L. (1961). The cerebrovascular response to reduction in arterial carbon dioxide tension. *J. clin. Invest.* **40**, 1297–303.

Wearn, J. T., Ernstene, A. C., Bromer, A. W., Barr, J. S., German, W. J. & Zchiesche, L. J. (1934). Normal behaviour of pulmonary blood vessels with observations of intermittence of flow in arterioles and capillaries *Am. J. Physiol.* **109**, 236–56.

References

Weber, E. (1908). Ueber die Selbstandigkeit des Gehirns in der Regulierung seiner Blutversorgund. *Arch. f. Physiol., Leipz.* 457–536.

Weber, E. H. (1831). *Hildebrandt's Lehrbuch d. Anat.* 4th edition. Vol. 3, p. 75. Braunschweig.

Wechsler, R. L., Dripps, R. L. & Kety, S. S. (1951). Blood flow and oxygen consumption of the human brain during anesthesia produced by thiopental. *Anesthesiology* 12, 308–14.

Wechsler, R. L., Kleiss, L. M. & Kety, S. S. (1950). The effects of intravenously administered aminophylline on the cerebral circulation and metabolism in man. *J. clin. Invest.* 29, 28–30.

Weed, L. H. (1914). The pathways of escape from the subarachnoid spaces with particular reference to the arachnoid villi. *J. med. Res.* 31, 51–91.

Weed, L. H. & Hughson, W. (1921). Intracranial venous pressure and cerebrospinal fluid pressure as affected by intravenous injections of solutions of various concentrations. *Am. J. Physiol.* 58, 101–30.

Weed, L. H. & McKibben, P. S. (1919). Pressure changes in the cerebral spinal fluid following intravenous injections of solutions of various concentrations. *Am. J. Physiol.* 48, 512–30.

Welch, K. & Sadler, K. (1965). Electrical potentials of choroid plexus of the rabbit. *J. Neurosurg.* 22, 344–9.

Wells, R. E., Denton, R. & Merrill, E. W. (1961). Measurement of viscosity of biologic fluid by cone plate viscometer. *J. Lab. clin. Med.* 57, 646–56.

Wendling, M. G., Eckstein, J. W. & Abboud, F. M. (1967). Cardiovascular responses to carbon dioxide before and after beta-adrenergic blockade. *J. appl. Physiol.* 22, 223–6.

Wentsler, N. E. (1936). Microscopic study of the superficial cerebral vessels of the rabbit by means of a permanently installed transparent cranial chamber. *Anat. Rec.* 66, 423–35.

West, J. B. (1962). Diffusing capacity of the lung for carbon monoxide at high altitude. *J. appl. Physiol.* 17, 421–6.

Weyne, J., Demeester, G. & Leusen, I. (1970). Effects of carbon dioxide, bicarbonate and pH on lactate and pyruvate in the brain of rats. *Pflügers Arch. ges. Physiol.* 314, 292–311.

Whelan, R. F. & Young, I. M. (1953). The effect of adrenaline and noradrenaline infusions on respiration in man. *Br. J. Pharmac. Chemother.* 8, 98–102.

White, J. C., Verlot, M., Silverstone, B. & Beecher, H. K. (1942). Changes in brain volume during anesthesia. The effects of anoxia and hypercapnia. *Archs Surg., Chicago* 44, 1–21.

Whitmore, R. L. (1968). *Rheology of the circulation.* Oxford: Pergamon.

Whittaker, V. P. (1965). The application of subcellular fractionation techniques to the study of brain function. *Prog. Biophys. Mol. Biol.* 15, 39–96.

Wienbeck, M., Golenhofen, K. & Lammel, E. (1968). Der Effeckt von CO_2 auf die Spontanaktivitat von isolierter glatter Muskulatur (Taenia coli des Meerschweinchens). *Pflügers Arch. ges. Physiol.* 300, R78.

Wierzuchowski, M. (1938). Isolated rabbit's head preparation for the study of the cervical sympathetic and the cephalic vascular reactions. *Archs int. Pharmacodyn. Thér.* 58, 47–60.

References

Wilkinson, I. M. S., Bull, J. W. D., Boulay, G. H. Du, Marshall, J., Ross Russell, R. W. & Symon, L. (1969). The heterogeneity of blood flow throughout the normal cerebral hemisphere. In *Cerebral blood flow*, ed. Brock, M., Fieschi, C., Ingvar, D. H., Lassen, N. A. & Schürmann, K., pp. 17–18. Berlin: Springer.

Williams, D. J. (1936). The origin of the posterior cerebral artery. *Brain*, **59**, 175–80.

Williams, D. J. & Lennox, W. G. (1939). The cerebral blood flow in arterial hypertension, arteriosclerosis, and high intracranial pressure. *Q. Jl Med.* **8**, 185–94.

Willis, T. (1664). *Cerebri anatome*. London: Martin & Allestry.

Wilson, W. P., Odom, G. L. & Shieve, J. F. (1953). The effect of carbon dioxide on cerebral blood flow, spinal fluid pressure and brain volume during Pentothal Sodium anesthesia. *Curr. Res. Anesth.* **32**, 268–73.

Winterstein, H. (1935). Über den Blutkreislauf im Kaninchershirn. *Pflüger. Arch ges. Physiol.* **235**, 377–88.

Wodick, R., Schwickardi, D. & Lübbers, D. W. (1966). Konzentration und kinetic der Atmungsfermente im Meerschweinchengehirn in vivo. *Pflügers Arch. ges. Physiol.* **291**, R25.

Wolfe, D. E., Potter, L. T., Richardson, K. C. & Axelrod, J. (1962). Localizing tritiated norepinephrine in sympathetic axons by electron microscopic autoradiography. *Science* **138**, 440–2.

Wolff, H. G. (1929). The cerebral circulation. xi*a*. The action of acetylcholine. xi*b*. The action of the posterior lobe of the pituitary gland. xi*c*. The action of amyl nitrite. *Archs Neurol. Psychiat., Chicago* **22**, 686–90.

—— (1936). The cerebral circulation. *Physiol. Rev.* **16**, 545–96.

—— (1938). The cerebral blood vessels – Anatomical principles. *Res. Publs Ass. Res. nerv. ment. Dis.* **18**, 29–68.

Wolff, H. G. & Blumgart, H. L. (1929). The cerebral circulation: vi. The effect of normal and of increased intracranial cerebrospinal fluid pressure on the velocity of intracranial blood flow. *Archs Neurol. Psychiat., Chicago* **21**, 795–804.

Wolff, H. G. & Forbes, H. S. (1928). The cerebral circulation. v. Observations of the pial circulation during changes in intra-cranial pressure. *Archs Neurol. Psychiat, Chicago* **20**, 1035–47.

Wolff, H. G. & Lennox, W. G. (1930). The cerebral circulation: xii. The effect on pial vessels of variations in the O_2 and CO_2 content of the blood. *Archs Neurol. Psychiat., Chicago* **23**, 1097–120.

Wollman, H., Alexander, S. C., Cohen, P. J., Chase, P. E., Melman, E. & Behar, M. G. (1964). Cerebral circulation in man during halothane anesthesia. *Anesthesiology* **25**, 180–4.

Woodward, D. L., Reed, D. J. & Woodbury, D. M. (1967). Extracellular space of rat cerebral cortex. *Am. J. Physiol.* **212**, 367–70.

Woollam, D. H. M. & Millen, J. W. (1954). Perivascular spaces of the mammalian central nervous system. *Biol. Rev.* **29**, 251–83.

—— (1955). The morphology of the blood–brain barrier. *Proc. 2nd Int. Cong. Neuropath. Lond.* **2**, 367–72.

Wortis, S. B. (1935). Respiratory metabolism of excised brain tissue; effects of

References

some drugs on brain oxidations. *Archs Neurol. Psychiat.*, Chicago **33**, 1022–9.

Wright, R. D. (1938). Experimental observations on increased intracranial pressure. *Aust. N.Z. J. Surg.* **7**, 215–35.

Wullenweber, R. (1965). Observations concerning autoregulation of central blood flow in man. *Acta neurol. scand.* **41**, Suppl. 14, 111–15.

Wyant, G. M., Merriman, J. E., Kilduff, C. J. & Thomas, E. T. (1958). The cardiovascular effects of halothane. *Canad. Anaesth. Soc. J.* **5**, 384–402.

Yamada, H. & Maie, S. (1954). Histological studies on the medullary pyramid of the dog. *Bull. Tokyo Med. Dent. Univ.* **1**, 99–104.

Yamakita, M. (1922). The gaseous metabolism and blood flow of the brain. 1. Under narcosis and hypnosis. *Tohoku J. exp. Med. Sendai* **3**, 414–95.

Yoshida, K., Meyer, J. S., Sakamoto, K. & Handa, J. (1966). Autoregulation of cerebral blood flow: electromagnetic flow measurements during acute hypertension in the monkey. *Circulation Res.* **19**, 726–38.

Zaimis, E. (1968). Vasopressor drugs and catecholamines. *Anesthesiology* **29**, 732–62.

Zierler, K. L. (1961). Theory of the use of arteriovenous concentration differences for measuring metabolism in steady and non-steady states. *J. clin. Invest.* **40**, 2111–25.

—— (1965). Equations for measuring blood flow by external monitoring of radioisotopes. *Circulation Res.* **16**, 309–21.

Zwetnow, N. (1968). Effects of intracranial hypertension: acid–base changes and lactate changes in CSF and brain tissue. *Scand. J. clin. Lab. Invest.* Suppl. 102, III: D.

Zwetnow, N., Kjällquist, A. & Siesjö, B. K. (1968). Elimination of autoregulation following a period of pronounced intracranial hypertension: is hypoxia involved? *Scand. J. clin. Lab. Invest.* Suppl. 102, V: F.

INDEX

acetylcholine
  action on cerebral vessels, 346
  evidence for cholinergic vasomotor pathways, 53
adrenaline
  action on vascular smooth muscle, 336
  and cerebral oxygen uptake, 335
  and pial vessels, 335
adrenergic blocking agents
  dichloroisoproterenol, 342
  D-INPEA, 342
  ergot compounds, 340
  phenoxybenzamine, 342
  propranalol, 342
afferent blood supply to brain
  cat, 6
  dog, 8, 111, 116
  man, 5
  ungulates, 7, 110
anastomoses
  arterial, extracranial, 3, 116; intracerebral, 4; intracranial, 8
  venous, 9, 111
angiography, 153
artificial perfusion of the brain, 80, 111, 112, 114
atropine, action on vascular smooth muscle, 209, 348
autoregulation (*see also* flow–pressure relation),
  definition, 80
  mechanisms, 82 ff

barbiturates
  and cerebral blood flow, 326
  and cerebral oxygen uptake, 115, 326
  and vascular smooth muscle, 327
barium, use as an experimental vasoconstrictor, 341
baroreceptors
  effect of denervation, 276, 278
  reflex control of cerebral blood flow, 166

bicarbonate
  evidence for active transport, 220, 222
  in c.s.f., 204, 217, 221
  in c.s.f.–plasma distribution, 205, 218
  in e.c.f., 205
  in electrochemical potential difference, 218, 222, 224
  in intracellular, 227
blood–brain partition coefficient
  for hydrogen, 147
  for nitrous oxide, 123
  for xenon-133, 134–6

capillaries
  cervical sympathetic ganglion, 28
  intracerebral, changes with age, 29; changes with hypoxia, 31; density, 24 ff; relation to oxygen$_2$ uptake, 28; ultrastructure, 19
  trigeminal ganglion, 28

carbon dioxide,
  cerebrovascular response, 177 ff
  effects of hypertension, 181
  effect of hypotension, 183
  effect of mid-brain stimulation, 269
  effects of spasm, 181
  in dog, 178
  in foetus and newborn, 180
  in man, 177
  in old age, 180
  in primates, 178
  interactions with hypoxia, 184
  neural contribution, 196
  speed of response, 185
  membrane conductance for chloride, 202
carotid artery
  as a measure of cerebral blood flow, 110
  effects of ligation, 6, 8

Index

cerebral blood flow
  alterations with: acceleration, 91
    adrenaline, 335
    barbiturates, 326
    e.c.f. or c.s.f. pH, 207, 210
    eclampsia, 79
    essential hypertension, 79
    halothane, 324
    hypercapnia, 177
    hyperoxia, 236
    hypocapnia, 178
    hypothermia, 318
    hypoxia, 232
    isoprenaline, 339
    mental activity, 311
    noradrenaline, 336
    perfusion pressure, 72, 161
    plasma pH, 201
    sleep, 314
    tilting, 91
    trichlorethylene, 325
    Valsalva manoeuvre, 90
  evidence for non-homogeneous flow, 137
  methods of measurement, 101 ff
    analysis of desaturation curves, 127
    angiography, 153
    autoradiography, 140
    cerebral mode transit time, 150
    clearance, heat, 118ff; hydrogen, 147; krypton-85, 126; nitrous oxide, 123; xenon-133, 126
    comparison of methods, 143
    compartmental analysis, 129
    computation of mean flow, 140
    indicator fractionation, 151
    non-diffusible indicators, 149
    regional flow, 140, 291
    rheoencephalography, 153
    venous outflow, 111
  relation to diameter of pial vessels, 160
  EEG, 328
  oxygen uptake of brain, 293
cerebral blood volume, 149, 150
cerebral oxygen uptake
  alterations with: adrenaline, 335
    anaesthetic agents, 322
    analeptic drugs, 319
    convulsions, 349
    hypercapnia, 309
    hypocapnia, 309

    hypothermia, 318
    hypoxia, 303
    isoprenaline, 328
    mental activity, 311
    noradrenaline, 339
    sleep, 314
  methods of measurement, *in vitro*, 29; *in vivo*, 288
  normal values in: adult, 33
    childhood, 299
    foetus and newborn, 294
    old age, 301
    parietal cortex, 34
  regional oxygen uptake, 290
  relation with blood flow, 284, 293
  EEG, 328
cerebrospinal fluid
  bicarbonate: concentration, 204
    equilibrium with e.c.f., 205, 218
    evidence for active transport, 220
    regulation: in acid–base disturbance, 221; at altitude, 208
  hydrogen ion: alterations with carbon dioxide, 206, 209
    content, 204
    evidence for active transport, 229
    regulation: in acid–base disturbance, 206; at altitude, 208; by cerebral blood flow, 215, by ventilation, 216
  lactate, 307
  potassium: content, 204
    regulation, 204
  pressure: effect of respiration, 90; of tilting, 91; of Valsalva manoeuvre, 90
    homeostatic function, 70, 89
    normal values, static measurements, 89; dynamic measurements, 90
    pulsations, 89, 90, 102
cervical sympathectomy, 54, 61, 65, 103
  effect upon cerebral blood flow, 266
  pial artery diameter, 254
  for vascular 'spasm', 264
cervical sympathetic ganglion
  capillary density, 28
  sympathetic pathways, 60
  vagal pathways, 60
chemoreceptors, arterial and cerebral blood flow, 276
chloroform, cerebral oxygen uptake, 323

Index

Circle of Willis, man (*see also* afferent blood supply to brain), 5
Cushing effect, 74
cyclopropane, effect on vascular smooth muscle, 325

desaturation curves (indicator)
  compartmental analysis, 129
  effect of period of injection, 130
  theoretical analysis, 127
  validity of H/A, 128
  validity of H_0, 128
diethyl ether, effect on cerebral oxygen uptake, 322
diffusion coefficient, for inert gases, 126
  for oxygen, 25, 33

electrical potential between c.s.f. and plasma, 219
  effect of acid–base disturbances, 219, 220
electrochemical potential
  for bicarbonate, 218
  for hydrogen ion, 218
  variations with: carbon dioxide, 224
    hydrogen ion, 225
electroencephalogram
  relation to: cerebral blood flow, 328
    cerebral oxygen uptake, 328
    stimulation of vagus and vasodilator pathways, 332
ergot, effect on diameter of pial arteries, 340
extracellular fluid
  bicarbonate, 218
  hydrogen ion: regulation, 213
    relation to vascular smooth muscle, 82, 212

Fick, Law of diffusion of gases, 25, 33
  principle, 123
flow–pressure relation, 161
  baroreceptors, 167
  effect of: hypercapnia, 164
    hypoxia, 165
    stimulation of sympathetic nerves, 170
    sympathectomy, 170
    surgery, 117
    vagotomy, 171

glial cells,
  astrocytes, 21

histiocytes, 22
microglial cells, 22
oligodendrocytes, 22
relation to capillaries, 20, 21
glucose, as brain substrate, 287
  lactate–glucose index, 308
  oxygen–glucose index, 308

haematocrit, blood viscosity, 96
  cerebral blood flow, 96, 97
halothane, cerebral blood flow, 324
  cerebral oxygen uptake, 324
  effect upon vascular smooth muscle, 325
heat clearance, analysis of errors, 120
  as a measure of cerebral blood flow, 118
histamine, 349
hydrogen clearance
  as a measure of: mean cerebral blood flow, 146
    regional cerebral blood flow, 147
hydrogen electrodes, 146, 148
hydrogen ion
  blood–brain barrier, 203
  cerebrospinal fluid, 207
  extracellular: plasma gradients, 205
  regulation, 213
  relation to smooth muscle tone, 211
  intracellular: measurements, 226
    DMO, 228
    estimates, 227
    effect of tissue buffers, 227
  plasma, effect on cerebral blood flow, 201
hypercapnia, effect upon: cerebral blood flow, 177
  pressure–flow relation, 164
hyperoxia, effect upon cerebral blood flow, 236
hypocapnia, effect upon cerebral blood flow, 177
lactate, c.s.f., 309
  tissue, 309
tissue hypoxia, 310
hypoxia, action on vascular smooth muscle, 248
  cerebral oxygen uptake, 303
  cerebrovascular response, 232
    at altitude, 239
    and capillary density, 31
    and carbon dioxide, 236, 238
    flow–pressure relation, 165

417

Index

hypoxia, *cont.*
    in foetus, 234
    neural contribution, 250
    old age, 234
    cortical potentials, 245
    lactate in tissues, 305

indicators
  diffusible
    hydrogen, 146
    krypton-85, 126
    nitrous oxide, 123
    xenon-133, 126
  fractionation techniques, 151
  non-diffusible, 149
    indocyamine green, 150
    iodinated albumin, 150
    labelled erythrocytes, 151
internal carotid artery plexus, 60
intracranial 'steal', 6
isoprenaline, cerebral blood flow, 339
  oxygen uptake, 340

jugular veins
  anastomoses
  blood for: measure of flow, 123
    measure of oxygen uptake, 288
    nitrous oxide method, 123
    outflow method, 111
    variations of drainage pattern, 11, 289

krypton-85, clearance, 126
  saturation method, 125
λ, (*see* blood–brain partition coefficient), 123

lactate, in c.s.f., 307
  lactate–glucose index, 308
  hypocapnia, 310
  hypoxia, 304
  tissue, 304

mental activity, cerebral blood flow, 312
  cerebral oxygen uptake, 311
mode transit time,
  relation to cerebral blood flow, 150
Monro–Kellie doctrine, 69, 72

NADH, increase in cerebral tissue hypoxia, 305, 306

nerves
  afferent, and headache, 63
    from dural sinuses, 65
    from pain receptors, 63
    from pial vessels, 62
    pathways, 64
  vasomotor, adrenergic, 53
    cerebral and other vascular beds compared, 66
    cholinergic, 53
    contribution of carotid sinus, 276
    contribution to resting smooth muscle tone, 254
    dilator, 48, 55
    fluorescent techniques, 54
    internal carotid artery plexus, 60
    origin, 45, 58
    reflex effects, 103
    staining methods, 50
    ultrastructure, 50
    vasoconstrictor, 273
    vertebral artery plexus, 58
nitrous oxide:
  effect upon cerebral blood flow, 323
  method of measuring cerebral blood flow, 123
  modifications of method, 126
noradrenaline:
  action on vascular smooth muscle, 338
  effect of α- and β-blocking agents, 340

oxygen
  diffusion in tissue, 25
  gradients, 40
  models, 37
  electrodes, 38, 40, 41
  glucose index, 308
  tension, as a measure of blood flow, 38
    capillary, 34
    during convulsions, 320
    mitochondrial, 41
    tissue, 38
    with mid-brain stimulation, 273
    threshold, venous P_{O_2}
    reaction, critical, lethal, 36

perivascular fluid, pH, 212
perivascular space, 16, 21

Index

pial arteries
　diameter
　　effects of: acetylcholine, 346
　　　adrenaline, 335
　　　barium, 341
　　　ergot, 340
　　　histamine, 349
　　　hydrogen ion, 212
　　　hypoxia and hypercapnia, 175
　　　hypo- and hypertonic solutions, 75
　　　propranalol, 342
　　　phentolamine, 342
　　　strychnine, 321
　　　sympathetic nerve stimulation, 107
　　　vagus stimulation, 107
　　　relation to: cerebral blood flow, 160; perfusion pressure, 72
phentolamine, action on pial arteries, 342
potassium concentration in c.s.f., 204
　depolarization of smooth muscle, 350
　efflux from smooth muscle, 189
　regulation in c.s.f., 204
precapillary cushions, 18
　sphincters, 17
pressure, arterial, 91
　effect of: acceleration, 92
　　tilting, 91
　cerebrospinal fluid, 70, 71, 73
　intracranial, 75
　　effect of hypo- and hypertonic solutions, 75
　　measure of cerebral blood flow, 104
　jugular venous, 73
　perfusion, 72
　　relation to cerebral blood flow, 76
　sagittal sinus, 73
　venous, 70, 71
pyruvate, and tissue hypoxia, 305

reflex control of cerebral vessels, 253
　effect of: carbon dioxide, 266
　　midbrain lesions, 273
　　stimulation, 271
　pathways, vasoconstrictor, 56, 254
　　vasodilator, 47, 55, 257
　　vertebral plexus, 58

resistance
　cerebro-vascular, mean, 76, 77
　effect of: baroreceptor denervation, 276
　　carbon dioxide, 79
　　hypertension, 79
respiration and control of cerebral blood flow, 216, 262
respiratory enzymes, distribution in brain, 40, 41
rete mirabile, 6
rheo-encephalography, 153

skull,
　compliance, 70
　window technique, 102, 107
sleep
　cerebral blood flow, 315
　cerebral oxygen uptake, 314
smooth muscle
　cerebral blood vessels
　　effects of: acetylcholine, 346
　　　adrenaline, 335
　　　adrenergic blocking agents, 340
　　　angiotensin, 351
　　　atropine, 87, 346, 348
　　　barbiturates, 326
　　　barium, 341
　　　bradykinin, 351
　　　cyclopropane, 325
　　　halothane, 324
　　　histamine, 349
　　　isoprenaline, 339
　　　noradrenaline, 336
　　　strychnine, 321
　　　tryamine, 344
　　　xanthines, 321
　intestinal
　　acetylcholine, 87, 347
　　action potentials, 87
　　adrenaline and hyperpolarization, 86, 345
　　atropine, 87
　　extracellular pH, 189; fluxes of ^{36}Cl, 189, 190; fluxes of ^{42}K, 191
　　neural control, 254ff
　　neuromuscular junctions, 52
　　responses, arterial pressure, 156
　　carbon dioxide, 173ff
　　hypoxia, 248
　　hyperoxia, 249
　　pH of perivascular fluid, 211

Index

smooth muscle, *cont.*
  systemic blood vessels
    action potentials, adrenergic blocking agents, 347; cholinergic drugs, 347; vasoconstrictor drugs, 345
    carbon dioxide, 191 ff
    depolarization, 345
    hyperpolarization, nitrites, 348
    neuromuscular junctions, 52
solubility coefficient for oxygen, 33
Stewart–Hamilton principle, 149
stellate ganglion blockade for vascular spasm, 264
strychnine
  cerebral oxygen uptake, 320
  pial artery diameter, 321

trichlorethylene and cerebral blood flow, 325
tyramine: smooth muscle, 344

ultrastructure
  adrenergic, 51
  capillaries, 19
  extracerebral arteries, 13
  glial cells, 21
  intracerebral arterioles, 16
  perivascular spaces, 21
  vasomotor nerves, 50
  venules, 20

vagus nerves:
  effect of section
    upon carbon dioxide, 199
    upon cerebral blood flow, 199
    upon hypotension, 267
    upon hypoxia, 252
  modification by hypoxia, 252
  effect of stimulation
    upon cerebral blood flow, 267
    upon modification by carbon dioxide, 268
    upon intracranial pressure, 106
    upon pial artery diameter, 56, 109, 158
Valsalva manoeuvre, 90
vascular impedance, 77
veins
  drainage, 70
  pressure, 71, 91
    effect of tilting, 91
    relation to c.s.f. pressure, 71, 91
  vertebral, 11
vertebral, plexus, 58
  veins, 11
viscosity
  blood, 96
  cerebral blood flow, 96
  measurement, 97
  size of blood vessels, 98

Xenon
  clearance, 126
    inhalation techniques, 143
Xanthines
  cerebral blood flow, 321